D1275700

**RECENT ADVANCES IN
ANIMAL NUTRITION — 2004**

Recent Advances in Animal Nutrition

2004

P.C. Garnsworthy, PhD

J. Wiseman, PhD

University of Nottingham

NOTTINGHAM
University Press

Nottingham University Press
Manor Farm, Main Street, Thrumpton
Nottingham, NG11 0AX, United Kingdom

NOTTINGHAM

First published 2005

British Library Cataloguing in Publication Data
Recent Advances in Animal Nutrition — 2004:
University of Nottingham Feed Manufacturers
Conference (38th, 2004, Nottingham)
I. Garnsworthy, Philip C. II. Wiseman, J.

ISBN   1-904761-00-3

Disclaimer

Every reasonable effort has been made to ensure that the material in this book is true, correct, complete
and appropriate at the time of writing. Nevertheless the publishers, the editors and the authors do not
accept responsibility for any omission or error, or for any injury, damage, loss or financial consequences
arising from the use of the book.

Typeset by Nottingham University Press, Nottingham
Printed and bound by The Bath Press, Bath

PREFACE

The 38th University of Nottingham Feed Conference was held at the Sutton Bonington Campus 14 – 16 September 2004. This was the first conference to be held in September rather than January, and the change of time proved to be popular with delegates. As indicated in the preface to the proceedings for 2003, this move to September has now become established in the Industry's 'Calendar of events'.

The first session was on nutrition of non-ruminants. Recent outbreaks of disease have led to increased biosecurity in the pig industry and to development of herds with high health status; Mike Varley considered the implications of health status on growth potential and nutrient requirements of growing pigs. High growth potential can only be utilised if feed intake is optimised; management and behavioural factors affecting feed intake of piglets around weaning were reviewed by Hubert van Hees. The EU has banned inclusion of used cooking oils in animal feed to avoid possible contamination and disease transmission issues. The removal of this relatively cheap constituent of fat blends means that new alternatives have to be sought; David Howells outlined the nutritional, physical, processing, availability and commercial factors that must be considered when choosing between these alternatives. Energy and nutrient requirements of companion animals have been updated recently; Kelly Swanson discussed the new recommendations in relation to life stage, reproductive status and body weight of cats and dogs.

The second session was on legislation. Miguel-Angel Granero-Rosell explained the progression of the new legislation on feed additives through the European Parliament, and described the Community Reference Laboratory for Feed Additives Authorisation. Keith Millar discussed the implications of legislation that will have a significant effect on the Feed Industry, particularly in the UK. Legislation designed to ensure food safety, traceability and authenticity can only be enforced if accurate methods of detecting infringements are available; Ian Murray reported the findings of a project that compared methods to detect and quantify mammalian tissues in feedstuffs; Ron Bardsley and Greg Tucker reviewed molecular techniques for analysing feed raw materials.

The third session was on general topics. Changes in the Asian livestock industries are predicted to have major implications for European producers and the Feed Industries; these implications were outlined by Paul Meggison. Manipulation of leptin has potential to improve production efficiency and modify body composition of animals, but the modes of action of the leptin axis are poorly understood, as highlighted in the review by Rod Hill.

The fourth session was on dairy cow nutrition. The transition period is critical for establishment of lactation and avoidance of metabolic problems in dairy cows; Tom Overton reviewed carbohydrate nutrition, and Bill Sanchez discussed cation-anion difference, in transition cows. Methionine is one of the first limiting amino acids in diets for dairy cows; Jean-Claude Robert compared products designed to supply protected methionine. Manipulation of milk fat remains an important target for dairy breeding and nutrition programmes; Richard Dewhurst reviewed strategies for achieving different targets. For many years, dairy nutritionists have been asking for a feed evaluation system that is based on nutrient responses rather than energy and protein; Harmen van Laar presented initial evaluations of a model that predicts responses in milk production and composition from changes in diet composition.

The organisers would like to thanks the Programme Committee for formulating such a stimulating programme, the speakers for presenting and writing their papers, and the Chairmen (Julian Wiseman, Brian Cooke, Mike Varley, Phil Garnsworthy) for keeping control of the sessions. We would also like to thank Sue Golds and her team of helpers for administrative duties that ensured the conference ran smoothly. The 38th Nottingham Feed Conference was sponsored by Trouw Nutrition.

P.C. Garnsworthy
J. Wiseman

CONTENTS

1

FEEDING THE HIGH HEALTH STATUS PIG

MIKE VARLEY
Provimi Ltd, SCA Mill, Dalton, Thirsk, N. Yorks

Introduction

Through the late 1990s and on into the present century the pig industry in the United Kingdom experienced very profound changes. The global price crash in 1997/98 was especially traumatic and the outbreaks of Classical Swine Fever (CSF) and Foot and Mouth Disease (FMD) were also particularly damaging. Post-weaning Multi-systemic Wasting Syndrome (PMWS) and its related Post-weaning Dermatitis Nephropathy Syndrome (PDNS) have also taken their toll on individual business but also on the whole industry supply chain. Partly as a direct consequence of these events, the UK industry went rapidly from a national herd of around 750,000 sows to about 500,000 sows. At the time of writing (October 2004) there has been a modest economic recovery from these problems and the industry is beginning to look forward once again with some cautious optimism.

One of the hard lessons learnt is the sheer importance of health status and the need for effective biosecurity measures both at the individual business level but also at the industry level. The UK Meat and Livestock Commission, through BPEX, have launched a new Health & Welfare Strategy to focus attention on health care and health status and to coerce more producers to progress towards better animal health.

Many individual producers also realise and understand the benefits that accrue from attaining a higher level of health status within their herds, and are prepared to invest in programmes of systematic de-population and re-population over a 5-8 year cycle to sustain a high level of health.

The alternative to high health status is operating a sow / grower / finisher unit with an ever decreasing health status propped up with ever increasing antibiotic and other drug inputs to maintain production in the face of spiralling costs. Estimates vary but, following a full or partial depopulation programme, herd performance in terms of feed intake, growth rates and FCRs can increase by around 30%. The investment therefore pays for itself quickly and the veterinary profession have been strongly suggesting these programmes for some time.

By applying good biosecurity measures following a de-population programme there is no reason why the improvements should not be maintained for between 3 and 5 years and by year 8 the whole exercise can be profitably repeated.

For the veterinarians involved, this brings a closer involvement with producers over time. For the nutritionists involved, there are new challenges. The provision of energy and nutrient requirements from weaning to slaughter for a given genotype may be on a totally new basis.

High health status pigs (HHSP) have significantly enhanced feed intake characteristics, and the indications are that they will respond to higher levels of essential amino acids. The danger is therefore that using the same feeding programmes as for a low health status pig (LHSP) may well lead to 'under-achieving pigs' that grow below their potential and that are also too fat.

The purpose of this chapter is to identify those aspects of nutrition where it is believed there are robust associations with health status and to provide some pointers where practical nutrition programmes might be influenced.

Defining health status

One of the problems in ultimately designing nutrition programmes for specific levels of health status is that it is not easy to define, evaluate and measure herd health status! Prevailing growth performance in the growing/finishing herd can be examined and this will provide a good insight when compared to existing industry benchmarks. The difficulty here is that herds vary significantly in genetics and management, and this clouds the assessment. Subsequently it is possible to collate all available data on lung scores from the abattoirs to provide an overview of the sub-clinical or chronic disease loading carried by the herd. This can also be useful information to build up the picture but again on its own does not give an accurate index of herd health status. The level of pharmaceutical inputs into a particular herd also adds to the story.

Attempts have been made to use specific analytical tools to yield a health status index, and the measurement of acute phase proteins is one of these methods. The idea is that when a group of animals experiences a chronic exposure to pathogens without necessarily having overt disease, they will also have highly activated immune systems and the production of an array of cytokines and other cellular material such as expanded lymphocyte lines will be evident. Acute phase protein production is also an integral component of these responses and could provide a tonic response. This means that the level of acute phase proteins is not a rapid and immediately responsive effect but the levels will be determined by the animal's recent immunological / pathogenic history over previous weeks. Amory *et al.* (2002) have explored this possibility but only saw modest correlation coefficients between acute phase protein measurements, lung scores and growth performance in a study including some 35 herds in the West Midlands.

There is obviously an urgent need for further research in this area but the answer probably will be the use of an index along similar lines to a genetic selection index. It may be possible to use all of the information available, statistically weighted for its importance and its economic value, to compute an overall index score for a particular herd within a given industry.

The equation at the end of the day may look something like this:

Health Score Index = Σ (ADG*k1 + LS*k2 + PhU*k3 + VE*k4)

where:

ADG = average herd daily gain over the previous 3 months
PhU = pharmaceutical use over the previous 3 months
VE = veterinary consultant subjective evaluation over the previous
3 months K1, k2, k3 - weighting factors

If this could be simplified to an index scale from 0 (poor) to 10 (excellent) then nutritionists would have a number upon which to base their own nutritional designs.

Growth potential in pigs

There have been significant and progressive changes in growth parameters in modern hybrid pigs and, in terms of average daily liveweight gains to slaughter weight and feed conversion efficiency, great strides have been made over the last 40 years and this process continues. This has emanated from the efforts of the breeding companies but technological developments in environmental control and nutrition have also played a role. The real genetic potential is probably in the order of between 120 and 130 days to slaughter at 100 kg when all the technologies are working in concert. This of course in commercial practice is hardly ever realised.

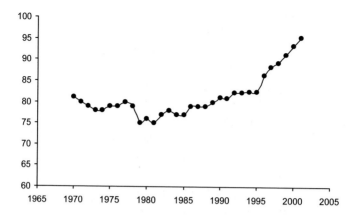

Figure 1.1 Changes in finishing weights (kg) of pigs in the UK 1970 to 2003 (MLC Yearbooks).

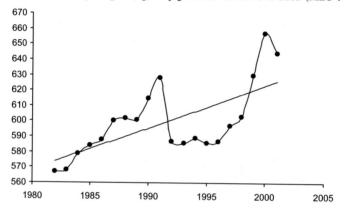

Figure 1.2 Changes in growth rates (g/d) of pigs in the UK 1970 to 2003 (MLC Yearbooks).

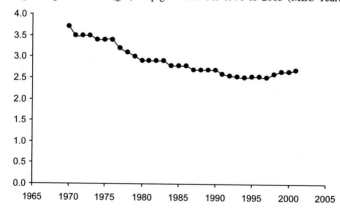

Figure 1.3 Changes in feed conversion ratio (kg feed/kg gain) of pigs in the UK 1970 to 2003 (MLC Yearbooks).

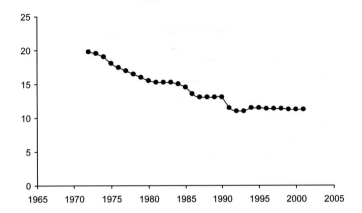

Figure 1.4 Changes in P2 fat depth (mm) pf pigs in the UK 1970 to 2003 (MLC Yearbooks).

Figures 1.1 to 1.4 give data from the UK Meat & Livestock Commission (Wiseman, *et al.*, 2004)) showing the phenotypic changes seen in growth and carcase performance in recent years. Clearly the current pig is profoundly different from that of 1970. The implications of these changes on nutritional requirements and management practices are of paramount importance to understand. Not only are genotypes different but health status is too.

Growth potential and health status

It is axiomatic that good growth is linked directly to good health status. Any producer who has carried out a depopulation-repopulation programme and begins a new production cycle with refreshed buildings and high health status pigs will have observed the very rapid growth rates that are achieved in the growing and finishing phases. This also brings a totally new cost structure to the business because of the increased throughput through the available building system but also because of significantly reduced feed costs. Experience suggests that this will continue for a considerable time after the de-population procedure has been carried out and, depending on the farm protocols in place for hygiene and general management, will give an advantage for between 5 and 8 years.

The data presented in figure 1.5 from the Australia Pig Health Scheme (2003) illustrate how an increased respiratory disease burden is related to growth performance to slaughter. With multiple diseases evident on a particular farm, then growth rates will be depressed.

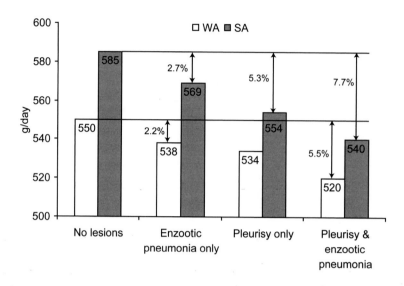

Figure 1.5 The relationship between growth performance and disease burden of pigs in Western Australia (WA) and South Australia (SA) (Australian Pig Health Scheme 2003).

Practical nutrition for high health status herds

In commercial practice, as herds hopefully move towards a higher level of health status (however this is defined), a change of strategy will be required. High health status pigs will have a higher level of voluntary feed intake and one of the consequences of this, even despite improved genetics for lean tissue growth, is that pigs may get too fat at the selected slaughter weight. What may be acceptable and correct energy/lysine ratios for a 'normal' pig will not be suitable for a HHSP. Energy requirements themselves also may need to be totally reviewed to accommodate changing feed intake patterns. Moreover, it is known that a profoundly challenged pig will mount a series of immune responses to counter this challenge and this in itself (Koutsos and Klasing, 2001) will divert up to 6% of available net energy to generate the necessary immune functions. The corollary of this is that a HHSP is not carrying this added burden to the energy budget and may need less energy to achieve the same lean tissue growth as FCR improves significantly.

It might be expected that the ratios of digestible amino acids required will be very similar at varying levels of health status but the work from Iowa State University   (Williams *et al.*, 1997) has shown that at different levels of immune stimulation (health status) growing pigs will respond differently to amino acid inputs and for higher health status pigs, they may reach their optimum growth at higher levels of amino acid inputs.

A further aspect of nutrition that is important in terms of health status is mineral and vitamin nutrition. If growing pigs have a 'daily requirement' for at least some of these macro and micro nutrients then, if higher levels of feed intake are reached, there may be a need to alter (reduce) the concentrations of these in both the premix formulation and the final finished feed formulation.

Conclusions

There is obviously a need for much more research in this area and current knowledge is far from complete. There can be speculation on changing requirements and animal responses from available data sets and knowledge but at this point in time a finished feed cannot be re-formulated with precision as health status changes. A priority in this and a basic first step is to establish a working index of health status in order to grade individual herds. This will not only benefit the producer for use in management protocols but also the veterinarian when identifying appropriate actions. This will also make the life of the nutritionist more easy in that a greater level of precision in on-farm formulation will be achieved.

Effective working health status-nutrition simulation models that can utilise health index information and facilitate adjustments in formulation parameters are required. For the moment qualitative changes to feeding programmes on the basis of our current understanding can be made. What is also clear is that once a higher level of health status is achieved, there are very significant cost savings to be had from the accrued growth and feed efficiency improvements.

References

Amory 2001. The effects of the environment on health status and welfare of growing pigs. PhD thesis. Harper Adams University College.

Koutsos, E.A. and Kirk C. Klasing, K.C.,2001. Interactions between the immune system, nutrition, and productivity of animals. In: *Recent Advances in Animal Nutrition. 2001*, pp 173-190. Ed. P.C. Garnsworth and J Wiseman. Nottingham University Press.

Wiseman, J., Varley, M.A. Waters, R, and Knowles, A., 2004. Livestock yields now, and to come: Case Study Pigs. In: *Livestock yields now and to come* , pp xxx, Ed. R. Sylvester-Bradley and J Wiseman, Nottingham University Press.

Williams, N.H. Stahly, T.S. and Zimmerman,D.R. 1997. Effect of chronic immune system activation on the rate, efficiency and composition of growth and lysine needs of pigs fed from 6 to 27 kg. *J. Anim. Sci.* **75**: 2463 - 2471.

2

MANAGING FEED INTAKE OF WEANED PIGLETS: INTERACTIONS BETWEEN NUTRITION, ETHOLOGY AND FARM MANAGEMENT

HUBÈRT VAN HEES[1], MIRJAM VENTE-SPREEUWENBERG[1] AND BERT VAN GILS[2]
[1]Nutreco Swine Research Centre (SRC), St. Anthonis, The Netherlands;
[2]Trouw Nutrition International, Putten, The Netherlands

Introduction

With pig prices within the EU currently at low levels there is only a small margin between making profit and losing money. Consequently, optimising farm management is the big challenge for pig producers. Feed is one of the most important factors with respect to the cost structure and profitability of a farm. The feed industry, in their role as consultants to the farmer, must take a leading role as a partner in business for farmers.

With respect to optimising piglet production, the industry currently has to deal with several challenges. The consolidation of the pig industry in major EU markets is a trend which has been apparent for many years. This results in larger units where hired stockmen and women assist a farm manager. In order to be profitable, these large units require working in accordance with strict standard operating procedures. With this size of operation flexibility is low. Large fluctuations in health status and growth performance between and within batches of piglets will result in a variable pig flow and will complicate farm management.

Furthermore, Figure 2.1 shows the success of new breeding technology. Sow productivity is increasing at a very high pace, especially since the introduction of BLUP-technology (Best Linear Unbiased Prediction) in the early 1990s. It is expected that the genetic potential, which is currently present at nucleus farms, will be expressed at sow farms within 3-5 years (Hypor, personal communication). According to Quiniou, Dagorn and Gaudré (2002) the improvements in genetic potential are not as great for more piglets weaned due to higher peri- and postnatal losses. Larger litter size will increase the number of underweight (< 1 kg life weight; Table 2.1) and supernumerary piglets. Larger litters will result in lengthier farrowing, which will increase the risk for intra-partum anoxia. As a result more weak and listless piglets will be born (Quiniou *et al.*, 2002). In addition, research has shown that later born piglets will receive colostrum with lower IgG content and thus less maternal protection (Le Dividich, Marineau, Thomas, Demay, Renoult, Homo, Boutin, Gaillard, Surel, Bouétard and Massard, 2004). Therefore, geneticists have increased the selection pressure for number of weaned piglets and piglet vitality (Hypor, personal communication).

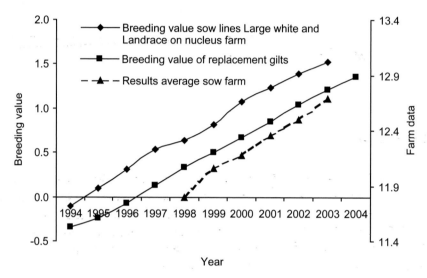

Figure 2.1 Genetic trends in sow productivity. Breeding values are shown on the left y-axis and average farm data on right y-axis. Both are expressed as total piglets born (Source: Hypor BV).

Table 2.1 Effect of litter size (excluding mummies) on piglet characteristics at birth (data from 965 litters at ITP, France: after Quiniou *et al.*, 2002).

Litter size class	≤ 11	12-13	14-15	≥ 16	p-value
Mean litter size	9.0	12.5	14.4	17.0	
Parity number	2.1	2.1	2.2	2.5	0.053
Birth weight (kg)					
- Mean	1.59 [a]	1.48 [b]	1.37 [c]	1.26 [d]	<0.001
- Within litter variation	0.26 [a]	0.27 [ab]	0.28 [b]	0.30 [c]	<0.001
Piglets < 1kg (%)	7 [a]	9 [a]	14 [b]	23 [c]	<0.001

a, b, c, d = significant differences between treatments.

The above-mentioned trends in farming and sow productivity will lead to more variability in piglet quality at weaning, both with respect to weaning weight and immune status (Milligan, Kramer and Fraser, 2002; Quiniou *et al.*, 2002).

Additionally, the ban on animal proteins (such as animal plasma and fish meal in some markets) in feed and the ban on feed antimicrobial growth promoters in the European Union from 1st January 2006 will force nutritionist to re-evaluate their formulations, with fewer nutritional tools available.

In a previous contribution to the Proceedings of this conference (Partridge and Gill, 1993) a list of principal factors assumed to be associated with growth and health after weaning were listed. Since then more insight has been generated into these topics. Although the importance of all these factors will not be disputed, there is currently a consensus, both in science as well as in practice, on the importance of early post-weaning feed intake.

In order to further optimise the post-weaning period, we believe that we have to focus on individual behavioural differences and take into account the interrelationships between the piglet voluntary feed intake (VFI) behaviour, farm conditions and nutrition (Figure 2.2). This chapter is a synthesis of own research data, practical experiences and recent literature.

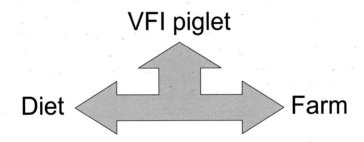

Figure 2.2 Scope of this paper: interaction between factors determining optimal nutrition of the piglet. VFI = voluntary feed intake.

It is outside the scope of this chapter to review in detail the physiological backgrounds of digestion in piglets and the impact of weaning, nor to discuss all aspects of feed intake behaviour in full detail. The reader is referred to recently published books (i.e. The Weaner Pig, edited by M.A. Varley and J. Wiseman, 2001, CABI; Weaning the Pig, edited by J.R. Pluske, J. le Dividich and M. W.A. Verstegen, 2003, Wageningen Academic Publishers) for further study of these topics.

Voluntary feed intake and the process of weaning

FEED INTAKE BEHAVIOUR IN NATURAL WEANING

In most mammals weaning is a process in which the young gradually become less dependent on the dam's milk. This is also observed in the domestic pig when it is kept under natural or semi-natural conditions. In a natural environment the sow will leave her social group to prepare a nest and give birth to her piglets. She will remain close to the nest for several days (Jensen and Recén, 1989). About 10 days after farrowing the sow and her piglets will leave the nest site to join the mother's social group. The piglets will experience big environmental and social changes, but all under the surveillance of their dam (Fraser, Milligan, Pajor, Philips, Taylor and Weary, 1998). In a natural environment the contact between the sow and her offspring decreases gradually. In the first week of their lives the piglets suckle 25-30 times per day. The nursing frequency declines steadily as the sow and the piglets separate themselves for increasing moments of time as the litter ages. The proportion of suckles, in which the litter is incomplete, increases also as the litter ages (Jensen and Récen, 1989). This is the start of the weaning process. Pigs are foragers and partly rely on learning by trial and error through sampling new feed sources. From the fourth week onwards, piglets will gradually decrease their intake of milk and, through exploratory behaviour and mimicking the sow, will include in their diets a small portion of the sow's diet (e.g. roots, seeds, small invertebrates) and even faeces and soil (Fowler and Gill, 1989). When piglets are 6 weeks of age they usually receive less than half of the dry matter intake from milk (English *et al.*, 1988 as cited by Fraser *et al.*, 1998). Moreover, the behaviour of the sow seems to promote this process of becoming independent of the sow's milk (Jensen and Récen, 1989; Bøe, 1991).

Thus, under natural conditions there is a gradual transformation from complete nutritional and behavioural dependence on the dam to independence (Fowler and Gill, 1989). The complete process will take on average 17 weeks (range 5-19 weeks; Bøe, 1991; Jensen and Récen (1989).

FEED INTAKE BEHAVIOUR IN COMMERCIAL PRACTICE

In commercial practice weaning has been turned into an event rather than a gradual process (Brooks and Tsourgiannis, 2003). The problems associated with weaning are mainly a consequence of commercial conditions.

Feed intake behaviour before weaning; some notes on creep feeding

Before weaning energy and nutrient intake is mainly determined by the sow's ability to produce milk. In a review by Pluske, Williams and Aherne (1995) it was shown that, on a litter basis, creep feeding did not contribute significantly to the energy intake before weaning and hence had little effect on litter weaning weights. The problem with evaluating these results is the high between- and within-litter variation with respect to creep feed intake. Furthermore, its potential effect is reduced with decreasing weaning age (Brooks and Tsourginanis, 2003). Indeed, in some markets creep feeding has been discarded by many sow farms (e.g. The Netherlands) because under their commercial practice its beneficial effects were hard to prove (e.g. Hoofs, 1993). However, the work of Bruininx (2002) has given creep feeding a new impulse. He has shown that creep feeding indeed can have its merits by quicker acceptance of dry feed after weaning, i.e. the time to first intake decreases (latency time) which results in a better post-weaning performance. A tendency for higher early VFI of piglets that ate >100 grams per day was also observed earlier by Delumeau and Meunier-Salaün (1995). If a farm decides that creep feeding is part of their standard operating procedures it should either be done in a proper way or not done at all. According to our own data, providing fresh feeds can almost double the intake when compared to feeds that were stale. Furthermore, providing ample feeder space (2 versus 6 and 1 versus 8 feeder places) is important because it stimulates more piglets to commence dry feed consumption as shown by Appelby, Pajor and Fraser (1991, 1992). We have observed that approximately 50% of the piglets ate little or no creep when a single space hopper was used. Furthermore, results of our own experiments showed that continuing creep feeding until 3 days after weaning improves ADFI during this critical period (178 versus 111 g/d; Nutreco SRC, 2003). Similar results were obtained by Bruininx (2002).

Feed intake behaviour after weaning; on group level

Several authors have listed the differences with the natural weaning process as briefly outlined before (Fraser *et al.*, 1998; Brooks and Tsourgiannis, 2003). Table 2.2 lists the changes a piglet faces at the weaning transition, focussing on the energy and nutrient intakes. In short, piglets are weaned at an age when they are still behaviourally, nutritionally and immunologically incompetent.

Indeed the most striking effect of weaning is a critical period of underfeeding during which the pig is acclimatising itself to the dry food (Le Dividich and Herpin, 1994). Energy requirements for maintenance are not met during at least the first three days after weaning (Bark *et al.*, 1986; Le Dividich and Herpin, 1994). To reach pre-weaning growth rates of between 200 and 280 g/d at 3 weeks of age, the piglets would need to consume between 320 and 475 g/d (Fowler and Gill, 1989; Pluske *et al.* 1995). These feed intakes are generally not being attained until the end of the second week following weaning (Figure 2.3). The impact on intestinal health and growth performance will be discussed subsequently.

Table 2.2 Differences in feed intake (VFI) characteristics before and after weaning.

VFI before weaning	ref.[a]	VFI after commercial weaning	ref.
- suckling process covers 15% of time	1	- eating and drinking accounts for only 1-3% of time	2
- high frequency of suckling (16-20 times per 24 h.)	3, 4, 5	- relatively low frequency of eating (8-9 times per 24 h.)	6
- small meals (suckles take 10-20 seconds; 40-50 ml/suckle)	3, 7	- occasionally high intakes	2, 8
- intake is synchronized by the sow's behaviour (grunting)	3	- in *ad libitum* systems there are initially no synchro-nizing cues.	
- little competition for nutrients	3	- depending on the feeding system competition occurs. Establishment of a social rank causes aggression at weaning	9, 10
- milk = solids + water		- unfamiliarity with method of food acquisition. Separate feelings of hunger and thirst and relate them to the feeder and water nipple or bowl.	11, 8
- high nutrient intakes		- first 14 days low intakes	12

[a] references: 1 Schouten, 1986; 2 Nutreco SRC, 2003; 3 Brooks and Burke,1998; 4 Bøe, 1991; 5 Jensen and Récen,1989.; 6 Bruininx, 2002; 7 Pluske and Dong, 1998; 8 Brooks and Tsourgiannis, 2003; 9 Held and Mendl, 2001; 10 Pluske and Williams, 1996; 11 Bark, Crenshaw and Leibbrandt, 1986; 12 Le Dividich and Herpin, 1994;

Feed intake behaviour after weaning; individual characteristics

Until now we have discussed development of VFI on a group level and on an average daily basis (ADFI). However, to understand more about the relation between VFI, diarrhoea and performance we have to go to the individual animal and look at the VFI pattern within a day.

Pigs prefer to eat during daytime (they are diurnal animals; Dalby, 1998). Moreover, the FI pattern of a mature pig follows a distinct circadian pattern,

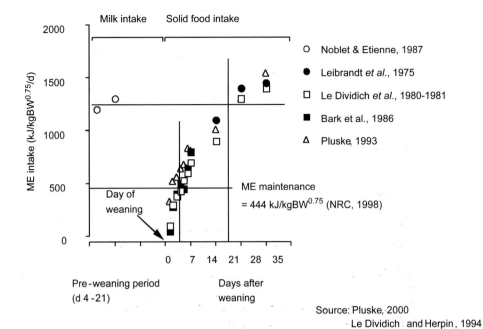

Figure 2.3 The voluntary feed intake of pigs before and after weaning. Lines cross at the time point when energy intake for maintenance and pre-weaning intake level is reached.

as shown by many authors (De Haer, 1992; Ramaekers, 1996; Augspurger, Ellis, Hamilton, Wolter, Beverly and Wilson, 2002). Most studies report two daylight-feeding peaks; one near the beginning and one near the end of the lighting period (reviewed by Dalby, 1996). This distinct pattern is often seen in animals living in groups and is for a large part socially facilitated. In farm practice other synchronizing cues are known (e.g. the sound of feeding equipment, the presence of the stockman). A weak circadian pattern is also seen in newly weaned piglets, but it becomes more distinct with increasing time post-weaning (Bruininx, 2002; Smith, Beaulieu, Patience, Gonyou and Boyd, 2004).

Electronic feeding stations have facilitated the study of VFI in more detail. Table 2.3 summarizes some feed intake characteristics of piglets in the immediate post-weaning period and of older animals. The general trend is that with increasing age pigs increase their rate of FI (RFI) and have larger intakes per visit. This was also concluded by Augspurger *et al.* (2002). Furthermore, pigs appear capable of adapting their feeding behaviour to maintain an adequate FI in situations where competition is limiting their optimal feeding pattern. This interaction will be discussed later.

Table 2.3 Feed intake (FI) characteristics of piglets and growing pigs.

	Piglets		Growing/fattening pigs [3]		Fattening pigs [4]
	day 0-8 [1]	*day 0-34* [2]	*Group*	*Individual*	
Visits to feeder					
- total	48.6	31.8	14.4	58.6	17
- with FI	7.5	12.2	nd [5]	nd	15
Per visit with FI					
Time per visit (min:s)	4:13	5:31	4:42	1:30	6:18
FI per visit (g)	19	44	159	38	294
Rate of FI (g/min)	4.5	8.0	32.0	27.2	47.0
Feeding time per day (min:s)	31:38	67:18	63:30	84:06	65:00

[1]Bruininx, 2002; 0-8 days post-weaning (at 4 weeks). 10-11 piglets/pen. Values are weighed averages of 3 trials (total visits to the feeder was an weighed average of 2 trials)

[2]Bruininx, 2002; 0-34 days post-weaning (at 4 weeks). Values are weighted averages of 3 trials (total visits to the feeder was an weighed average of 2 trials)

[3]De Haer, 1992; Chapter 2. *Ad libitum* feeding of 8 pigs/pen, 30-100 kg BW.

[4]Ramaekers, 1996. 12 pigs/pen, 56 -110 kg BW.

[5]nd = not determined

When measuring the course of ADFI on group level one would observe a gradual increase, with typically a small plateau on days 3-5 and than a further increase. However, data obtained with electronic feeding stations show that between individuals a large variation exists. For example, Bruininx, Van der Peet-Schwering, Schrama, Vereijken, Vesseur, Everts, Den Hartog and Beynen (2001) found that within 4 hours after weaning, approximately 50% of the pigs started eating, whereas it took about 50 hours before 95% of all pigs had started eating. Also subsequent patterns are often very irregular, for example with peak intakes to 300 g/day followed by intakes below maintenance level (e.g. Brooks and Tsourgiannis, 2003). In individual housing Vente-Spreeuwenberg, Verdonk, Bakker, and Verstegen (2004) reported that VFI on any day during the first four days post-weaning had very little correlation with VFI of the subsequent or previous day. From day four onwards these correlations became stronger.

Each piglet seems to develop its own feed intake pattern dependent on individual characteristics (e.g. social rank) and external factors (e.g. group dynamics, competition for feed or feeding times). Table 2.4 summarizes some individual characteristics and its effect on VFI during the first few days post-weaning. Creep feeding has been discussed before, but the other characteristics will be discussed in more detail below. However, one should realize that some are interrelated, e.g. a higher weight at weaning (BW) can be related with teat order, sex, creep feed intake (gilts ate more creep compared to castrates, Delumeau and Meunier-Salaün, 1995) and coping style (Van Erp-Van der Kooij, Kuijpers, Schrama and Ekkel, 2000).

Table 2.4 Consequences of individual characteristics on early (first few days) post-weaning feed intake behaviour and overall impression on the impact of this characteristic.

Individual characteristic	Consequence	Overall impression [1]
Gilt versus boars/castrates	ADFI 0 / ↑	0
Higher body weight at weaning	Latency time ↑ ADFI ↓	+
Genetics	Little piglet data	0 / + (potentially high)
Coping style	Latency time 0 ADFI 0	0
Suckling a front teat /dominant pig	Latency time ↑ ADFI ↓	0
Creep feed eater	Latency time ↓ ADFI ↑	
	Numbr. meals ↑	++

[1] 0 = little impact. + moderate impact, ++ = high impact

There seems to be little difference between sexes (catrates and gilts; boars and gilts). Bruininx *et al.* (2001) used electronic feeding stations to monitor individual VFI of group housed weanling pigs. Gilts had only a numerically higher ADFI during the immediate post-weaning period. This is in agreement with Delumeau and Meunier-Salaün (1995). No differences were detected with respect to other FI characteristics (e.g. number of visits, time per visit, FI per visit). Also in more mature animals there are little or no differences reported between sexes for feeding patterns (De Haer and De Vries, 1993; Gonyou, Chapple and Frank, 1992; Augspurger *et al.*, 2002).

In the same study of Bruininx *et al.* (2001), heavier piglets at weaning (mean 9.3 kg) appeared to have a delayed onset (increased latency time) of eating compared to lighter piglets (mean 6.7 kg). Lighter piglets had more visits to the feeder than heavy piglets (34 and 27 respectively) but the number of visits with FI and time per visit were comparable (approx. 12 per 24 hours and 6 minutes, respectively). Additionally, lighter piglets tended to eat more during the first 3-4 days post-weaning but heavy piglets ate more thereafter. Heavy piglets are assumed to be more dominant piglets and presumably more involved in aggressive behaviour when establishing a social hierarchy (Bruininx, 2002). There are also indications that they occupy the more productive front teats and are less prone to eat creep feed which makes them less accustomed to dry feed intake compared to lighter weaned piglets (Algers, Jensen and Steinwall, 1990; reviewed by Held and Mendl, 2001). In addition, heavy piglets have a higher rate of feed intake compared to light piglets (7.8 and 6.9 g/min, respectively; calculated from data from Bruininx *et al.*, 2001). Although heavier pigs are at risk immediately post-weaning, they outperform lighter pigs when evaluated at the end of the rearing period or even at slaughter (reviewed by King and Pluske, 2003).

In growing-fattening pigs, De Haer and De Vries (1993) reported differences in feeding pattern between Dutch Landrace and Yorkshire. Hyun, Wolter and Ellis (2001) reported substantial genetic variation in feed intake pattern between pig genotypes (Meishan, York or Yorkshire*Meishan cross). The same group reported a higher feed consumption rate and larger meals for a synthetic line compared to pigs with a Pietrain ancestry (207 versus 172 g and 37 versus 29 g/min respectively; Augspurger *et al.*, 2002). Bruininx (2002) did not detect genetic differences, but this might be related to the fact that his piglets had a similar genetic background from a rotational breeding scheme. Thus, according to the literature, genetic effects on FI patterns might be of importance. Indeed, number of meals has a high heritability (h^2 of 0.45; De Haer and De Vries, 1993) and is an animal specific trait. Moreover, there is circumstantial evidence that piglets from genetic crosses of more vigorous eaters have higher initial VFI at weaning.

It was hypothesised that the way individual piglets cope with stressful situations, e.g. in a 'backtest' (Hessing, Hagelsø, van Beek, Wiepkema, Schouten and Krukow, 1993; Van Erp-Van der Kooij, Kuijpers, Eerdenburg and Tielen, 2001), could be related to its VFI behaviour after weaning. However, in a recent experiment we were not able to demonstrate this relationship (Nutreco SRC, 2003), neither did we find a correlation between teat order, post-weaning behaviour and growth performance, which is in agreement with results obtained by Delumeau and Meunier-Salaün (1995). There were however, some behavioural parameters that correlated with body weight gain during the first week post-weaning as shown in Table 2.5.

These data indicated that well-growing piglets are those that frequently visit the feeder, have a long total duration of meals and have a high frequency of meals.

Relationship between voluntary feed intake, incidence of diarrhoea and performance

INCIDENCE OF POST-WEANING DIARRHOEA

During the first two weeks post-weaning, piglets frequently develop diarrhoea.

Table 2.5 Mean and standard deviations (Std) of voluntary feed intake characteristics during the first 36 hours after weaning and their Pearsons correlation coefficient (R) with growth during the first seven days post-weaning. Piglets were weaned at 27 days, 4 pens with 7 both barrows and gilts were observed (Nutreco SRC, 2003).

Variable	Average value	Std	R	p-value
Time until 1st feeder visit (min)	78	73	-0.34	0.075
Duration 1st visit (sec)	25	30	0.44	0.019
Number of feeder visits	36	16	0.61	<0.001
Total duration of visits (min/36 hrs)	27.5	18	0.68	<0.001
Time until 1st meal (min) [1]	304	440	-0.21	0.280
Number of meals	24.5	13	0.67	<0.001
Total duration of meals (min/36 hrs)	26.5	18	0.70	<0.001
Number of visits to waterer nipple	29.4	9	0.38	0.045

[1] a meal was defined as a visit lasting >15 seconds

Reported incidences of diarrhoea range from 32 to 55% (Ball and Aherne, 1982; Hampson, 1986; Nabuurs, 1991). Madec, Bridoux, Bounaix and Justin (1998) looked at digestive disorders in weaned piglets in 106 farms describing 616 pens. Seventy six percent of the pens showed diarrhoea, with 20% of the pens showing diarrhoea for the duration of 10 days or more (median = 3.6 days).

THE RELATION BETWEEN FEED INTAKE AND DIGESTIVE DISORDERS

FI pattern, performance and health can be related through different pathways. First of all there may be a relation through digestibility. Secondly, it appears that FI pattern can affect protein utilization and tissue accretion, as shown in growing and fattening pigs (De Haer, 1992; Ramaekers, 1996). The former will be discussed below in more detail

Makkink (1993) proposed a theory in which two VFI patterns are distinguished after weaning. In the first case, a piglet almost immediately starts to consume its diet in sufficient amounts (i.e. well above its energy requirements for maintenance). This piglet subsequently increases its feed intake and adapts its digestive system gradually to the new feeding routine and feed composition (Makkink, 1993). In the second case, piglets refrain from consuming sufficient amounts of feed. This occurs frequently as described in the previous section. The consequence of this is that the digestive and absorptive capacity is decreased and the resistance to gastrointestinal diseases impaired (as reviewed by Vente-Spreeuwenberg and Beynen, 2003). Recent data demonstrate that, within 24 hours, jejunal mucosal atrophy occurs and that the process of villus atrophy is completed within 72 h after the withdrawal of food (Niinikoski Stoll, Guan, Kansagra, Lambert, Stephens, Hartmann, Holst and Burrin, 2004).

In agreement with this, Madec *et al.* (1998) showed in a meta analysis that low feed intake immediately after weaning is a risk factor for the occurrence of digestive disorders. Piglets eating less than 140 g / day during the first week were more likely to develop digestive disorders (odds ratio 33.6). On the contrary piglets eating more than 190 g per day were not likely to develop diarrhoea (odds ratio 1.1) (Madec *et al.*, 1998). Vente-Spreeuwenberg, Verdonk, Beynen and Verstegen (2003) showed that when piglets showed numerous inconsistent faeces – i.e. 72 % of the piglets showed a faeces score of at least 2 on a scale from 0 being normal faeces to 3 being diarrhoea - there is a positive correlation (R^2 = 46%) between the average feed intake for days 3 to 7 and the villus length at day 7. Additionally, the number of days with inconsistent faeces (score $\geq$ 1) was negatively correlated (R^2 = 25%) with the villus length (Vente-Spreeuwenberg *et al.*, 2003).

A period of low energy and nutrient intake is often compensated for by a period with high intakes, at least when *ad libitum* feeding is applied (Kamphues, 1987). These high feed intakes, as observed in some individuals, will result in lower total tract digestibility as suggested by Seve (1979) and observed by Kamphues (1987) and increases the risk of for diarrhoea. The latter researcher showed that the effect of the level of VFI on digestibility was more evident in younger pigs compared to older pigs as shown in Table 2.6. In this table, the large reduction in crude fibre digestibility may indicate a reduced fermentative capacity or an increased rate of passage in the hindgut. Moreover, the same research group found decreased ileal digestibilities of starch (-15 to –18%) and protein (-13%) in overfed piglets when compared to piglets that were not overfed. As a result, an increasing proportion of feed will escape digestion and absorption in the small intestine and will enter into the lower gut. Consequently, the risks for unwanted fermentation and eventually fermentative or pathogenic diarrhoea will increase.

Furthermore, Kamphues (1987) studied the effects of fasting and re-feeding on the digestive process in the so-called Karenz model. In this model piglets are allowed to recover from weaning first. On day 4-7 feed is withdrawn and they are fasted for 18-24 hours. Subsequently, they are allowed *ad libitum* access to feed again. During the first six hours of re-feeding FI will be very high, i.e. typically more than 17 g/kg BW (Figure 2.4). The digestive consequences of this peaked FI are described in detail by Kamphues (1987).

Table 2.6 **Coefficient of apparent faecal digestibilities in relation to age and dry matter intake level (comparing *ad libitum* feeding (ad lib) with restrictive feeding (restrict); after Kamphues, 1984, cited by Kamphues, 1987).**

Age		*6 weeks*			*18 weeks*	
	Restrict	*Ad lib*	*Difference*	*Restrict*	*Ad lib*	*Difference*
DM intake[1]	24.4	52.5	+28.1	334	590	+25.6
Organic matter (OM)	0.880	0.851	-0.029	0.905	-0.892	-0.013
Crude protein	0.841	0.791	-0.050	0.868	0.860	-0.008
Crude fat	0.715	0.668	-0.047	0.855	0.852	-0.003
Residual OM	0.926	0.910	-0.016	0.943	0.936	-0.007
Crude fibre	0.446	0.341	-0.105	0.536	0.444	-0.092

[1] Dry matter intake in g/kg body weight

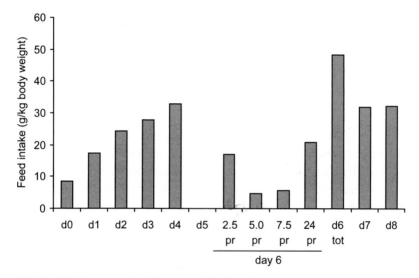

Figure 2.4 Feed intake pattern in a Karenz- model. d0 = day of weaning. pr = period post re-feeding, e.g. 5.0 pr = between 2.5-5.0 hours. Data: University of Hohenheim and Nutreco SRC, 1996.

In short, the peaked FI first affects stomach function. Especially in the young pig, this organ will not be able to properly acidify and mix large quantities of feed. Furthermore, its function as a storage compartment and its capacity to regulate the outflow to the duodenum is exceeded. Indeed, it was shown that in a fasted animal the greatest portion of the stomach contents is emptied by 2h post feeding (Casteel, Brown, Lattimer and Dunsmore, 1998). Conversely, pigs with many meals of limited size have a more continues flow of digesta through the intestine (Ruckebush and Bueno, 1976). As suggested by De

Haer and De Vries (1993) this even intestinal transit will allow for better digestion by increasing the time for contact between food and enzymes and between digestion products and absorptive surface (see also Rerat and Corring, 1991). Another explanation of an increased digestibility could be that, with a more optimal FI pattern pancreatic secretions are stimulated (Hee, Sauer and Mosenthin, 1988).

Thus, the malfunctioning of the stomach will reflect on the processes in the small but also in the large intestine. In the latter, a lower net water resorption will ensue (Kamphues, 1987). Clinically, transient diarrhoea is observed which typically starts between 8-12 hours after re-feeding. The faeces are then characterized by a reduced DM% (<200g/kg), a low pH (<6.2) and increased lactate levels (1-5 g/kg faeces). The colour of the faeces changes to 'feed-like' and a sour smell is observed. This data indicates that, besides VFI on a daily basis, the intake pattern within a day is also related to intestinal health and performance.

NUTRITIONAL IMPLICATIONS

From what has been described above it can be concluded that a sub-optimal FI pattern in piglets will aggravate the effect of the impaired digestive capacity caused by the initial period of under feeding after weaning. The question which arises is what options there are in diet formulation to counteract overfeeding? This question becomes even more relevant in markets where pharmacological levels of zinc and the preventive use of (high levels of) antibiotics are not allowed.

Diet composition

One option we propose is to reduce the rate of feed consumption by including fibrous ingredients. Although not always significant, from literature data and in-house research, it can be concluded that increasing the dietary fibre level in general decreases ADFI. However, to the authors' knowledge, there are no literature sources addressing the relation between fibrous ingredients and VFI patterns. It can by hypothesised that fibre can increase the chewing action necessary to ingest the feed pellet or meal and consequently slows down the rate of feed consumption. In piglets the Karenz-model was used to investigate the effect of fibre sources on overfeeding. Piglets were weaned at 4 weeks of age (range 6.0-11.6 kg BW) and fed one of three diets *ad libitum* before and after day five, the day of feed withdrawal. The control diet was based on wheat, barley, soya bean meal and native cornstarch with a vitamin and mineral premix. In the two test diets either sugarbeet pulp or wheat bran replaced the cereals. During the pre-fasting period, ADFI of the control animals tended to be higher compared to the test diets (27.7 versus 23.9 g/kg BW; p<0.10). On day five piglets were fasted for 24 hours. After re-feeding VFI showed a similar pattern with typically high initial intakes. Faeces samples, taken directly after re-feeding, 8, 24 and 48 hours after re-feeding, showed a lower pH drop, lower lactate and higher dry matter levels in the piglets fed the diet containing wheat bran. Incidence of diarrhoea was decreased in these animals as shown by Table 2.7.

Table 2.7 Effect of fibre sources on the incidence of diarrhoea (% of animals) in overfed piglets. Data: University of Hohenheim and Nutreco SRC, 1996.

	Control	*Sugar beet pulp*	*Wheat bran*
Crude fibre content (g/kg)	31	51	48
Normal	0	0	0
8 hrs after re-feeding	58[a]	35[ab]	16[b]
24 hrs after re-feeding	26[a]	25[a]	0[b]
48 hrs after re-feeding	21	25	5

[a,b] = significant differences between treatments, p<0.05.

In conclusion, the data indicate that some fibrous raw materials, i.e. wheat bran, can ameliorate the effect of overfeeding. To what extent a changed RFI played a role in this protective effect needs to be elucidated. In-house data in growing pigs indicate that fibre indeed influences RFI. In gilts, feed intake and average daily gain were decreased (P<0.05) with the high fibre diets (69 g versus 31 g crude fibre originating from sugar beet pulp, citrus pulp and soybean hulls/kg). These animals responded to high fibre with a decreased RFI (P<0.05) and smaller meals (P<0.06) while the number of meals were not affected. In castrates, feeding a high or a low fibrous diet did not significantly affect the feed intake characteristics, but values pointed in the same direction (Nutreco SRC, 2002).

These data suggest that increasing the fibre level of the diet can decrease the risk of overfeeding. Conversely, lowering the fibre content could be a nutritional tool to ameliorate the effect of limited access to feed. This is discussed subsequently.

Another option in diet formulation could be the selection of protein and carbohydrate sources which are easily digestible. It is our contention that, in relation to what has been outlined before, this means not only a high end-point digestibility, but also rate of digestion should be taken into account. For example, Savoie (1992) has shown, both during *in vitro* as well as in *in vivo* studies, that protein sources can differ in this respect. Probably the feasibility of using expensive milk protein sources in piglet diets is also based on the fact that dairy protein has a high rate of digestion, and is therefore almost completely digested even in piglets with a sub-optimal feeding pattern. For starch sources a similar approach could be taken (i.e. Fledderus, Bikker and Weurding, 2003). Although starch digestion seems to be almost complete within 2 weeks post-weaning, selecting starch sources may be more critical in the immediate post-weaning period under poor conditions.

In addition, enzyme supplementation (ß-glucanase and a-amylase) to a wheat-barley-soya based diet decreased the incidence of diarrhoea in overfed piglets as was reported by Hogenkamp, Hayler, Drochner and Mosenthin (1997).

The question remains as to how relevant over-feeding is in the practical situation. In order to test the hypothesis of over-eating, piglets were selected that had an early ($\leq$ 2 h.) or late ($\geq$ 23 h.) start of feed intake combined with a relatively fast ($\geq$ 18.0 g/kgBW $^{0.75}$) or slow ($\leq$ 2 g/kgBW $^{0.75}$) daily increase in feed intake. Growth performance and small intestinal morphology were measured. However, results indicated that the time between weaning and first feed intake was only very weakly related to overall performance and intestinal integrity. In addition, branched chain volatile fatty acids and ammonia-N (i.e. indicators of protein fermentation) were only numerically higher in late starters with a fast increase in FI (Bruininx, Schellingerhout, Lensen, Van der Peet-Schwering, Schrama, Everts, den Hartog and Beynen, 2002). Diarrhoea incidence could not be determined on an individual level because of the experimental set-up. Seemingly, these data do not substantiate the relevance of over-eating. However, there are many farm factors that determine the occurrence of diarrhoea. Besides the situation of over-feeding, the presence of pathogenic bacteria and the colonisation resistance of the intestinal microflora determines whether pathogens will get the opportunity to multiply and cause diarrhoea. One could hypothesize that, under conditions of low hygiene, over-eating indeed results in a higher incidence of diarrhoea. Sub-optimal climate conditions might even aggravate this process. Some examples of diet by sanitary conditions interactions will be dealt with below.

Feed form

Besides adjusting diet composition, VFI characteristics are affected by feed form. When comparing the rate of feed intake between piglets fed pellets or the same diet as an agglomerated meal (by steam treatment) the time spent eating increased by almost 50% while total FI did not differ. The meal diet appeared to be ingested at a slower rate (Table 2.8). Furthermore, piglets ate more meals while average meal size decreased. These results are in agreement with Laitat, Vandenheede, Désiron, Canart, and Nicks (1999) who investigated the effect of group size and feed form in weaned piglets and observed a higher rate of feed intake with pellets compared to meal. Pellets are thus eaten much quicker then meal resulting in less time at the feeder. Moreover with increasing group size, feeder space became more limited and piglets were forced to shift their FI to the night. This effect was more pronounced with meal.

Table 2.8 Effect of feed form on intake behaviour of piglets between 35-60 days of age (weaning at 21 days; Nutreco SRC, 1996).

	Pellet	*Agglomerated meal*	*Meal vs. pellets*
Rate of intake (g/min)	16.4	11.0	- 33%
Average meal size (g)	24.6	15.7	- 36%

There are two possible explanations for the higher rate of FI in pellets. Firstly, piglets seem to prefer pellets over meal (Laitat, Vandenheede, Désiron, Canart and Nicks, 2000). In addition, in the older work of Seerley, Miller and Hoefer (1962) a higher average intestinal retention time with a meal diet compared to a pelleted diet was estimated, thus indicating an association with passage rate through the intestinal tract.

The data imply that the use of a pelleted diet, through its effect on feed consumption rate and intestinal transit time, can be a risk factor for intestinal disorders. This was demonstrated by Guerrero, López, Brito and García (1991) who found a higher mortality rate in small-for-age piglets with pellet diets compared to meal. Conversely, in general, pelleted and also crumbled diets will improve growth performance, even though incidence of diarrhoea was slightly increased (Table 2.9).

Table 2.9 The effect of feed form on growth performance, diarrhoea incidence and water intake of group housed weaned piglets (0-28 days post-weaning; Nutreco SRC, 1995).

	Meal	*Pellet*	*Crumb*	*p-level*
N	16	16	16	
Average daily feed intake (g/d)	638	644	620	ns
Average daily gain (g/d)	365	398	394	<0.05
Incidence of diarrhoea (%)	8.9	13.2	13.4	<0.05

Farm factors in relation to feed intake behaviour and diet composition

INTERACTION BETWEEN FARM HYGIENE STATUS AND THE DIET

The immune system is often activated under poor hygiene conditions. The immune response to infection, inflammation or trauma is mediated by a combination of cytokines and is associated with increased concentrations of plasma proteins produced by the liver (Gruys, Toussaint, Landman, Tivapasi, Chamanza and Van Veen, 1999). Those cytokines have direct and indirect effects on growth performance and metabolism, including a decreased feed intake (Johnson, 1997). At our research facilities we have clearly found that the effect of diet composition and additives is dependent on the hygiene status of the piglets. At our research facility we found that under normal conditions, 40 ppm of the antimicrobial growth promoter avilamycin showed an improvement in growth performance between –2 and 13% from weaning to 4 weeks post-weaning (Nutreco SRC, 1999a; 1999b). However after an oral infection with pathogenic E. coli K88, avilamycin addition led to an improvement of growth performance of 30% (Nutreco SRC, 1999). In general, the effect of alternatives for antimicrobial growth promoters is more pronounced in malnourished/ diseased piglets when compared to apparently healthy weanling pigs.

In order to investigate the relationship between dietary factors and sanitary conditions we developed two models. In the first model the piglets are orally infected with pathogenic E. coli: the COMO-model (coli model. Meijer, Van Gils, Bokhout, De Geus, Harmsen, Nabuurs and Van Zijderveld., 1997). The second model mainly affects management factors to create sub-optimal conditions: the MAMO-model (management model). In the MAMO-model, the rooms are not cleaned before the piglets enter, the temperature is lowered by 2 degrees, the rate of ventilation is increased, the amount of dust is increased and manure of sows is put in each pen. Figure 2.5 shows the relative effect of MAMO compared to clean conditions. Results show that the effect of MAMO conditions are greatest during the first 2 weeks post-weaning and decrease thereafter. During the first 2 weeks post-weaning, the feed intake is decreased by 23 %. The effect of MAMO conditions is largest on the incidence of diarrhoea. The same MAMO-model is also used in growing pigs at the SRC. Results show that, the impact of low sanitary conditions on growth performance depends on the health status of the animals at the start of the experiment. Healthy, well growing pigs at the starter phase are hardly affected by sanitary conditions (Nutreco SRC 1997).

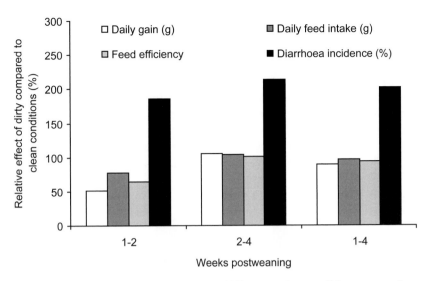

Figure 2.5 The relative effect of sub-optimal (MAMO) versus clean conditions on growth performance of weaned piglets in different periods after weaning (n=24; Nutreco SRC, 1999).

At weaning protein digestion is impaired. Makkink (1993) found an increased ratio of precipitable protein to total crude protein in the jejunal digesta on 3 and 6 days after weaning, indicating that protein digestion was impaired. Caine (1997) showed that the apparent ileal protein digestibility was low 7 days after weaning. Furthermore, the production of aminopeptidase in the brush border is decreased after weaning (Vente-Spreeuwenberg, Verdonk, Koninkx, Beynen and Verstegen, 2004). In addition, we hypothesised that under conditions of over-feeding, time for proper protein digestion seems to be limited and more protein might escape digestion and enter the hindgut. Abundant presence of proteins or non-absorbed amino acids in the ileum and large intestine may result in a proliferation of proteolytic flora, producing a large amount of breakdown products such as biogenic amines, ammonia, phenols and indoles. Some catabolites are potentially toxic and thereby affect growth or differentiation of intestinal epithelial cells. Furthermore the breakdown products are also associated with an increased incidence of diarrhoea (Gaskins, 2000).

Under high sanitary conditions, increased protein levels in general support body weight gain. However under poor conditions, high protein levels may lead to an increased incidence of diarrhoea, especially when no antibiotics are used. Therefore, it was investigated whether the addition or not of an alternative for an anti-microbial growth promoter (AMGP), based on medium chain fatty acids (Bactoil), at two different protein levels in the diet affected the growth performance and the incidence of diarrhoea in piglets infected with pathogenic E. coli (COMO-model). The incidence of diarrhoea was higher for piglets fed the high protein level (Figure 2.6). Furthermore, the effect of the alternative for the AMGP was most pronounced at the high protein level, when the piglets were most susceptible to diarrhoea (Nutreco SRC, 2004). It also indicates that it depends on the whole feed matrix whether dietary supplements are effective in preventing diarrhoea.

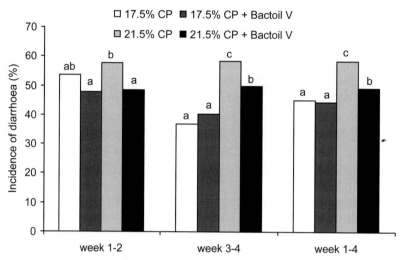

Figure 2.6 The effect of protein level (CP: 17.5 vs. 21.5%) and the addition or not of an alternative for an anti-microbial growth promoters based on medium chain fatty acids (Bactoil V) on the incidence of diarrhoea (%) in E. coli infected piglets (Nutreco SRC, 2003).

INTERACTION BETWEEN STOCKING DENSITY, FEEDER SPACE ALLOWANCE AND VOLUNTARY FEED INTAKE IN GROUP HOUSED PIGLETS

Group dynamics

VFI of pigs will adapt to social interaction between animals in a group. Two kinds of interactions can be distinguished, i.e. social facilitation and agonistic behaviour. The first stimulates FI and causes the peaked distributions of FI during the day (see section 2.2.3). Agonistic behaviour and competition can result in less daily eating time, less frequent visits and reduced ADFI for subordinate animals. With respect to early post-weaning VFI, mixing piglets of different body weights seems favourable. Partly, this is due to the lower rate and lower vigour of agonistic behaviours in mixed-weight groups (reviewed by Held and Mendl, 2001). However, there is a risk that large differences in body weight may cause heavy dominant pigs to guard the feeder and prevent others from eating. This risk can be reduced when ample feeder space is provided.

The VFI of gilts seems to be more influenced by the group in which they are living. Bruininx *et al.* (2001) reported a higher FI during the second half of the rearing period in gilts in mixed body-weight groups compared to homogeneous body weight groups. In fattening pigs a higher FI is observed when gilts are housed together with barrows and fed on a long trough (Van der Peet-Schwering and Binnendijk, 1994). In older pigs, Gonyou *et al.* (1992) showed that group-housed pigs ate less than individually penned pigs. Additionally, the reduction in feed intake of group- versus individual-housed pigs was larger in gilts (-9.3%) than in barrows (-3.1%). In this trial a double space dry feeder was used.

It is important to note that in experiments the effect of group size, pen space (m^2/pig), feeder space allowance and feeder adjustment are often confounded (e.g. Laitat *et al.*, 1999; Smith *et al.*, 2004). Furthermore, it seems that these aspects become relevant in the second halve of the rearing period with the increase in body dimensions (mainly the head; e.g. Smith, *et al.*, 2004).

Group size

In general, literature data indicate that an increased group size is related to reduced growth rates in weanling pigs (e.g. Wolter, Ellis, Curtis, Parr and Webel, 2000). However, the effect might be small, i.e. ADFI was reduced by 0.51 grams per piglet extra, as reviewed by Madec, Le Dividich, Pluske and Verstegen (2003). The same group reported no effect on within-group variation. Also, Smith *et al.* (2004) concluded that space allowance has more impact on nursery pig performance than group size per se.

Feeder space requirements and feeder adjustment

The assumption that piglets should be given the opportunity to eat simultaneously, as was the case when they were still suckling, cannot be substantiated with available literature data. For example, although a single space wet/dry feeder does not perform as well, no differences in growth performance was detected between a linear trough and a single space dry-feeder (Pluske and Williams, 1996). Also Beattie, Weatherup and Kilpatrick (1999) reported no effect on performance with increased feeder space allowance. In contrast, feeder space affected FI and ADG (Figure 2.7) in a recent experiment at our facilities. However, the effect was only significant during the early-post-weaning period. In this trial piglets of mixed sex were grouped according to body weight. Coefficient of variation for body weight increased from 3.5 % at weaning until 11.9 % at 28 d post-weaning and did not differ between feeder space. The apparent disagreement between studies might be related to other factors influencing feeder space requirements, like group composition, breed, crowding and maybe also sanitary conditions.

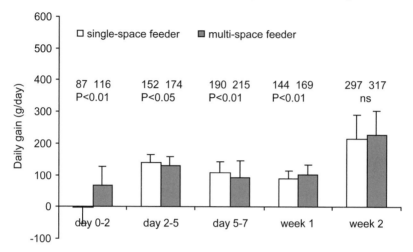

Figure 2.7 Effect of feeder space allowance on average daily weight gain (depicted by bars: g/d: ± std) and average feed intake (separate values). 10 piglets per pen; multi-space = 8 feeding places. a, b = p<0.05 (Nutreco SRC, 2004).

Although feeding activity in groups is usually synchronised, observation in studies with growing pigs (Gonyou *et al.*, 1992) and weaned piglets (Spicer and Aherne, 1987) point to the fact that animals do not usually eat simultaneously from the same feeder, despite the fact that feeder dimensions would seem to allow them to do so. In addition, it was shown earlier that breeds can differ in FI characteristics. This can be of importance when designing a pen and feeding system. For example, in fattening pigs Augspurger *et al.* (2002) reported a difference of 14 minutes feeder occupation time per animal between a synthetic line and a pig of mainly Pietrain genetics. These data indicate that social interactions and breed can also aggravate a lack of feeder space allowance.

In a recent experiment in piglets, reported by Smith *et al.* (2004), the effect of feeder adjustment and group size-space allowance was studied. In this study the piglet to feeder space ratio was fixed at 4. There was a significant interaction between feeder adjustment and group size. Too tight settings resulted in a decreased growth performance (ADG -10%), mainly in the last part of the rearing period. The effect of tighter feeder settings was more pronounced with increased crowding (0.35 versus 0.23 m²/pig). Also more body injuries were observed with increased crowding and smaller feeder gap opening but were not associated with individual growth performance. Not space allowance but tighter feeder gap opening was associated with higher within group variation at 42 d post-weaning (Smith *et al.*, 2004).

Adaptive responses of pigs

Pigs are able to adapt their FI behaviour if circumstances force them to do so, even overriding circadian rhythms (Dalby, 1998). Circadian feed intake pattern changes depending on the ease with which feed can be accessed. In a trial with group housed piglets fed *ad libitum* feeder visits showed typical peaks in the morning and the afternoon when a long trough was used. With a single space wet/dry feeder, feeder occupation time was more spread over the 24 hours of the day (Figure 2.8).

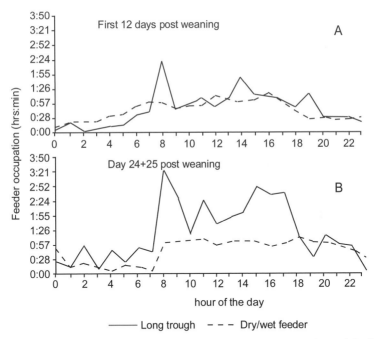

Figure 2.8 Daily feed intake patterns of piglets in relation to time post-weaning and feeding system. (Nutreco SRC, 1996).

In older pigs, work by Ramaekers (1996) showed that FI pattern also varies with the feeding system. With restricted feeding FI pattern shifted towards the early hours of the day coinciding with the start of a new feeding cycle. The circadian pattern of FI in this study was more pronounced than in the study of De Haer (1992), probably because of a higher number of animals per feeder (12 versus 8) as was also observed by Hyun and Ellis (2001). Furthermore, in hot climate conditions FI can shift to the cooler nightly hours (Ramaekers, 1996). It is not clear whether these changes are relevant for welfare or nutrient utilisation.

It can be hypothesised that, if crowding occurs and access to the feeder becomes more limiting, RFI becomes more critical. Pigs who eat more rapidly need to spend less time at the feeder. The work of Hyun and Ellis (2001, 2002) showed that finisher pigs were able to maintain growth rate and ADFI by having fewer but longer visits to the feeder, eating larger meals and also by increasing their RFI. In contrast, this adaptive capacity was not observed in younger growing pigs (Table 2.10). There are to the authors' knowledge no data available in piglets. However, it may be reasonable to expect that piglets will also have only a limited ability to increase their RFI in case of limited access to the feeder. Previously we discussed that feed form and dietary fibre content can facilitate improved rate of feed intake.

Table 2.10 Effect of increased number of animals per pen and feeder on feed intake characteristics as affected by age (after Hyun and Ellis, 2001, 2002).

	Group size (pigs/pen-feeder)				
Trait	2	4	8	12	SEM
Growing pigs, 26-50 kg					
Number of visits	21	22	18	19	0.52
Time per visit (min:sec)	6:20	6:11	6:43	5:53	0.166
FI per visit (g)	99	87	102	95	2.67
Rate of FI (g/min)	16	15	16	17	0.33
Total feeder time (min/day)	123	128	112	99	2.11
Finishing pigs, 80-115 kg					
Number of visits	16	12	9	10	0.43
Time per visit (min:sec)	7:36	8:54	10:48	9:24	0.60
FI per visit (g)	217	268	341	321	13.7
Rate of FI (g/min)	29	31	33	35	1.66
Total feeder time (min/day)	109	99	98	89	5.65

Concluding remarks

Pig production is becoming more and more efficient and industrialised. On the one hand the number of piglets per litter increases which may result in more weak piglets and increased variability between piglets. On the other hand production units become larger and pig farmers cannot accept that health problems and variability between piglets reduce the efficiency of the production process. In a natural environment, weaning is a gradual process. However, in the modern pig industry, weaning is an abrupt event, causing problems for young piglets. Feed intake around weaning is very important, because insufficient energy intake is a risk factor for health problems. A high feed intake during the first day(s) after weaning results in less illness, fewer production losses and a better efficiency of the pig farm. Therefore, feed intake around weaning is regarded as a critical control point in the modern production process of profitable pig farming. In order to optimise piglet rearing and to accommodate variability, feed intake and particularly feed intake behaviour of the piglet should given central importance. In this chapter factors affecting feed intake behaviour of piglets have been described in order to assess feed and environmental factors that are critical in the interaction between feed intake behaviour, feeding management and farm management. These 3 aspects are summarised below.

Feed intake behaviour varies considerably between piglets. In group-housed piglets, eating takes time and occurs during more meals per day. Young piglets eat, generally, more slowly and more frequently than older pigs. Moreover, it seems that piglets are less capable in adapting their feed intake pattern to different circumstances. Therefore, sub-optimal conditions will result more likely in a depressed feed intake in piglets than in older pigs. Furthermore, a group of piglets is not homogenous and sub-optimal feed intake of the group will result in under-nutrition for a number of piglets within that group. In order to stimulate postweaning feed intake as soon as possible, creep feed is essential.

Whether or not an irregular feed intake will cause diarrhoea is determined by a large number of feed and environmental factors, and feed management. Overfeeding and irregular feed intake may lead to a diminished digestibility, especially in young piglets. This may result in intestinal disorders and diarrhoea. Feed factors are, among others, the fibre source and fibre content. Fibre affects the total feed intake, feed intake pattern and intestinal absorption kinetics. The protein content of the feed plays a role on the level of digestive disorders, especially under sub-optimal conditions. Furthermore, feed form has a direct effect on feed intake behaviour and thereby determines the risk for overfeeding and digestive disorders.

Hygiene status, being part of farm management, is a clear risk in itself on digestive disorders, but also interacts with feed composition. Group size, pen space and feeder space all affect feed intake, but are often confounded. It seems that feeder space is especially important during the first days after weaning.

Overall, it can be concluded that feed composition, feeding system, feeding management and farm circumstances all affect feed intake behaviour and should not be separated from each other. Managing the social behaviour of a group by those factors may both stimulate a good feed intake of the individual pig and decrease variability between piglets. The interaction between farmer and feed supplier needs to be further optimalised in order to achieve more profitable pig production.

References

Algers, B., Jensen, P. and Steinwall, L. (1990) Behaviour and weight changes at weaning and regrouping of pigs in relation to teat quality. *Applied Animal Behaviour Science,* **26**, 143-155.

Appleby, M., Pajor, E. and Fraser, D. (1991) Effects of management options on creep feeding by piglets. *Animal Production,* **53**, 361-366.

Appleby, M., Pajor, E. and Fraser, D. (1992) Individual variation in feeding and growth of piglets: effects of increased access to creep food. *Animal Production,* **55**, 147-152.

Augspurger, N. R., Ellis, M., Hamilton, D. N., Wolter, B. F., Beverly, J. L. and Wilson, E. R. (2002) The effect of sire line on the feeding patterns of grow-finishing pigs. *Applied Animal Behaviour Science,* **75**, 103-114.

Ball, R. O. and Aherne, F. X. (1982) Effect of diet complexity and feed restriction on the incidence and severity of diarrhea in early-weaned pigs. *Canadian Journal of Animal Science,* **62**, 907-913.

Bark, L. J., Crenshaw, T. D. and Leibbrandt, V. D. (1986) The effect of meal intervals and weaning on feed intake of early weaned pigs. *Journal of Animal Science,* **62**, 1233-1239.

Beattie, V. E., Weatherup, R. N. and Kilpatrick, D. J. (1999) The effect of providing additional feed in a highly accessible trough on feeding behaviour and growth performance of weaned piglets. *Irish Journal of Agricultural and Food Research,* **38**, 209-216.

Bøe, K. (1991) The process of weaning in pigs: when the sow decides. *Applied Animal Behaviour Science,* **30**, 47-59.

Bruininx, E. M. A. M., Van der Peet-Schwering, C. M. C., Schrama, J. W., Vereijken, P. F. G., Vesseur, P. C., Everts, H., Den Hartog, L. A. and Beynen, A. C. (2001) Individually measured feed intake characteristics and growth performance of group-housed weanling pigs: Effects of sex, initial body weight, and body weight distribution within groups. *Journal of Animal Science,* **79**, 301-308.

Bruininx, E. M. A. M. (2002) *Individually measured feed intake characteristics in group-housed weanling pigs.* PhD Thesis. University of Utrecht, Utrecht, NL.

Bruininx, E. M. A. M., Schellingerhout, A. B., Lensen, E. G. C., Van der Peet-Schwering, C. M. C, Schrama, J. W., Everts, H, Den Hartog, L. A., and Beynen, A. C. (2002) Associations between individual food intake characteristics and indicators of gut physiology of group-housed weanling pigs differing in genotype. *Animal Science*, **75 (1)**, 103-113.

Brooks, P. and Burke, J. (1998) Behaviour of sows and piglets during lactation. In *The lactating sow*, pp 301-338. Edited by M. Verstegen, P. Moughan and J. Schrama. Wageningen Academic Publishers, Wageningen, NL.

Brooks, P. H. and Tsourgiannis, C. A. (2003) Factors affecting the voluntary feed intake of the weaned pig. In *Weaning the pig; concepts and consequences,* pp 81-116. Edited by J. R. Pluske, J. Le Dividich and M. W. A. Verstegen. Wageningen Academic Publishers, Wageningen, NL.

Caine, W. R. (1997) *Ileal recovery of endogenous amino acids in pigs.* PhD Thesis. Wageningen University, Wageningen, NL.

Casteel, S. W., Brown, L. D., Lattimer, J. and Dunsmore, M. (1998) Fasting and feeding effects on gastric emptying time in juvenile swine. *Contemporary Topics,* **37(5)**, 106-108

Dalby, J. A. (1998) Behaviour and choice feeding. In *Progress in pig science,* pp 183-207. Edited by J. Wiseman, M. A. Varley and J. P. Chadwick. Nottingham University Press, Nottingham, UK.

De Haer, L. C. M. (1992) *Relevance of eating pattern for selection of growing pigs.* Ph.D. thesis. Wageningen University, Wageningen, NL.

De Haer, L. C. M. and De Vries, A. G. (1993). Feed intake patterns of and digestibility in growing pigs housed individually or in groups. *Livestock Production Science,* **33**, 277-292.

Delumeau, O. and Meunier-Salaün, M. C. (1995) Effect of trough familiarity on the creep feeding behaviour in suckling piglets and after weaning. *Behavioural Processes,* **34**, 185-196.

Fledderus, J., Bikker, P. and Weurding, R. E. (2003) In vitro assay to estimate kinetics of starch digestion in the small intestine of pigs. In *Proceedings of the 9th International Symposium on Digestive Physiology in Pigs – volume 2,* pp 4-6. Edited by R. O. Ball. University of Alberta, Canada.

Fowler, V. R. and Gill, B. P. (1989) Voluntary food intake in the young pig. In *The voluntary food intake of pigs,* pp 51-60. Edited by J.M. Forbes, M. A. Varley and T. L. J. Lawrence. Occasional publication of the BSAS, Edinburgh, UK.

Fraser, D., Milligan, B. N., Pajor, E. A., Philips, P. A., Taylor, A. A. and Weary,

D. M. (1998) Behavioural perspectives on weaning in domestic pigs. In *Progress in pig science,* pp 121-138. Edited by J. Wiseman, M. A. Varley and J. P. Chadwick. Nottingham University Press, Nottingham, UK.

Gaskins, H. R. (2000) Intestinal bacteria and their influence on swine nutrition. In *Swine Nutrition,* pp 537-587. Edited by R. L. Mackie, B. A. White and R. E. Isaacson. Chapmann & Hall, New York, USA.

Guerrero, J. L., López, O., Brito, P. C. and García, A. (1991). Influence of the housing system and the type of feeding on the performance of pigs with low weaning weights. Zootechnica de Cuba, **1**, 7-16.

Gonyou, H. W., Chapple, R. P. and Frank, G. R. (1992) Productivity, time budgets and social aspects of eating in pigs penned in groups of five or individually. *Applied Animal Behaviour Science,* **34**, 291-301.

Gruys, E., Toussaint, M. J. M., Landman, W. J. M., Tivapasi., M., Chamanza, R. and Van Veen, L. (1999) Infection, inflammation and stress inhibit growth. Mechanism and non-specific assessment of the processes by acute phase proteins. In: *Production Diseases in Farm Animals. 10th international conference 1998,* pp. 72-84. Edited by Th. Wensing. Wageningen Academic Publishers, Wageningen, NL.

Hampson, D. J. 1986. Influence of creep feeding and dietary intake after weaning on malabsorption and occurrence of diarrhoea in newly weaned pig. *Research in Veterinary Science,* **41**, 63-69.

Held, S. and Mendl, M. (2001) Behaviour of the young weaner pig. In: *The weaner pig: Nutrition and Management,* pp. 273-298 Edited by M. A. Varley and J. Wiseman. Papers from the occasional Meeting of the British Society of Animal Science at Notthingham University in Sept. 2000. Cabi Publishing, Wallingford, UK.

Hee, J., Sauer, W. C., Mosenthin, R. (1988) The effect of frequency of feeding on the pancreatic secretions in the pig. *Journal of animal physiology and animal nutrition,* **60** (**5**), 249-256.

Hessing, M., Hagelsø A., van Beek, J., Wiepkema, P., Schouten W. and Krukow, R. (1993) Individual behavioural characteristics in pigs. *Applied Animal Behaviour Science,* **37**, 285-295.

Hogenkamp, D., Hayler, R., Drochner, W. and Mosenthin, R. (1997) The effect of enzyme treatment of piglet diets on the incedence of post weaning diarrhea. In *Digestive physiology in pigs,* pp 466-469. Edited by J.P. Laplace, C. Fevrier and A. Barbeau. EAAP Publication no. 88, Wageningen Academic Publishers, Wageningen, NL

Hoofs, A. (1993) *Creep feeding of suckling piglets as a farm management practice.* Report P1.97. Institute for Practical Pig Husbandry, Sterksel, NL (in Dutch).

Hyun Y. and Ellis, M. (2001) Effect of group size and feeder type on growth performance and feeding patterns in growing pigs. *Journal of Animal Science,* **79**, 803-810.

Hyun Y. and Ellis, M. (2002) Effect of group size and feeder type on growth performance and feeding patterns in finishing pigs. *Journal of Animal Science,* **80**, 568-574.

Hyun, Y. Wolter, B. F. and Ellis, M. (2001). Feed intake patterns and growth performance of purebred and crossbred Meishan and Yorkshire pigs. *Asian-Australian Journal of Animal Science,* **14**, 837-843.

Jensen, P. and Recén, B. (1989) When to wean - Observations from free-ranging domestic pigs. *Applied Animal Behaviour Science,* **23**, 49-60.

Johnson, R. W. (1997) Inhibition of growth by pro-inflammatory cytokines: an integrated view. *Journal of Animal Science,* **75**, 1244-1255.

Kamphues, J. (1987) *Untersuchungen zu Verdauungsvorgängen bei Absetzferkeln in Abhängigkeit von Futtermenge und –zubereitung sowie von Futterzusätzen.* Habilitationsschrift, Tierärztliche Hochschule Hannover, D. (In german).

Laitat, M., Vandenheede, M., Désiron, A., Canart, B. and Nicks, B. (1999) Comparison of performance, water intake and feeding behaviour of weaned pigs given either pellets or meal. *Animal Science,* **69**, 491-499.

Laitat, M., Vandenheede, M., Désiron, A., Canart, B. and Nicks, B. (2000) Granulés our farine en post-sevrage: le choix des porcelets. *Journées Recherche Porcine en France,* **32**, 157-162.

Le Dividich, J. and Herpin, P. (1994) Effects of climatic conditions on the performance, metabolism and health status of weaned piglets: a review. *Livestock Production Science,* **38**, 79-90.

Le Dividich, J., Martineau, G., Thomas, F., Demay, H., Renoult, H., Homo, C., Boutin, D., Gaillard, L., Surel, Y., Bouétard, R. and Massard, M. (2004) Acquisition de l'immunité passive chez les porcelets et production de colostrum chez la truie. *Journées Recherche Porcine en France,* **36**, 451-456. (In French).

King, R. H. and Pluske, J. R. (2003) Nutritional management of the pig in preparation for weaning. In *Weaning the pig; concepts and consequences,* pp 37-52. Edited by J. R. Pluske, J. Le Dividich and M. W. A. Verstegen. Wageningen Academic Publishers, Wageningen, NL.

Madec, F., Bridoux, N. Bounaix, S. Justin, A. (1998) Measurement of digestive disorders in the piglet at weaning and related risk factors. *Preventive Veterinary Medicine,* **35**, 53-72.

Madec, F., Le Dividich, J., Pluske, J. R. and Verstegen, M. W. A. (2003) Environmental requirements and housing of the weaned pig. In *Weaning the pig; concepts and consequences,* pp 337-360. Edited by J. R. Pluske, J. Le Dividich and M. W. A. Verstegen. Wageningen Academic Publishers, Wageningen, NL.

Makkink, C. A. (1993) *Of piglets, dietary proteins and pancreatic proteases.* PhD Thesis. Wageningen University, Wageningen, NL.

Meijer, J. C., Van Gils, B. Bokhout, B., De Geus, M., Harmsen, M., Nabuurs, M., and Van Zijderveld, F. (1997) Weaned pigs orally challenged with E. *coli* following stress as a model of postweaning diarrhoea. *Proceedings symposium Gastro-intestinal disorders in juveniles,* ID-DLO, Lelystad, NL

Milligan, B. N., Fraser, D. and Kramer, D. L. (2002) Within-litter birth weight variation in the domestic pig and its relation to pre-weaning survival, weight gain and variation in weaning weights. *Livestock Production Science,* **76**, 181-191.

Nabuurs, M. J. A. (1991) *Etiologic and pathogenic studies on postweaning diarrhea.* PhD Thesis. University of Utrecht, Utrecht, NL

Niinikoski, H., Stoll, B., Guan, X., Kansagra, K., Lambert, B.D., Stephens, J. Hartmann B., Holst, J.J. and Burrin, D.G. (2004) Onset of small intestinal Atrophy is associated with reduced intestinal blood flow in TPN-fed Neonatal piglets. Journal of Nutrition, **134**, 1467-1474.

Partridge, G. G. and Gill, B. P. (1993) New approaches with pig weaner diets. In *Recent advances in Animal Nutrition – 1993,* pp 221-248. Edited by P. C. Garnsworthy and J. C. A. Cole. Notthingham University Press, Nottingham.

Pluske, J. R. (2000) The small intestine of the young pig: Structure and function and the relationship with diet. In *Book of Abstracts of the 51st Annual Meeting of the European Association for Animal Production - No6,* pp167. Edited by J.A.M. van Arendonk, A. Hofer, Y van der Honing, F. Madec, K. Sejrsen, D. Pullar, L Bodin, J. A. Fernandez and E. W. Bruns. Wageningen Academic Publishers, Wageningen, NL.

Pluske J. R. and Dong, G. Z. (1998) Factors influencing the utilisation of colostrum and milk. In *The lactating sow,* pp 45-70. Edited by M. Verstegen., P. Moughan. and J. Schrama . Wageningen Academic Publishers, Wageningen, NL.

Pluske, J. R. Le Dividich, J. and Verstegen, M. W. A. (2003) *Weaning the pig; concepts and consequences.* Wageningen Academic Publishers, Wageningen, NL.

Pluske, J. R. and Williams, I. H. (1996) The influence of feeder type and the method of group allocation at weaning on voluntary food intake and growth in piglets. *Animal Science,* **62**, 115-120.

Pluske, J. R., Williams, I. H. and Aherne, F. X. (1995) Nutrition of the neonatal pig. In *The Neonatal Pig: Development and Survival,* pp 187-235. Edited by M. A. Varley. CAB International, Wallingford. UK.

Quiniou, N., Dagorn, J. and Gaudré, D. (2002) Variation of piglet' birth weight and consequences on subsequent performance. *Livestock Production Science,* **78**, 63-70.

Ramaekers, P. J. L. (1996) *Control of individual daily growth in group-housed pigs using feeding stations.* PhD Thesis. Wageningen University, Wageningen, NL.

Rerat, A. and Corring, T. (1991) Animal factors affecting protein digestion and absorption. In *Digestive physiology in pigs. Proceedings of the 5th international symposium on digestive physiology in pigs,* EAAP publication nr 54, pp 5-34. Wageningen Academic Publishers, Wageningen, Netherlands.

Ruckebusch Y. and Bueno, L. (1976) The effect of feeding on the motility of the stomach and small intestine in the pig. *British Journal of Nutrition,* **35**, 397-405.

Savoie, L. (1992) In vitro simulation of protein digestion: an integrated approach. In *In Vitro Digestion for Pigs and Poultry* pp 146-161. Edited by M.F. Fuller. CAB International, Wallingford, UK.

Schouten, W. G. P. (1986) *Rearing conditions and behaviour in pigs.* PhD thesis, Wageningen University, Wageningen, NL.

Seerley, R. W., Miller E. R. and Hoefer, J. A. (1962) Rate of food passage studies with pigs equally and ad libitum fed meal and pellets, *Journal Animal Science,* **21**, 834-837.

Seve, B. (1979) Bien concevoir les aliments des porcelets. *Revue Francaise Production Animales,* **82** , 19-29. (in French)

Smith, L. F., Beaulieu, A. D., Patience, J. F., Gonyou, H. W. and Boyd, R. D. (2004) The impact of feeder adjustment and group size-floor space allowance on the performance of nursery pigs. *Journal of Swine Health and Production,* **12**, 111-118.

Spicer, H. M. and Aherne, F. X. (1987) The effects of group size/stocking density on weanling pig performance and behaviour. *Applied Animal Behaviour Science,* **19**, 89-98.

Van der Peet-Schwering, C. M. C. and Binnendijk, G. P. (1994) *Split-sex-feeding of barrows and sows.* Report P1.107. Institute for Practical Pig Husbandry, Rosmalen, NL. (in Dutch)

Van Erp-van der Kooij E., Kuijpers A., Schrama, J. W. and Ekkel, E. D. (2000) Individual behavioural characteristics in pigs and their impact on production. *Applied Animal Behaviour Science,* **66**, 171-185.

Van Erp-Van der Kooij E., Kuijpers A., Van Eerdenburg F. and Tielen M. (2001) A note on the influence of starting position, time of testing and test order on the back test in pigs. *Applied Animal Behaviour Science,* **73**, 263-266

Varley, M. A. and Wiseman, J. (2001) *The weaner pig: Nutrition and Management.* Papers from the occasional Meeting of the British Society of Animal Science at Notthingham University in Sept. 2000. Cabi Publishing, Wallingford, UK.

Vente-Spreeuwenberg, M. A. M. and Beynen, A. C. 2003. Diet-mediated modulation of small intestinal integrity in weaned piglets. In *The Weaner pig: Concepts and Consequences*, pp 145-199. Edited by J. Pluske, J. Le Dividich and M. W. A. Verstegen. Wageningen Academic Publishers, Wageningen, NL

Vente-Spreeuwenberg, M. A. M., Verdonk, J. M. A. J. Bakker, G. C. M. and Verstegen, M. W. A. (2004) Effect of dietary protein source on feed intake and small intestinal morphology in newly weaned piglets. *Livestock Production Science,* **86 (1/3)**, 169-177.

Vente-Spreeuwenberg, M. A. M., Verdonk, J. M. A. J. Koninkx, J. F. J. G. Beynen, A. C. and Verstegen, M. W. A. (2004) Dietary protein hydrolysates versus the intact proteins do not enhance mucosal integrity and growth performance in weaned piglets. *Livestock Production Science,* **85(2-3)**, 151-164.

Vente-Spreeuwenberg, M. A. M., Verdonk, J. M. A. J. Beynen, A. C. and Verstegen, M. W. A. (2003) Interrelationship between gut morphology and faeces consistency in newly weaned piglets. *Animal Science,* **77**, 85-94

Wolter, B. F., Ellis, M., Curtis, S. E., Parr, E. N., and Webel, D. M. (2000) Group size and flour-space allowance can affect weanling-pig performance. *Journal of Animal Science,* **78**, 2062-2067.

3

IMPLICATIONS OF NEW FAT SOURCES FOR NON-RUMINANT NUTRITION

DAVID HOWELLS
Advanced Liquid Feeds LLP, Alexandra House, Regent Road, Liverpool, L20 1ES

Background

The combination of the PCB-contamination in Used Cooking Oils (UCO) at Verkest in Belgium and the subsequent Foot and Mouth outbreak in the United Kingdom (UK) resulted in the banning by the European Union of UCO for inclusion in Compound Feed (CF). Because the UK already had in place a robust and independent testing, traceability and auditing regime, the European Commission was able to grant the UK a 2-year derogation beyond the ban in the rest of Europe and this derogation comes to an end at the end of October, 2004. UCO has been a major constituent of fat blends for CF with the UK producing 80-90,000 tonnes per annum out a total inclusion of fat in CF in excess of 250,000 tonnes per annum. The removal of this relatively cheap by-product has significant implications for formulation of new blends and at the time of writing this chapter (September 2004), the industry has still to decide amongst the various options that have been presented to them.

The areas of concern that have to be addressed are:

• Nutritional	-	Growth performance following the use of the blends in diets.
• Physical acceptability	-	Carcass quality, wet-litter considerations,
	-	Palatability.
• Mill processing characteristics	-	Pellet quality, dust suppression, die lubrication
• Raw material availability		
• Commercial		

Nutritional considerations

At the Feed Conference, there will be as many different expert opinions as there are delegates. Whether people are an advocate of animal fat blends, or follow the approach outlined by Wiseman et al. (1998), or have a different model, none of the solutions that present themselves will satisfy all of the above criteria. If animal fat blends are ignored (for obvious reasons) there are three distinct approaches:

1.   **Straight Vegetable Oil** CF formulators have used blends for many years without really re-visiting which fat blend to use, but the new blends required post-October 2004 have obviously focused their thoughts, they are naturally concerned about a "leap into the unknown" and have turned to prediction equations. The equations of Wiseman et al. (1998) give more or less equal weighting to many of the main world liquid vegetable oils (e.g. soya, sunflower, rape and palm) but the former 3 have distinct disadvantages in terms of carcass quality and price.

      Detailed in Table 3.1 are the gas liquid chromatography data for the four main world oils.

Table 3.1 Fatty acid profiles (g/100g oil) of the 4 major global vegetable oils

	Palm	*Rape*	*Soya*	*Sunflower*
C12[1]	0.5	0	0.5	0
C14	1.2	0.5	0.5	0
C16	45	7	12	6
C18	3	2	3	4
C18-1	37	55	25	20
C18-2	10	23	54	65
C18-3	0	9	6	1
C20 and over	0	3	1	2

[1]Length of carbon chain followed by number of double bonds

2.   **Current Copy** As an industry, Fat Blenders have tended to shun the various nutritional models in favour of the pragmatically successful approach of constructing the feed fat blends to mimic the basic nature of the carcass fat of the appropriate species with some variation to allow for the requirement for "essential" fatty acids. Throughout the chapter, fats for broiler grower diets will be considered but the same

considerations apply to pig diets as well. Detailed below in Table 2 are the fatty acid profioles Chicken Fat and the main fat blend used by the sector; it is no accident that the two are very similar .

It would be trivial for us to reconstruct an identical specification using Mixed Acid Oils, Palm Fatty acid Distilate (PFAD), Soya Acid Oil, Palm Oil, Rapeseed Oil and Soya Oil; the disadvantages are that it is "theoretically" less valuable than straight Vegetable Oils because of the high free fatty acid (FFA) and higher saturated fatty acid content, and it is also relatively expensive to use Rapeseed Oil as a building block.

Table 3.2 Fatty acid profiles (g/100g oil) of 'chicken fat' and a commercial blend for a feed-grade fat.

	Chicken fat	*20 Linoleic fat blend*
C14[1] and below	3.5	Max 5
C16	22.5	20 – 25
C16-1	6	2
C18	5.5	4 – 8
C18-1	40	35 – 42
C18-2	19	20 – 22
C18-3	1	1 – 5
C20 and above	1	2 –5
FFA[2]		Max 40

[1]Length of carbon chain followed by number of double bonds
[2]Free Fatty acids

3. **UCO/Palm Switch** The third approach advocated as a commercially-viable solution is to accept a higher Saturated Fat content than the above blends, but retaining the C18-2 and FFA specifications. This effectively replaces the current UCO with the cheapest (and likely to be so in the long-term) of the world oils, Palm Oil. This is virtually a Vegetable Oil equivalent to the Tallow / Soya blends which were used by many formulators prior to the removal of animal fats from blends and merely replaces some saturated stearic with another saturated fatty acid, palmitic. The disadvantage of this approach is that, like Tallow/ Soya blends, they are theoretically less desirable than the straight vegetable oils. There are reservations about the ability of a simple two or three-parameter equation to predict adequately the behaviour of a complex chemical mixture containing over a thousand distinct moieties.

Trial results for very high FFA contents and high saturated fat contents have produced results which suggest that a more complex model may be required. One thing is certain; the blends used by different growers in the next few months will test the various models. Table 3.3 shows the specification of the Blend in which linoleic acid is maintained at the level of current blends.

Table 3.3 Fatty acid profiles (g/100g oil) of UCO (Used Cooking Oils) and a commercial blend based on palm oil for a feed-grade fat.

	Current UCO 20 linoleic blend	*Palm Oil 20L blend*
C14[1] and below	Max 5	Max 3
C16	20 – 25	30 – 32
C16-1	2	1
C18	4 – 8	4
C18-1	35 – 42	35
C18-2	20 – 22	20 – 22
C18-3	1 – 5	1 – 4
C20 and above	2 –5	2
FFA[2]	Max 40	Max 40
Typical MFA[3]	91	93

[1]Length of carbon chain followed by number of double bonds
[2]Free Fatty acids
[3]Monomeric fatty acids

As can be seen the only significant change would be a higher palmitic level at the expense of oleic and an improved monomeric fatty acid level.

Physical acceptability

There are three areas in this section which traditionally would have been of concern to the commercial sector – Carcass quality, Wet-litter, Palatability.

- Palatability is rarely a problem for broiler diets but obviously is so for pig diets. The main problem area formerly was changing from an Animal Fat diet to Vegetable Oil base – *or vice versa.* With the disappearance of Tallow (which by its nature could indeed smell quite strongly, especially in summer) as a Feed Fat ingredient in the last 9 years, the palatability problem has virtually disappeared and none of the three different Blend-types above should encounter any significant problems.

• Commercial growers frequently blame Wet-Litter problems on the added-fat part of the diet. It comes as no surprise at this time of the year (September) when new crop grains are coming in that customers ask if anything has changed in fat blends. Conclusions over the year are that the biggest sources of scouring/wet litter are:

Dramatic changes in diet – including the fat
Mycotoxins
Husbandry

Change from the present UCO blends to any of the regimes suggested above is unlikely to cause problems since Soya Oil, Rapeseed Oil and Palm Oil are already being incorporated, albeit after several cooking cycles. Nonetheless users are encouraged users to phase in the new fat regime in two or three stages just to minimise the possibilities.

• The only area in which any significant differentiation between the blends is expected is in carcass quality . Of the three blends the only one which will give rise to a problem is the Crude Vegetable Oil Blend where oily carcasses is a major issue. The incorporation of the more saturated Palm Oil or PFAD would obviously overcome this problem but would compromise the theoretical nutritional value. It should be noted that the saturated fat content of Blends 2 and 3 (Table 3.3) actually exceed that of "normal" chicken depot fat.

Mill processing characteristics

There is nothing to choose between any of the blends offered in terms of die lubrication but, for dust suppression and pellet quality, the blends containing higher quantities of saturated fats (Blends 2 and 3; Table 3.3) and particularly those containing PFAD would perform much better. The liquid vegetable oils will soak into the pellet very quickly and both pellet quality and dust would be a problem. PFAD, whose melting point characteristics are such that solid fat will be deposited rapidly on the surface of the pellet, scores well in this particular area. The tryglceride Palm Oil is significantly softer and slower to solidify (this is not a paradox but the kinetics of solidification for the smaller fatty acid molecule are obviously faster than a large triglyceride molecule) and therefore not quite as efficient as PFAD in this respect..

Raw material availability

Naturally the only raw materials included in the above formulations are those which will be available in the long term and in the appropriate tonnages which are commercially sensible. Those buyers who would like to use just Mixed Acid Oils (which are the cheapest of the raw materials) or Factory UCO (which may still be used legally throughout Europe) must understand that the availability of these ingredients is very limited. Any attempt to distort the demand for these particular products will, by normal Adam Smith economics, drive the prices of these "cheap" ingredients up to the price of the next equivalent, i.e. the Crude Oils .

It is not recommended to "cherry pick" the best and cheapest ingredients locally. Market forces will inevitably mean that those undertaking this are following a route that will be so competitive that yesterday's cheap solution will be totally unavailable tomorrow.

Commercial

Commercial reality must now be brought into this discussion, albeit that the Feed Industry survives with ever-tightening regulatory control and constraint is harder and harder to survive in. The solution that the industry chooses must be one that is commercially viable. For those who think fat is too expensive, it should be pointed out that this is in part because it was the easiest target for industrial burners to pick on first; in future anything which is an evidently cheap source of energy for the Feed Industry will also be cheap for them.

The following points must be taken into account:

- The UK currently is barely self-sufficient in Soya Oil. Any large increase in demand will inevitably be imported and at a price which will be at a premium to Dutch markets.
- The EEC's illogical obsession with biodiesel has committed the UK to using approximately 1.5m tonnes of Rapeseed Oil within the next two decades which is 4 times the existing production. Under those circumstances it is inevitable that Rapeseed Oil will stay at substantial premium to Soya Oil. At the moment of writing (September 2004) that premium is approximately £30/ t.
- The Palm Oil price has discounted Soyabean Oil consistently for many years and, apart from times of extreme crop failures, it will continue to do so. The differential today is approximately £100 /t.

- PFAD will always trade at a discount to Palm Oil

In addition it is important to take into account of the UK's viability as a meat producer in the context of competition:

- Countries that still use cheap UCO/Yellow Grease
- Countries that still use cheap Tallow and Meat and Bone Meal
- Countries which have access to cheaper Soya Oil and Meal

Table 4 below summarises the advantages and disadvantages for the distinct philosophies.

Table 3.4. Relative values of fat / oil sources

	Nutritional	*Physical*	*Mill processing*	*Availability*	*Commercial*
Liquid vegetable oils	Theoretically best	Worst	Worst	OK	Worst
Current look-alike	Medium	OK	OK	OK	Not good
Palm oil based	Theoretically worst	OK	OK	OK	Best

Table 3.4 shows that there will have to be some compromise, and the practical solution may well by somewhere between these choices. It is inevitable that commercial considerations will prevail in a sector which is currently and will continue to be under pressure from imports. With Palm Oil 25 cheaper at the moment (September 2004) than Soya Oil it will clearly feature heavily in the final decision. The financial benefits and process benefits from the inclusion of PFAD may well justify its continued usage in Feed Fats.

References

Wiseman, J. Powles, J. and Salvador, F. (1998). Comparison between pigs and poultry in the prediction of the dietary energy value of fats. *Animal Feed Science and Technology* **71**, 1-9.

4

AN UPDATE ON THE ENERGY AND NUTRIENT REQUIREMENTS OF COMPANION ANIMALS

KELLY S. SWANSON AND GEORGE C. FAHEY, JR.

Department of Animal Sciences, University of Illinois, Urbana, IL 61801

Introduction

According to fossil records, dogs and cats have lived with humans for over 10,000 years. Although dogs and cats were originally domesticated for practical purposes (e.g., herding, sport, hunting, rodent control), their primary role in today's society is that of companionship. Just as their role in society has changed over time, their lifestyle also has changed. They are a member of the family in many households, and commonly live under the same roof as their human owners. This intimate relationship between companion animals and humans has probably been a major catalyst for research efforts to enhance dog and cat health and longevity through nutritional and medical interventions. In addition to improved veterinary medical practices, advances in nutritional sciences and pet food manufacturing have been highly successful in the realization of these goals.

Until ~150 years ago, companion animals were expected to be self-sufficient, occasionally receiving scraps or bones from the human food table. However, that practice began to change in 1860 when James Spratt of London developed the first commercial dog food in the form of a biscuit (Lazar, 1990). Given his success in Europe, commercial pet foods arrived in the U.S. in the early 1900s (Case *et al.*, 2000). Although very few commercial pet foods were initially available from a small number of companies, it is estimated that over 300 pet food manufacturers and over 3,000 different brands of pet foods currently exist in the world (Case *et al.*, 2000). Because little was known regarding energy and nutrient requirements of dogs and cats, the first pet foods were marketed for consumption by both species. Due to the extensive research performed with dogs and cats over the past 50 years, we now know that they are very different metabolically and have different energy and nutrient requirements.

Dog and cat energy and nutrient requirements

The following is a brief overview of the basic nutrient requirements of healthy dogs and cats. For a more extensive review on these topics, refer to Hussein (2003a, 2003b). For specific nutrient requirements or recommendations, refer to NRC (2003) or AAFCO (2004), respectively.

ENERGY AND NUTRIENT REQUIREMENTS OF THE DOG

Energy. In many respects, feeding an adult dog for optimal health is similar to that of an adult human. Feeding a balanced diet that meets all the energy and nutrient requirements but prevents obesity and associated disorders is the overall goal. Although energy is not a nutrient per se, it is required for normal metabolic and physiological functions. Atwater factors tend to overestimate metabolizable energy (ME) values of most pet foods (Case *et al.*, 2000). Therefore, modified Atwater factors (3.5, 3.5, and 8.5 kcal/g protein, carbohydrate, and fat, respectively; 14.64, 14.64 and 35.56 kJ respectively) are used to estimate ME content of dog diets. Three formulas are commonly used to calculate ME requirements (MER) of adult dogs at maintenance:

1) $MER = K \times BW\ kg^{0.67}$
   with K = 99, 132, and 160 kcal for inactive, active, and very active dogs, respectively
   with K = 414, 552 and 669 kJ for inactive, active, and very active dogs, respectively

2) $MER = 100 \times BW\ kg^{0.88}$
   [418 kJ]

3) $MER = 132 \times BW\ kg^{0.75}$ (Case *et al.*, 2000).
   [552 kJ)

Energy requirements are dependent on numerous factors (e.g., life stage, physiological state, genetics, and activity level). Therefore, these equations provide an energy intake from which to start, but body condition should be regularly monitored and used as a gauge to adjust energy intake on an individual basis. Due to their tendency to overeat and become obese on the highly palatable and highly digestible commercial dog foods now available, energy intake is probably the most important concern in feeding adult dogs. The fact that up to 40% of dogs presented to veterinarians in the U.S. are now overweight underlined the prominence of canine obesity and the importance of closely monitoring energy intake of dogs (Sunvold and Bouchard, 1998).

Water. In addition to energy, several key components (water, proteins, carbohydrates, fats, vitamins, and minerals) are required to support normal bodily functions. Although it is usually the nutrient given the least attention, water is the most important dietary nutrient for dogs. Being the major component of most bodily tissues and fluids, water serves as the solvent for most intra- and extra-cellular chemical processes. The dog's water requirements are met primarily from drinking water. The contribution of food water and metabolic water (produced during energy metabolism) depend on the water and energy content of foods, respectively. While urine production is the greatest source of water loss, respiratory and faecal losses are other primary contributors. In dogs, the amount of water consumed usually reflects the amount of water lost (Adolph, 1939). As with humans and other animals, food water and salt content, environmental temperature, physiological status (e.g., lactation), and activity level influence water requirement and consumption. Due to its important role in the body, fresh drinking water should be available at all times.

Protein. Dietary protein provides essential and nonessential amino acids (AA) needed for maintenance, growth, gestation, and lactation. The essential AA, ones that cannot be produced by the body (or not in sufficient amounts needed for optimal performance) and must be included in the diet, include: arginine, histidine, isoleucine, leucine, lysine, methionine, phenylalanine, threonine, tryptophan, and valine. If provided with enough nitrogen and carbon sources, sufficient amounts of nonessential AA can be synthesized by the body. Amino acids are the building blocks for numerous body proteins including hair, skin, nails, connective tissue, muscle, enzymes, hormones, and other biologically active compounds. Because body proteins are continuously subjected to degradation and synthesis, it is critical to assure a continuous supply of essential AA through the diet to avoid deficiency (Hussein, 2003a). Protein requirements are dependent on protein quality (e.g., AA content, AA balance), digestibility, energy intake, nutritional status, feeding pattern, age, growth rate, reproductive status, and protein:energy ratio (Pond *et al.*, 1995). Excess AA are not stored in the body and are catabolised in energy-yielding reactions or converted to fat or glycogen as energy stores.

Carbohydrates. Although dogs do not have an obligate requirement for carbohydrates in the diet, this diverse group of compounds is the primary energy source of omnivores (Guilbot and Mercier, 1985) and plays important roles in the body. Functionally, carbohydrates may be classified into four groups including absorbable, digestible, fermentable, and nonfermentable sources (NRC, 2003). While absorbable and digestible carbohydrate sources are primarily used as sources of energy, fermentable carbohydrates (e.g., dietary fibres) play an important role in maintaining intestinal health. Although numerous definitions of dietary fibre have been published over the past few decades, the Panel on the Definition of Dietary Fiber and the Standing Committee on the Scientific Evaluation of Dietary Reference Intakes, Food and Nutrition Board, Institute of Medicine, have recently proposed a new definition of dietary fibre (Institute of Medicine, 2001). According to this panel, the following fibre definitions were proposed:

1)   *Dietary fibre* consists of nondigestible carbohydrates and lignin that are intrinsic and intact in plants

2)   *Added Fibre* consists of isolated, nondigestible carbohydrates that have beneficial physiological effects in humans;

3)   *Total Fibre* is the sum of *Dietary Fibre* and *Added Fibre*.

Along with cellulose, hemicelluloses, pectin, and gums, which are usually considered to be dietary fibre, nondigestible oligosaccharides and resistant starches naturally occurring in plants, are also included under this definition. Potential health benefits of consuming adequate dietary fibre include improved colonic health and reduced incidence of obesity and related disorders.

Fats. Dietary fats provide a concentrated source of energy and supply essential fatty acids (FA) required for several structural and metabolic functions. Fats are required for the absorption of fat-soluble vitamins, are important in cell membrane integrity and fluidity, are precursors of metabolically active compounds such as prostacyclins, prostaglandins, leukotrienes, and thromboxanes, and play an important role in the regulation of the immune system. Dietary fat is primarily found in the form of triglycerides (three FA linked with a glycerol unit) with varying degrees of saturation, which is highly dependent on source. Animal fat sources have a higher proportion of saturated FA, while plants sources may have considerable amounts of the unsaturated forms.

Although dogs can synthesize adequate amounts of saturated FA de novo starting with glucose (Salati and Goodridge, 1996), they lack the enzymes necessary to produce omega-3 and omega-6 FA. The enzymes, $\Delta 12$ desaturase [needed to produce a double bond in linoleic acid (18:2n-6)] and $\Delta 15$ desaturase [needed to produce a double bond in α-linolenic acid (18:3n-3)], are not synthesized by dogs. Given sufficient amounts of linoleic acid, dogs can synthesize γ-linolenic (C18:3n-6) and arachidonic (C20:4n-6) acids by desaturating and elongating this precursor. Desaturation and elongation also result in the synthesis of the long chain omega-3 FA, eicosapentaenoic (EPA, 20:5n-3), docosapentaenoic (DPA, 22:5n-3), and docosahexaenoic (DHA, 22:6n-3) acids from α-linolenic acid. Therefore, while linoleic and α-linolenic acids are considered to be essential FA in dog diets, their derivatives may be regarded as "conditionally essential."

Vitamins. In contrast to the other organic components in the diet (i.e., protein, carbohydrates, fat), vitamins cannot be used as an energy source or as structural compounds. Moreover, vitamins are a mixed group of chemical compounds and are not related to one another as are proteins, carbohydrates, and fats. Although required in trace amounts, vitamins often act as coenzymes and are important participants in a variety of metabolic pathways. Vitamins are classified as being either fat-soluble (i.e., A, D, E, K) or water-soluble [i.e., ascorbic acid (vitamin C), thiamin (B_1), riboflavin (B_2), niacin, pyridoxine (B_6), pantothenic acid, biotin, folic acid, cobalamin (B_{12}), and choline]. While this classification system does not provide any information pertaining to function, it does have relevance to the mechanisms of absorption from the gastrointestinal tract and excretion from the body. In general, animals have the ability to store fat-soluble vitamins (e.g., fatty tissue, liver) for use in times of low intake. In contrast, animals have limited capacity to store water-soluble vitamins which, thus, must be consumed on a regular basis. For an extensive review of individual vitamin function and signs of deficiency or toxicity, refer to McDowell (2000). Although the functions and deficiency and toxicity signs for most vitamins have been known for decades, quantitative requirements have been poorly defined in dogs and this area still needs extensive study.

Minerals. Minerals, commonly divided into macrominerals and trace minerals, are inorganic compounds required in trace amounts by the body. Macrominerals, required in gram amounts per 1,000 kcal ME, include calcium (Ca), phosphorus (P), magnesium (Mg), sodium (Na), chlorine (Cl), and potassium (K). Trace minerals, required in milligram amounts per 1,000 kcal ME, include iron (Fe), zinc (Zn), copper (Cu), manganese (Mn), Iodine (I), selenium (Se), molybdenum (Mo), boron (B), and chromium (Cr). Minerals are involved in numerous physiological functions, playing an important role in acid-base balance, skeletal structure, nerve impulse transmission and muscle contraction, cell signaling, DNA and RNA metabolism, and as metalloenzymes. For an in depth review of mineral function and deficiency and toxicity signs, refer to McDowell (2003). Although minerals such as arsenic (As), silicon (Si), nickel (Ni), vanadium (V), and others are present in the environment, insufficient information exists to make dietary recommendations for these minerals in dogs.

METABOLIC IDIOSYNCRASIES OF THE CAT

Although domestic dogs and cats are both members of the class Mammalia and order Carnivora, they have diverged metabolically over time. Dogs are members of the Canidae family and are omnivores, while cats are members of the Felidae family and are strict carnivores, requiring several nutrients only found in animal sources. This section will not repeat information already mentioned regarding energy and nutrient requirements of the dog, but will focus on the unique metabolic requirements of the cat.

High protein requirement. Cats have substantially higher protein requirements for maintenance than other species. Many species have the ability to adapt to changes in dietary protein intake by increasing or decreasing hepatic enzyme activity responsible for AA catabolism. However, cats have high concentrations of hepatic AA catabolic enzymes and urea cycle enzymes, which result in high obligatory nitrogen loss in urine (Rogers *et al.*, 1977). Due to their lack of enzymatic control, consistent feeding of a high protein diet is required to maintain normal health of cats (Rogers and Morris, 1980). Cats also have a greater sulphur AA requirement than dogs because they synthesize and excrete felinine, a sulfur-containing compound thought to function as a territorial marker in urine.

High sensitivity to arginine deficiency. Although arginine is not considered to be an essential AA in most adult animals because they have the ability to produce enough to meet metabolic requirements, it is essential in both adult dogs and cats. However, cats are much more sensitive to arginine deficiency than are dogs. In most adult animals, ornithine is produced from glutamate and proline in the intestinal mucosa and from arginine in the liver. Cats lack the ability to synthesize ornithine in the intestine and rely solely on arginine for its production throughout life. Because ornithine is a primary participant in the urea cycle, arginine deficiency in cats results in a rapid and severe hyperammonemia (Morris and Rogers, 1978). If not quickly corrected, arginine deficiency leads to vomiting, muscle spasms, ataxia, coma, and death (Morris and Rogers, 1978).

Taurine requirement. Taurine, a ß-amino sulphonic acid, plays an important role in bile acid conjugation, retinal function, female reproduction, and normal function of the myocardium. Taurine is synthesized from sulphur-containing AA (e.g., methionine, cysteine) in sufficient amounts in most animals. However, because of low hepatic activity levels of cysteine dioxygenase and cysteine sulphonic acid decarboxylase, cats synthesize small amounts of taurine and cannot meet their metabolic requirements (Hardison *et al.*, 1977; Morris and Rogers, 1992). This problem is exacerbated by the fact that cats use only taurine for bile salt formation and cannot conjugate bile acids with glycine as other animals can when taurine supply is low. Food processing has been shown to influence taurine requirements of cats, with higher requirements for cats consuming canned vs. dry food (Hickman *et al.*, 1990). The heat processing that takes place with canning results in the formation of Maillard reaction products, which results in lower taurine digestibility (Morris *et al.*, 1994). Because plants do not contain taurine, animal sources rich in the AA such as meat, poultry, fish, and shellfish must be included in cat food diets.

Arachidonic acid requirement. In the dog, linoleic acid is essential while dihomo γ-linolenic and arachidonic acids are conditionally essential. That is, if provided with sufficient amounts of linoleic acid, the dog has the ability to synthesize adequate amounts of the other two by desaturation and elongation mechanisms. However, due to low Δ5 desaturase and absence of Δ6 desaturase activity in the liver, cats cannot synthesize enough arachidonic acid to meet their metabolic needs and it must be provided in the diet (Rivers *et al.*, 1975). Again relating back to the strict carnivorous nature of the cat, arachidonic acid is only present in animal sources.

Preformed vitamin A. ß-carotene and other carotenoids, the yellow and orange pigments found in foods such as carrots and green vegetables, are vitamin A precursors in most herbivorous and omnivorous species, including the dog. Although cats have the ability to absorb ß-carotene, they lack the 15,15'-dioxygenase enzyme required to convert ß-carotene to vitamin A in the small intestinal mucosa, and require preformed vitamin A in the diet (Schweigert *et al.*, 2002). Because the liver is the primary vitamin A storage organ, fish liver oils and liver and other organ meats from animal sources are rich in preformed vitamin A (e.g., retinyl palmitate, retinyl acetate). Caution should be taken, however, as long-term feeding of liver, organ meats, or fish may lead to vitamin A toxicity.

Preformed vitamin D. When exposed to sunlight, two pro-vitamin D compounds [ergosterol in plants; 7-dehydrocholesterol (7-DHC) in animals] are transformed into vitamin D. In plants, sunlight transforms ergosterol into vitamin D_2 (ergocalciferol), which has vitamin D activity in most animals. In animals, sunlight converts 7-DHC into vitamin D_3 (cholecalciferol). Ergocalciferol and cholecalciferol are both precursors of the active form of vitamin D (1,25-dihydroxycholecalciferol) in most animals. Dogs have an adequate supply of 7-DHC in the skin and produce enough vitamin D to meet metabolic needs if exposed to sunlight. In contrast, cats have a limited capacity to synthesize cholecalciferol (How *et al.*, 1994) and require preformed vitamin D (ergocalciferol and cholecalciferol are both utilized by the cat) in the diet. Their limited capacity to synthesize cholecalciferol is due to the low levels of 7-DHC present in the skin (Morris, 1999). Cats have extremely high 7-DHC-Δ7 reductase concentrations in the skin, rapidly converting 7-DHC into cholesterol and leaving little precursor available for cholecalciferol formation.

Niacin. The AA tryptophan is a precursor for the synthesis of niacin in the liver and intestine. Animal species differ in their ability to convert tryptophan to niacin. Similar to the vitamin D example, cats have all of the necessary enzymes to synthesize niacin, but produce very little because a competing pathway limits the amount of precursor available. Conversion efficiency is largely due to liver concentrations of picolinic acid carboxylase, the enzyme that diverts one of the intermediates (2-amino-3-acroleyl fumaric acid) toward the glutaryl-CoA pathway instead of allowing this compound to condense with quinolinic acid, the immediate precursor of niacin (McDowell, 2000). Cats have the highest picolinic acid concentrations of any animal studied and require niacin in the diet (Ikeda *et al.*, 1965).

Energy and nutrient requirements/recommendations

Numerous agencies and organizations regulate the production, marketing and sale of commercial pet foods in the United States (Table 4.1; Case *et al.*, 2000). Although it has no regulatory responsibilities to the pet food industry, the National Research Council (NRC) plays an important role in companion animal nutrition because it is responsible for collecting and evaluating research and making nutrient recommendations.

Another important agency closely associated with the pet food industry is the Association of American Feed Control Officials (AAFCO), which is responsible for setting standards for substantiation claims and provides an advisory committee for state legislation. Prior to the development and acceptance of the AAFCO Nutrient Profiles, the NRC publications were the recognized authorities for pet food formulation and substantiation of nutritional adequacy claims on the labels of commercial pet foods (Case *et al.*, 2000). Due to some significant flaws associated with the 1985/1986 NRC publications (e.g., lack of safety margins, recommendations made from purified or semi-purified diets), the 1974/1978 NRC versions were still used as the standard by pet food manufacturers. In the early 1990s, the AAFCO Nutrient Profiles were published and quickly replaced the NRC recommendations as the "gold standard" for the pet food industry. The AAFCO recommendations not only provided their suggested nutrient inclusion levels (rather than minimum requirements provided by NRC), but also included maximum levels for some nutrients.

Table 4.1. Governing agencies for commercial pet foods[1].

Agency	Function
Association of American Feed Control Officials (AAFCO)	Sets standards for substantiation claims and provides an advisory committee for state legislation
Canadian Veterinary Medical Association (CVMA)	Administers voluntary product certification
Environmental Protection Agency (EPA)	Regulates use of pesticides in raw materials and feeds; regulates processing plant discharges
Federal Trade Commission (FTC)	Regulates trade and advertising
Food and Drug Administration (FDA)	Specifies permitted ingredients and manufacturing procedures
National Research Council (NRC)	Collects and evaluates research and makes nutrient recommendations
Pet Food Association of Canada	Trade organization that represents pet food manufacturers
Pet Food Institute (PFI)	Trade organization that represents pet food manufacturers
State Feed Control Offices	Enforces Commercial Feed Law within states
United States Department of Agriculture (USDA)	Regulates pet food labels and research facilities

[1]Adapted from Case *et al.*, 2000.

Prior to the 2003 NRC publication, the most recent NRC publications for dog and cat energy and nutrient requirements were 1985 (NRC, 1985) and 1986 (NRC, 1986), respectively. The 2003 NRC publication is much improved from the 1985/1986 versions in both format and content, reflecting the considerable amount of published research performed in the area of companion animal nutrition over the past 20 years. The remainder of this chapter will provide an overview of the most recent NRC publication for dog and cat energy and nutrient requirements (NRC, 2003) with special emphasis placed on the most significant changes from the previous NRC publications.

While the difference in sheer volume of text between NRC versions is obvious (~450 pages in 2003 vs. 79 and 78 pages in 1985 and 1986 versions, respectively), changes in format and content are also very apparent. Because of the negativity associated with the previous versions, the new NRC publication was completely revamped, producing a publication of much higher quality and usefulness for companion animal nutritionists. Given its importance as a scientific document, the authors of the 2003 NRC document used mostly peer-reviewed publications to establish energy and nutrient recommendations and did not consider much unpublished information. The recent version also was developed using information from the study of practical diets rather than semi-purified diets. In contrast to the 1985/1986 versions, which did not include any margin of safety with recommendations, energy and nutrient recommendations in the new version were adjusted when sufficient information was available. In addition to the minimal energy and nutrient requirements provided in previous versions, the 2003 version also provides an adequate intake (AI), recommended allowance (RA), and safe upper limit (SUL) for nutrients where sufficient data exist. Format of tables in the 2003 version is much more informative, but more complicated as well. In addition to providing a minimum, AI, RA, and SUL for each nutrient (when sufficient data exist), separate recommendations have been provided for the life stages of growth and late gestation/lactation, and recommendations are provided in several formats, all of which will be discussed below. Finally, the list of energy and nutrient recommendations has expanded from the previous NRC versions that focused only on classically essential nutrients, to include classically and conditionally essential nutrients.

As Table 4.2 demonstrates, several topics (chapters) have been added or expanded in the recent version that were either overlooked or incompletely covered in the 1985/1986 versions. Briefly, Comparative Digestive Physiology of Dogs and Cats (chapter 1), Feeding Behavior in Dogs and Cats (chapter 2), Special Considerations for Laboratory Animals (chapter 10), Physical Activity and Climate (chapter 11), and Other Food Constituents (chapter 14) were all new additions to the 2003 NRC publication. In addition to adding several new chapters, existing chapters were significantly improved. Unlike the previous versions that discussed energy and all nutrients (except water) in one chapter (26 and 37 total pages in the cat and dog versions, respectively), the 2003 publication has developed a chapter for energy and each nutrient class separately (~250 total pages). Finally, Diet Formulation and Feed Processing (chapter 12) and Nutrient Composition of Ingredients Used in Dog and Cat Foods (chapter 13) sections have been expanded from ~5 pages in previous versions to ~40 pages in the current version. The ingredient composition list has been greatly expanded and is now a very useful database for companion animal nutritionists.

Table 4.2. Comparison of material presented in National Research Council (NRC) nutrient requirements of dogs and cats publications.

Chapt.	2003 (Dogs and Cats) Title	Pages	1986 (Cats) Title	Pages	1985 (Dogs) Title	Pages
1	Comparative Digestive Physiology of Dogs and Cats	1-20	Introduction	1-2	Introduction	1
2	Feeding Behavior in Dogs and Cats	21-27	Nutrient Requirements	3-28	Nutrient Requirements and Signs	2-38
3	Energy	29-51	Water	29	Water	39
4	Carbohydrates and Fiber	55-90	Formulated Diets for Cats	30-33	Composition of Ingredients of Dog Foods	40-41
5	Fat and Fatty Acids	91-123	Other Food Constituents	34-36	Formulated Diets for Dogs	42-43
6	Protein and Amino Acids	125-162	Composition of Feeds	37-38	Tables	44-62
7	Minerals	163-216	Tables	41-68	References	63-75
8	Vitamins	217-278	References	69-76	Index	77-79
9	Water	279-285	Index	77-78		
10	Special Considerations for Laboratory Animals	287-293				
11	Physical Activity and Climate	295-355				
12	Diet Formulation and Feed Processing	357-363				
13	Nutrient Composition of Ingredients Used in Dog and Cat Foods	365-400				
14	Other Food Constituents	401-411				
15	Nutrient Requirement Tables	413-438				
	Acronyms	439-443				
	About the authors	445-447				

Just as the extent to which energy and nutrients have been discussed in the text has been greatly expanded or changed, so has the format, number, and complexity of the energy and nutrient requirement tables. For Example, Table 4.3 presents a list of the energy and nutrient requirement tables provided in the 2003 NRC publication. Although the tables in the 2003 NRC are more complicated (i.e., cannot simply pull a number from a table), they are more informative and should result in more accurate dietary recommendations for dogs and cats. According to NRC (2003), the minimum requirement (MR) is defined as the minimal concentration or amount of a maximally bioavailable nutrient that will support a defined physiological state. Adequate intake is defined as the concentration or amount of a nutrient demonstrated to support a defined physiological state when no MR has been demonstrated. Recommended allowance is based on the MR with consideration for the normal variation in bioavailability of the nutrient in feed ingredients of typical quality. If no MR is available for a given nutrient, the RA is based on the AI value. The NRC RA is comparable to AAFCO's minimum recommendation. The SUL is the maximal concentration or amount of a nutrient that has not been associated with adverse effects (NRC, 2003) and is comparable to AAFCO's maximum level. The 2003 NRC for dogs and cats is the first NRC publication to use the SUL term. The values provided for each category are dependent on the refereed data available. Therefore, some nutrients do not have values provided for each category. All nutrients do, however, have an RA associated with them.

While the MR, AI, and RA appear relatively straightforward as regards interpretation, the SUL used in the 2003 NRC may be misinterpreted, and its use is not accepted by many in the pet food industry. The controversy surrounding the use of the SUL stems from several issues. Safe upper limit is defined as the maximal concentration or amount of a nutrient that has not been associated with adverse effects. Therefore, the SUL is completely dependent on concentrations that have been tested in published experiments and thus is different than that proven to be toxic or harmful. A similar issue is whether the actual value tested in the experiment is listed as the SUL or if a safety factor has been included. Finally, the nature of adverse effects when an SUL is exceeded is different depending on nutrient. Consider the adverse effects of overconsuming a carbohydrate such as raffinose that may lead to loose stools, frequent defecation, and reduced nutrient digestibility, but is a reversible condition, as compared with overconsuming Ca, which may result in permanent skeletal disorders and (or) deformities. Although inclusion of SUL is controversial, the NRC publication is a scientific as compared to a regulatory document. The NRC defined the most accurate SUL values possible, given the published data available. This information may or may

Table 4.3. Nutrient requirement tables provided in the 2003 NRC publication.

Table	Title
15-1	Change in Requirements and Allowances Expressed Relative to ME (per 1,000 kcal[4.184MJ]) if Requirements Vary Directly with Body Weight
15-2	Daily Metabolizable Energy Requirements for Growth of Puppies
15-3	Dietary Nutrient Concentrations and Daily Nutrient Amounts for Growth in a 5.5-kg, 3-Month Old Puppy Consuming 1,000 kcal [4.184MJ]ME•day^{-1} (Energy Concentration of 4 kcal [16.74kJ]ME•g^{-1} DM) After Weaning
15-4	Daily Metabolizable Energy Requirements for Adult Dogs at Maintenance
15-5	Dietary Nutrient Concentrations and Daily Nutrient Amounts for Maintenance in a 15-kg Adult Dog Consuming 1,000 kcal [4.184MJ] ME•day^{-1} (Energy Concentration of 4 kcal [16.74kJ]ME•g^{-1} DM)
15-6	Daily Metabolizable Energy Requirements for Bitches in Late Gestation
15-7	Daily Metabolizable Energy Requirements (Maintenance Plus Milk Production) for Lactating Bitches Based on Number of Puppies and Weeks of Lactation
15-8	Dietary Nutrient Concentrations and Daily Nutrient Amounts for Late Gestation or Peak Lactation in a 22-kg Bitch with Eight Puppies Consuming 5,000 kcal [20.92MJ]ME•day^{-1} (Energy Concentration of 4 kcal [16.74kJ]ME•g^{-1} DM)
15-9	Daily Metabolizable Energy Requirements of Growth in Kittens
15-10	Dietary Nutrient Concentrations and Daily Nutrient Amounts for Growth in an 800-g Kitten consuming 180 kcal [753kJ]ME•day^{-1} (Energy Concentration of 4 kcal [16.74kJ] ME•g^{-1} DM)
15-11	Daily Metabolizable Energy Requirements for Adult Cats at Maintenance
15-12	Dietary Nutrient Concentrations and Daily Nutrient Amounts for Maintenance for a 4-kg Adult Cat Consuming 250 kcal [1046kJ] ME•day^{-1} (Energy Concentration of 4 kcal 16.74kJ]ME•g^{-1} DM)
15-13	Daily Metabolizable Energy Requirements for Lactating Queens
15-14	Dietary Nutrient Concentrations and Daily Nutrient Amounts for Late Gestation or Peak Lactation in a 4-kg Queen with Four Kittens Consuming 540 kcal [2259kJ]ME•day^{-1} (Energy Concentration of 4 kcal [16.74kJ]ME•g^{-1} DM)

not be a resource for regulatory agencies such as AAFCO to adjust maximal concentrations used in pet food formulation. The controversy surrounding use of SUL underpins the great need for additional research in this area.

Several assumptions regarding energy concentration of the diet and animal BW and energy intake are required to standardize the data presented in tables. These assumptions are listed as footnotes in tables in the 2003 NRC document. Energy concentration of foods for both dogs and cats was assumed to be 4.0 kcal [16.74 kJ]ME/g dry matter (DM). This energy concentration assumption is different than that used by AAFCO (3.5 kcal [14.64kJ] ME/g DM), the 1985 Dog NRC (3.67 kcal [15.36kJ] ME/g DM), and 1986 Cat NRC (5.0 kcal [20.92kJ] ME/ g DM), and must be considered when comparing values from these documents. Nutrient requirements for growing puppies in the 2003 NRC are based on a 5.5-kg puppy consuming 1,000 kcal [4.184 MJ]ME/ d. For crude protein (CP) and AA, two requirements for growth are provided, depending on age of puppy (4-14 wk old; >14 wk old). This is different than the 1985 NRC that used a 3-kg puppy consuming 600 kcal [2510kJ]ME/d, and the AAFCO Nutrient Profiles that combines growth and reproduction into one profile and uses the higher value for either life stage. The 15-kg adult dog consuming 1,000 kcal [4.184MJ]ME/d for adult maintenance used by the 2003 NRC is also different than that of AAFCO (no body size or caloric intake assumed) and the 1985 NRC (10-kg dog consuming 742 kcal [3105kJ]ME/d). The 2003 NRC publication also provides adjustment factors (% change) for nutrient requirements that are based on body size of both dogs and cats. Unlike the previous NRC and AAFCO (2004) publications, the 2003 NRC also provides separate energy nutrient requirements for gestating and lactating bitches, based on a 22-kg bitch with eight puppies in late gestation or peak lactation consuming 5,000 kcal [20.92MJ]ME/d. Metabolizable energy requirements for lactating bitches, which are based on number of puppies in the litter and week of lactation, also are provided. The previous NRC publication singled out late pregnancy and lactation for MER, but not nutrient intake. Again, the AAFCO Nutrient Profiles combine growth and reproduction into one profile and choose the greater of the two values and thus do not differentiate gestation or lactation from growth.

Nutrient requirements for growing kittens in the 2003 NRC publication are based on an 800-g kitten consuming 180 kcal [753kJ]ME/d. Similar to that of dogs, AAFCO profiles for cats combine growth and reproduction into one profile. The 1986 Cat NRC based requirements on a 10-20 wk old kitten, but did not specify BW or energy intake. Nutrient requirements for adult cats at maintenance are based on a 4-kg adult cat consuming 250 kcal [1046kJ]ME/d in the 2003 NRC. Nutrient requirements for gestating and lactating cats are based on a 4-kg queen with four kittens in late gestation or peak lactation consuming 540 kcal [2259kJ]ME/d. Metabolizable energy requirements for lactating queens, which are based on number of kittens in the litter and week of lactation, are provided. As with dogs, AAFCO does not specify body size or energy intake for adult cats at maintenance. The 1986 Cat NRC does not provide a separate table for adult cats, but makes mention of changes needed for gestation and lactation (adjust arachidonic acid, zinc, vitamin A, and taurine concentrations compared to kitten requirements) and reproduction (adjust vitamin A and taurine concentrations compared to kitten requirements) in adult cats in a footnote of the table providing requirements for growing kittens.

For each category of nutrient requirement (MR, AI, RA, SUL), units of expression are presented as: 1) amount per kg of dietary DM; 2) amount per 1,000 kcal [4.184MJ]ME per day; 3) amount per kg BW; and 4) amount per kg of $BW^{0.75}$ for dogs and of $BW^{0.67}$ for cats (NRC, 2003). While this system is more complicated than the previous Cat (per kg DM only) and Dog (per kg BW and 1,000 kcal [4.184MJ]ME only) NRC publications and AAFCO profiles (per kg DM and 1,000 kcal [4.184MJ]ME only), it will provide the reader with several options for calculating nutrient requirements. Because the dietary ME content determines amount of food consumed, nutrient requirements expressed relative to ME are probably the most appropriate.

In contrast to AAFCO Nutrient Profiles that provide percentage values recommended in the diet, the 2003 NRC publication presents fat, protein, and some mineral requirements in units of g/kg. Some vitamins and minerals are in units of mg or μg. The most obvious unit difference from previous NRC versions occurred with the fat-soluble vitamins. In the 1985 Dog NRC, vitamins A, D, and E were expressed as international units (IU) (no requirement was listed for vitamin K). The 1986 Cat NRC expressed the fat-soluble vitamins in terms of IU as well as mg (vitamins A and E) and mg (vitamins D and K). The 2003 version presents requirements of vitamin A as retinol equivalents (RE; 1 IU = 0.3 RE), vitamin D as μg (1 IU = 0.025 μg), vitamin E as mg (1 IU = 1 mg, but depends on isomer), and vitamin K as mg.

Finally, the tables in the 2003 NRC now include requirements of some "conditionally essential" nutrients not included in previous versions, and for some nutrients that previously had too little data to make an accurate recommendation. The FA requirements now include linoleic, α-linolenic, arachidonic, and eicosapentaenoic and docosahexaenoic acids. The 1986 Cat NRC only provided requirements for linoleic and arachidonic acids, while the 1985 Dog NRC listed only linoleic acid. The 2003 NRC also added RAs for phenylalanine (dogs), vitamin K (dogs), and arachidonic acid (puppies), which were not provided in previous versions.

Conclusion

Much new data have been generated in the area of companion animal nutrition since the last dog and cat NRC documents were published (1985 and 1986 for the Dog and Cat NRC publications, respectively). Authors of the 2003 NRC publication on dog and cat nutrition have thoroughly analyzed these data and revised and (or) developed new energy and nutrient requirements for dogs and cats according to life stage, reproductive status, and BW. These authors have dramatically overhauled the previous NRC documents into one highly informative and useful document, adding discussion on numerous new topics, greatly expanding the content of previous documents, and presenting nutrient requirements in a variety of units useful to companion animal nutritionists. Although some of the formatting changes have led to more complexity, this level of precision is required for the wide diversity among domestic dog and cat populations. Although it is a scientific, not regulatory, document, the 2003 NRC Nutrient Requirements of Dogs and Cats will be a valuable resource for students, teachers, researchers, pet food industry personnel, and regulatory agents.

Editor's note; Dietary energy in the original chapter was expressed in kcal which has been retained. However data have also been expressed in MJ using the conversion factor: 1000kcal = 4.184 MJ (4184kJ)

Literature cited

AAFCO (2004) *Official publication.* Association of American Feed Control Officials Inc., Oxford, IN.

Adolph, E. F. (1939) Measurement of water drinking in dogs. *Am. J. Physiol.* **125**, 75-86.

Case, L. P., Carey, D. P., Hirakawa, D. A. and Daristotle, L. (2000) Canine and Feline Nutrition (2nd ed.). Mosby Inc., St. Louis, MO.

Guilbot, A. and Mercier, C. (1985) Starch. In *The Polysaccharides*, pp 209-282. Edited by G. O. Aspinall. Academic Press, New York, NY.

Hardison, W. G., Wood, C. A. and Proffitt, J. H. (1977) Quantification of taurine synthesis in the intact rat and cat liver. *Proc. Soc. Exp. Biol. Med.* **155**, 55-58.

Hickman, M. A., Rogers, Q. R. and Morris, J. G. (1990) Effect of processing on the fate of dietary [14C] taurine in cats. *J. Nutr.* **120**, 995-1000.

How, K. L., Hazewinkel, H. A. and Mol, J. A. (1994) Dietary vitamin D dependence of cat and dog due to inadequate cutaneous synthesis of vitamin D. *Gen. Comp. Endocrinol.* **96**, 12-18.

Hussein, H. S. (2003a) Basic nutrient requirements for healthy adult dogs. In *Petfood Technology*, pp 2-13. Edited by J. L. Kvamme and T. D. Phillips. Watt Publishing, Mt. Morris, IL.

Hussein, H. S. (2003b) Basic nutrient requirements for healthy adult cats. In Petfood *Technology*, pp 14-28. Edited by J. L. Kvamme and T. D. Phillips. Watt Publishing, Mt. Morris, IL.

Ikeda, M. H., Tsuji, H., Nakamura, S., Ichiyama, A., Nishizuki, Y. and Hayaishi, O. (1965) Studies on the biosynthesis of nicotinamide adenine dinucleotides. II. Role of picolinic acid carboxylase in the biosynthesis of NAD from tryptophan in mammals. *J. Biol. Chem.* **240**, 1395-1401.

Institute of Medicine (2001) *Proposed definition of dietary fiber.* A Report of the Panel on the Definition of Dietary Fiber and the Standing Committee on the Scientific Evaluation of Dietary Reference Intakes, Food and Nutrition Board. National Academy Press, Washington, DC.

Lazar, V. (1990) Dog food history. *Pet Food Ind.* Sept/Oct, pp.40-44.

McDowell, L. R. (2000) Vitamins in Animal and Human Nutrition (2nd ed.). Iowa State University Press, Ames, IA.

McDowell, L. R. (2003) Minerals in Animal and Human Nutrition (2nd ed.). Elsevier Science Ltd., Amsterdam, The Netherlands.

Morris, J. G. (1999) Ineffective vitamin D synthesis in cats is reversed by an inhibitor of 7-dehydrocholesterol-delta7-reductase. *J. Nutr.* **129**, 903-908.

Morris, J. G. and Rogers, Q. R. (1978) Ammonia intoxication in the near-adult cat as a result of a dietary deficiency of arginine. *Science* **199**, 431-432.

Morris, J. G. and Rogers, Q. R. (1992) The metabolic basis for the taurine requirement of cats. Taurine: Nutritional value and mechanisms of action. *Adv. Exp. Med. Biol.* **315**, 33-34.

Morris, J. G., Rogers, Q. R., Kim, S. W. and Backus, R. C. (1994) Dietary taurine requirement of cats is determined by microbial degradation of taurine in the gut. *Adv. Exp. Med. Biol.* **359**, 59-70.

National Research Council (NRC) (1985) *Nutrient Requirements of Dogs.* National Academy Press, Washington, DC.

National Research Council (NRC) (1986) *Nutrient Requirements of Cats.* National Academy Press, Washington, DC.

National Research Council (NRC) (2003) *Nutrient Requirements of Dogs and Cats.* National Academy Press, Washington, DC.

Pond, W. G., Church, D. C. and Pond, K. R. (1995) Dogs and cats. In *Basic Animal Nutrition and Feeding*, pp 531-545. John Wiley & Sons, Inc.,

New York, NY.

Rivers, J. P. W., Sinclair, A. J. and Crawford, M. A. (1975) Inability of the cat to desaturate essential fatty acids. *Nature* **258**, 171-173.

Rogers, Q. R., Morris, J. G. and Freedland, R. A. (1977) Lack of hepatic enzymatic adaptation to low and high levels of dietary protein in the adult cat. *Enzyme* **22**, 348-356.

Rogers, Q. R. and Morris, J. G. (1980) Why does the cat require a high protein diet? In *Nutrition of the Dog and Cat*, pp 45-66. Pergamon Press, New York, NY.

Salati, L. M. and Goodridge, A. G. (1996) Fatty acid synthesis in eukaryotes. In *Biochemistry of Lipids, Lipoproteins, and Membranes*, pp 101-128. Edited by D. E. Vance and J. E. Vance. Elsevier Science, Ltd., Amsterdam, The Netherlands.

Schweigert, F. J., Raila, J., Wichert, B. and Kienzle, E. (2002) Cats absorb ß-carotene, but it is not converted to vitamin A. *J. Nutr.* **132**, 1610S-1612S.

Sunvold, G. D. and Bouchard, G. F. (1998) Assessment of obesity and associated metabolic disorders. In *Recent Advances in Canine and Feline Nutrition, Vol. II*, pp 135-148. Edited by G. A. Reinhart and D. P. Carey. Orange Frazer Press, Wilmington, OH.

5

NEW EUROPEAN LEGISLATION ON FEED ADDITIVES

M.A. GRANERO-ROSELL
European Commission, Health & Consumer Protection Directorate-General, Animal Nutrition Unit, Brussels

This chapter has been prepared from a transcript of the presentation given at the Nottingham Feed Conference in September 2004. Views expressed here are those of the author and not necessarily those of the Commission

Introduction

This chapter presents the new Regulation 1831/2003 of the European Parliament and of the Council on additives used in animal nutrition. The areas to be discussed are:

* Background: Steps in the preparation of the Regulation
* Main objectives of the Regulation
* Main provisions
* Other provisions
* Feed additives: the new definition, requirements and categories
* The Community Reference Laboratory for Feed Additives Authorisation

Preparation of the Regulation

The steps in the institutional decision-making process of Regulation 1831/ 2003 were:

1. The Commission's Initial Proposal was made in March 2002
2. The European Parliament (EP) delivered its opinion in first reading in November 2002

3. The Council of Ministers agreed its "Common Position" on the proposal in March 2003
4. The EP had the second reading in April 2003
5. The final text of the Regulation, incorporating amendments made during the previous steps, was adopted by the EP and Council in September 2003 and published in October 2003

This is the usual procedure for all European Union legislation. The Regulation then entered a transitional period from November 2003 and full entry into application in October 2004. Some aspects of the Regulation were already in force in September 2004, in particular the notification procedure for existing products currently authorized.

Interpretation of the intentions of legislation is always subjective and sometimes it is difficult to understand the precise goals of details of particular provisions included in a piece of legislation. European Legislation arises from complex interactions between the Commission, the European Parliament and the Member States, who all have significant input into the legislation. The end result for which no single body or institution feels entirely satisfied, and which might not exactly meet, for example, the initial objectives of the Commission. Nevertheless, it is legislation.

The Main objectives in the initial proposal of Regulation 1831/2003 were:

• To align rules for feed additives with General Food Law Regulation 178/2002, as announced in the White Paper on Food Safety in 2000
• To produce clear separation between risk assessment and risk management
• To simplify procedures for authorization

The main provisions of Regulation 1831/2003 are:

• Phasing out the use of antibiotics (other than coccidiostats and istomonostats) by 31 December 2005
• Some categories of feed additives were redefined, and the categories are now:

  Technological additives
  Sensory additives
  Nutritional additives
  Zootechnical additives
  Coccidiostats

- The scope of the Regulation was extended to include amino-acids and their analogues, urea and urea derivatives (which were in Directive 82/471/EEC), and silage additives (which had not been reflected previously in Community legislation)
- The Regulation takes into account that additives may be used by adding them to feed, or may be supplied through drinking water or other routes.

The main provisions of Regulation 1831/2003 concerning procedures are:

- Each authorisation is time-limited (ten years)
- Authorisation is granted using a Commission Regulation
- The European Food Safety Authority (EFSA) carries out the risk assessment of the applications entirely. (There is no longer a Member State rapporteur)
- The EFSA evaluation period is 6 months

Other provisions of Regulation 1831/2003 useful to highlight are:

- The regulation provides for re-evaluation of all additives already authorised and this has to be completed by 2010
- It introduces the possibility of post-market monitoring requirements, for example maximum residue limits, on a case-by-case basis
- There is a transition period for existing products currently on the market, which must follow a notification system
- Notifications are verified by EFSA and authorisations for products not notified will be withdrawn
- There is a possibility of clarification of existing authorisations given to groups of products, such as flavourings and some vitamins
- A Register of Feed Additives will be established to clarify the status of products that are authorised
- To increase transparency of the authorisation process, data in dossiers compiled during the evaluation procedure will be made available. These include a summary of the application, detailed opinion of the EFSA, and a more detailed authorisation that will include items such as maximum residue limits and methods of analysis.
- Authorisations will be given to specified authorisation holders in the case of zootechnical additives and products consisting, containing or derived from genetically modified organisms (GMOs)
- Development of updated, comprehensive and flexible guidelines for applicants. These include general guidelines produced by the

Commission in the form of legal requirements and additional detailed guidance produced by EFSA

The Regulation contains the following new definition of Feed Additives:

Substances, micro-organisms or preparations intentionally added to feed in order to perform one or more of the following functions:

a) Favourably affect the characteristics of *feed*,
b) Favourably affect the characteristics of *animal products*,
c) Favourably affect the colour of ornamental fish and birds,
d) Satisfy the *nutritional needs* of animals,
e) Favourably affect the *environmental consequences of animal production*,
f) Favourably affect *animal production, performance or welfare*, particularly by affecting the gastro-intestinal flora or digestibility of feedingstuffs, or
g) Have a *coccidiostatic or histomonostatic effect*

The requirements for Feed Additives are:

A feed additive shall not:

a) Have an adverse effect on animal health, human health or the environment,
b) Be presented in a manner which may mislead the user,
c) Harm the consumer by impairing the distinctive features of animal products or mislead the consumer with regard to the distinctive features of animal products.

The new categories of feed additives are:

- Technological additives, e.g. preservatives, acidity regulators, antioxidants, emulsifiers, anti-caking agents
- Sensory additives, e.g. colorants, flavourings
- Nutritional additives, e.g. vitamins, trace elements
- Zootechnical additives, e.g. micro-organisms (including probiotics), enzymes (including digestibility enhancers)
- Special zootechnical additives, e.g. coccidiostats and others

Characteristics of the Regulation that are similar to previous regulations are:

- Positive list principle: only products on the list can be used, and under the conditions of use specified in the authorisation
- Authorisation is given at Community level
- Conditions of use: used only for specific categories of animal species, with defined limitations in doses, concentrations, handling, and other conditions
- *Scientific evaluation* is conducted by EFSA on the data submitted to demonstrate that the additive has no harmful effects: for *human health*, for *animal health*, for *the user*, or for *the environment*,
- and that the additive performs the claimed functions
- Certain authorisations are linked to a legally responsible person (a specified authorisation holder); others are a general authorisation, not linked to a specific person

Community Reference Laboratory on Feed Additive Authorisations

Regulation 1831/2003 introduced a Community Reference Laboratory (CRL) for the evaluation of methods of analysis of feed additives and for storage of reference samples for the whole EU. The CRL is the Joint Research Centre, located at the Institute of Reference Materials and Testing in Geel, Belgium.

The main tasks of the CRL are:

- Reception of reference samples
- Evaluating the methods of analysis described in the application dossier
- Submitting an evaluation report to EFSA
- Validation of methods in certain cases
- Testing also possible
- Other tasks

The CRL may be assisted by a Consortium of national reference laboratories. Requirements and roles are defined for laboratories participating in the Consortium

It is likely that a flat fixed fee will be charged to applicants in order to support running costs of the CRL. The CRL may provide part of this fee to the Rapporteur laboratory in charge of the evaluation. The participant laboratories will not receive financial support from the Commission for this task.

6

LEGISLATION THAT WILL HAVE A SIGNIFICANT IMPACT ON THE FEED INDUSTRY, PARTICULARLY IN THE UK

KEITH MILLAR
Head of Animal Feed Unit, Food Standards Agency, London

Introduction

This chapter has a slightly narrower focus than ones published on previous occasions. However, it is intended to give a clear snapshot of European Community (EC) controls adopted in the last eighteen months, and those subject to on-going negotiations in Brussels, that have a bearing on the UK feed industry and other relevant stakeholders. Officials from the Animal Feed Unit of the Food Standards Agency (FSA) represent the United Kingdom at most EC animal feed negotiations and implementation into domestic law falls to the Agency, the Department for Environment, Food and Rural Affairs (Defra) and the Veterinary Medicines Directorate.

As in previous years, there has been a steady stream of animal feed legislation flowing from Brussels that has been agreed at either Ministerial or official level. Most legislation is of a highly technical nature and can be introduced by the European Commission following positive qualified majority votes by government expert officials. Wider ranging controls with greater political ramifications are subject to agreement under co-decision procedures requiring joint clearance by the Council of Ministers and the European Parliament.

General food and feed law

Council and European Parliament Regulation 178/2002 sets out the general framework for Community food law and most feed law will flow from it in years to come. The relevant provisions on feed apply from 1 January 2005. The Regulation also established the European Food Safety Authority (EFSA), which acts as the European Commission touchstone regarding risk assessment. EC Regulations are directly applicable in all Member States and cannot be repeated in national legislation, but it is necessary to link them to enforcement powers.

National legislation will make it an offence to contravene or fail to comply with specific articles in Regulation 178/2002. These are set out mainly in Articles 15, 16, 18 and 20 of the Regulation. Article 15 contains a prohibition on the placing on the market, or the feeding to food-producing animals, of unsafe feed. Article 16 prohibits the presentation of feed in such a way as to mislead consumers. Article 18 requires feed business operators to have traceability systems in place in relation to input products and to products supplied by that business. Article 20 lays down the responsibilities and obligations of feed business operators, particularly with regard to feed that does not, or may not, satisfy feed safety requirements.

Feed additives regulation

Updating EC rules on feed additives was one of the priority actions outlined in the Commission's White Paper on Food Safety issued in 2000. It was a measure specifically promised by Commissioner Byrne during his inauguration hearings in the European Parliament in September 1999. The new Regulation 1831/2003 represents both a strengthening and a streamlining of the laws on safety evaluation and marketing authorisation of feed additives.

Under the Regulation only additives that have been through an assessment and authorisation procedure can be put on the market, used or processed. Authorisations would be for specific animal species and with a maximum rate of inclusion as appropriate. Authorisations of new feed additives would be for a ten-year (renewable) period only. The rules require that companies demonstrate the additive's positive effect for the animal (efficacy), that it is of suitable quality, and the absence of significant risks for human health, animal health and the environment (safety). The European Food Safety Authority (EFSA) will be responsible for conducting these evaluations; this includes recommendations for maximum residue levels (MRLs).

The Regulation covers a wide range of additives, which can be divided into five broad categories:

- technological additives (e.g. preservatives);
- sensory additives (e.g. flavours, colorants);
- nutritional additives (e.g. vitamins);
- zootechnical additives (e.g. gut flora improvers)

• coccidiostats and histomonostats (additives to prevent certain diseases in poultry and rabbits).

Maximum residue limits in livestock products will be established for some feed additives where it proves necessary. A post-monitoring system as well as regular testing of foodstuffs, as is already common practice, will help to ensure that these are observed. In the case of coccidiostats, stricter measures will be introduced if they also have antibiotic activity. In such cases, a new dossier for re-evaluation within a four-year period will have to be presented and MRLs could be set to minimise risks to human or animal health. Reflecting existing statutory requirements, feed additives will need to be clearly labelled.

If the additive complies with safety and other requirements the Commission, within three months of receipt of the EFSA opinion, will generally propose a draft Regulation authorising the additive for a ten-year period and setting MRLs for the active substance in the additive where appropriate. If the draft Regulation is not in line with the opinion from EFSA, the Commission has to explain its reasoning. All authorisations will be renewable for ten-year periods; application for renewal should be made to the EFSA at least one year before the expiry date.

Authorisations in certain categories will be linked to an authorisation holder who will be responsible for the implementation of a post-monitoring plan. The authorisation holder will have to communicate to the Authority any new information concerning the safety of the product.

The Regulation also requires the re-evaluation of all existing feed additives. In order to ensure continued authorisation of these additives, companies marketing them or other interested parties have to notify the Commission and EFSA by November 2004 with details of the current authorisation of the additive. EFSA will verify this information and inform the Commission. Within another six years dossiers supporting re-authorisation based on safety, quality and efficacy must be submitted to EFSA for assessment. If an existing authorisation is the subject of an expiry date, a relevant dossier must be submitted at least one year before that date.

Official feed and food controls

Council and European Parliament Regulation 882/2004 was adopted earlier this year and sets out general requirements for the competent authorities, or enforcement bodies, that are responsible for checking that businesses are complying with feed and food legislation (and animal health and welfare rules). This includes controls of products produced and sold within the EU as well as controls on imports from and exports to third countries. It also includes a framework for financing of official controls.

The Regulation consolidates and extends existing sector specific rules, the aim being to improve the consistency and effectiveness of controls across the Community and thereby raise standards of food safety and consumer protection and also facilitate the functioning of the internal market. The objective is to create a more comprehensive and integrated, risk-based, EU-wide, 'farm to table' approach to official controls.

With regard to the feed sector, the key new provisions are as follows:

Imports - the Regulation introduces harmonised rules at EU level for import controls on products of non-animal origin (non-POAO). The provisions include mandatory requirements for prior notification of consignments of 'high risk' non-POAO, including 'high risk' non-POAO feed, and for controls on such products to be undertaken at specified designated ports. This is the best way to ensure targeting of controls and effective management of risks and is justified in the interests of safeguarding public and animal health. The 'high risk' products will be identified at EU level through comitology procedures and will be subject to a greater level of control. It is also worth noting that for other ('low risk') feed, the option to require notification and to carry out checks at specified ports is retained.

Financing – in general terms, the provisions require mandatory fees in those sectors where this is a requirement under existing Community legislation, with the option for Member States to charge in other sectors. In terms of the feed sector, the important elements are:

- mandatory fees for approvals of feed establishments (as now);
- recovery of costs from businesses for 'excess' controls required following detection of non-compliance. The Commission has confirmed, however, that this applies only in the case of significant issues (e.g. major dioxin incidents) which are not foreseen in national control plans;
- fees may be established for import controls on 'high risk' non-POAO, including 'high risk' non-POAO feed, when such products are identified through the EU comitology procedure; and
- the Commission will review the charging arrangements within three years of the Regulation coming into force with a view to extending the range of sectors subject to mandatory fees.

The Regulation is directly applicable and Food Standards Agency officials are considering whether any legislative or administrative measures need to be revoked, amended or introduced to supplement the EC Regulation and will be consulting with stakeholders in due course.

GM Food and Feed Regulation

Regulation (EC) No 1829/2003 dealing with authorisation procedures and labelling issues for genetically modified food and feed became law in the European Union on 18 April 2004. The Regulation replaces the previous approval procedures for GM foods, as contained in Regulation 258/97, and introduces for the first time rules for the labelling of GM animal feed and a harmonised procedure for the scientific assessment and authorisation of GMOs and GM food and feed. Products from animals fed GM animal feed will continue to be exempt from the labelling requirements.

The Regulation requires the labelling of all food and feed ingredients produced from GM sources. However foods produced with GM processing aids or enzymes and products from animals fed with GM feed will not require labelling. A threshold of 0.9% for the adventitious presence of GM material will apply, below which labelling of food or feed is not required. There will also be a 0.5% threshold for the adventitious presence of GM material yet to be authorised in Europe, provided it has a favourable safety assessment from the EU's scientific committees. This latter threshold will last for three years. There will be a centralised authorisation procedure, as set out in the original Commission proposal, which will require applications to be sent to a Member State to forward on to the European Food Safety Authority for assessment.

The UK supports genuine consumer choice and has consistently argued for a policy based on sound science that was practicable and enforceable. The Agency believes the compromise text failed to meet these objectives. The UK could not support threshold values less than 1% nor the revised centralised authorisation procedure without the Article 308 of the Treaty of Rome as a legal base. For these reasons the UK voted against the compromise and submitted a declaration for the Council minutes explaining why the UK could not agree with the proposal.

Regulation on Traceability and Labelling of Genetically Modified Organisms

At the EU Environment Council in July 2003 a qualified majority of Member States reached a political agreement on the EC proposal on the traceability and labelling of GMOs. Key elements of the agreement reached included:

(a)   a threshold of 0.9% for the adventitious (accidental) presence of GMOs for labelling purposes;

(b)   a 0.5% threshold for the adventitious presence of GMOs with a favorable risk assessment from an EU scientific committee but not yet authorised in the EU;

(c)   the labelling of all foods derived from GMOs irrespective of whether there is DNA or protein of GM origin in the final product;   and

(d)   identification of bulk shipments of agricultural commodities requiring operators to provide a list of all viable GMOs which are in a shipment on its arrival into the EU.

Feed hygiene

Agriculture Ministers adopted the Feed Hygiene Regulation in April 2004 which fulfils a commitment in the Commission's White Paper on Food Safety (January 2000) and contains provisions to strengthen feed safety, including the traceability of feed in the case of incidents. Its provisions apply from 1 January 2006. However, for feed businesses that require approval or registration, the conditions have to be complied with by 1 January 2008.

All feed businesses, involved in making or marketing feeds and feed products will have to be approved or registered and comply with certain standards concerning facilities, storage, and record keeping. This includes farms and food manufacturers selling co-products for feed. In addition, businesses (but not farms) will have to ensure that all potential hazards are identified and controls put in place under HACCP principles. Feed businesses will only be permitted to import feed from non-EU country establishments, if these establishments comply with the requirements of the Regulation. The Regulation exempts from its scope, farms which sell small quantities of feed to other local farms, although they will continue to be subject to existing feed safety legislation.

During negotiations a provision that would have made it mandatory for businesses to have financial guarantees to cover the withdrawal of feed and food and animals produced therefrom was removed. The European insurance industry indicated that it did not have the capacity to provide such cover and costs were likely to be prohibitively expensive. Such guarantees will now only be introduced, if appropriate, after a Commission feasibility study has been conducted.

Dioxins

When Directive 2002/32/EC on undesirable substances in animal feed was adopted its Annex omitted the dioxin MPLs for a range of feedingstuffs and feed materials introduced by way of Directive 2001/102/EC. An amending Commission Directive (2003/57/EC) links these MPLs to Directive 2002/32 and the general conditions on undesirable substances which it contains. This amending Directive also introduced a new, higher limit for fish protein hydrolysates containing more than 20% fat, of 2.25 ng WHO-PCDD/F-TEQ/kg; these products were previously subject to the same MPL as for fishmeal, of 1.25 ng WHO-PCDD/F-TEQ/kg. The Directive also increased the maximum permitted level for dioxins for fresh fish used for pets, zoos and circus animals to equate it with the 4 ng/kg WHO-TEQ level for human consumption; raised the provisional dioxin MPL for kaolinitic clays used as binders from 0.75ng/kg to 1.00ng/kg; and applied this new MPL to a range of other minerals also authorised for use as binders, anti-caking agents and coagulants.

Other undesirable substances

The blending down of feed materials with levels of contamination above specified MPLs was previously permitted, under specific conditions. Directive 2002/32, which has applied since 1 August 2003, prohibits this practice. Industry and some Member States made representations about the impact of this measure on certain feed materials, and in response the Commission held discussions to consider whether there should be provisional adjustments (increases) to particular MPLs.

At the Commission's request, the Scientific Committee on Animal Nutrition (SCAN) commenced a review of the annex to the Directive to establish whether the existing MPLs should be revised and any new substances added to it. However, this work was overtaken by the Commission bringing forward its own proposals for temporary upward revision of the MPLs for arsenic, lead and fluorine which in some cases differed from those suggested by SCAN. Revised MPLs for these substances were subsequently adopted as Directive 2003/100/EC. The opportunity was also taken to rationalise the MPLs for aflatoxin B_1, to introduce an MPL for free gossypol in whole cottonseed.

Formal risk assessments for these and other undesirable substances are now being carried forward by a scientific panel of EFSA which took over SCAN's responsibilities. There is, as yet, no fixed date for the completion of this work.

Trace elements

Commission Regulation (EC) No 1334/2003 amended the permitted maximum inclusion rates in feed for a number of trace elements (iron, cobalt, copper, manganese and zinc). Many of the inclusion levels were set many years ago, and required updating in the light of scientific knowledge regarding the nutritional needs of animals, having regard to the potential impact on human health of high levels of certain additives and the effects on the environment.

As a result, a number of maximum rates of inclusion have been reduced, but some key applications of trace elements remained broadly the same as before, e.g. 170 mg copper / kg in piglet diets for animals up to 12 weeks of age and 35 mg/kg for 'other bovines'. As a Regulation, the measure is directly applicable in all Member States and is enforced in England by SI 2004 No. 1301, which came into force on 1 June 2004.

Percentage ingredient declaration

Provisions requiring percentage ingredient declaration on the labels of compound feed for food producing animals were transposed into law in England by the Feeding Stuffs, the Feeding Stuffs (Sampling and Analysis) and the Feeding Stuffs (Enforcement) (Amendment) (England) Regulations 2003 (SI 2003 No. 1503), and were due to come into force on 6 November 2003.

On 8 September 2003 a number of UK feed manufacturers lodged an application in the High Court in England for judicial review of these provisions. The applicants sought to have the relevant aspects of Directive 2002/2/EC referred to the European Court of Justice (ECJ) for a determination of its validity, arguing that it was not a public health measure and would breach their intellectual property rights through the requirement to disclose details of their feed formulations. The applicants also sought interim relief via the suspension of the provisions implementing the Directive pending the adjudication of the ECJ.

On 6 October 2003 the High Court ruled in favour of the applicants, ordering the suspension of the provisions implementing the Directive in England pending the adjudication of the European Court. In consequence, percentage ingredient declaration for compound feedingstuffs in England did not come into force on 6 November 2003. Similar provisions in the legislation of Northern Ireland, Scotland and Wales have also been suspended. The Food Standards Agency is aware that there are similar applications, either pending or in progress, in some other Member States for suspension of percentage declarations. The ingredients of compound feedingstuffs are required to be listed in descending order by weight on the labels of feed, with the use of categories for ingredient listing purposes no longer permitted.

At the request of one of the parties involved, a hearing was fixed for 30 November 2004 for them to make final oral submissions to the Court. However, there is no indication of when a final judgement will be made by the ECJ.

Positive list of feed materials

In the Spring of 2003, the Commission issued a report to the European Parliament surveying the benefits and drawbacks inherent in the creation and maintenance of a positive list of feed materials. It pointed out that the benefits could include consistent labelling, traceability and the prevention of fraud. Drawbacks included the difficulty in arriving at clear definitions and criteria for authorisation, the implications for trade and innovation, and the impact on national and regional diversity.

The Commission's report concluded that the existence of a positive list would not in itself be a guarantee of feed safety, and examined a number of alternative means of enhancing this, such as legislation regarding manufacturing processes and codes of practice to be followed by feed operators. The report suggested that better feed safety can be attained through provisions to ensure feed hygiene, the improvement of existing controls on feed, the extension of both the list of prohibited ingredients and the non-exclusive list of feed materials, and the recasting of feed labelling rules.

Animal by-products regulation

The EU Animal By-Products Regulation (1774/2002) has applied in Member States since 1 May 2003. However, transitional and implementing measures were not agreed at EU level until mid-April 2003 and introduction of the national legislation was delayed so that it could also give effect to the additional measures. The Animal By-Products Regulations 2003 were laid before Parliament on 9 June and came into force in England on 1 July 2003. Equivalent legislation in Scotland, Wales and Northern Ireland followed shortly thereafter.

The Regulation tightens earlier rules on the processing, use, disposal, trade and import of animal by-products. Its purpose is the protection of animal and public health and it results from a review by the European Commission and a number of opinions of the EU Scientific Steering Committee. These were largely driven by developments in our knowledge of BSE and of suitable control measures, but also take account of the risk to animal and public health from other pathogens that may potentially be introduced or spread by animal by-products. Trade and import rules are broadly unchanged but the Regulation introduces some other changes which have a considerable impact in the UK. In particular, it:

- bans the routine burial of fallen stock;
- allows the treatment of animal by-products in approved composting or biogas plants;
- maintains the UK ban on swill feeding;
- requires the treatment of previously uncontrolled animal by-products such as blood and feathers.

The Feeding Stuffs (Sampling and Analysis) (Amendment) (England) Regulations 2004 came into force on 13 September 2004. This instrument implements an EC measure on the identification by microscopy of ingredients of animal origin in animal feeds. It replaces previous guidelines on this subject and reflects technical advances to improve the identification of very small quantities (less than 0.1%) of animal material. The new measure also specifies the procedures to be followed in conducting the analysis and calculation of results.

Guidelines for the microscopic identification of ingredients of animal origin in feed were first established by Commission Directive 98/88/EC and implemented in the Feeding Stuffs (Sampling and Analysis) Regulations 1999. However, these methods were optional and in practice resulted in a wide variation in the methods used.

These controls were introduced to help protect consumers of animal products in relation to transmissible spongiform encephalopathies (TSEs) and it is therefore important that the measures in place to check compliance with prohibitions on the use of animal material in feeds are applied consistently.

The new EC method, which reflects technical advances in methods of detection, is mandatory and will therefore permit the identification of lower levels of animal material. The new method should ensure a harmonised approach throughout the EU, which should improve determination rates.

Fishmeal

Following the original ban in 1988 on feeding ruminants with ruminant protein, it has been illegal to feed ruminants with mammalian protein - with certain exceptions - in the UK since November 1994, and to feed any farmed livestock, including fish and horses, with mammalian meat and bone meal (MMMB) since 4 April 1996. EU-wide controls, in the form of a 'temporary' ban on the feeding of processed animal protein to animals kept, fattened or bred for the production of food, were implemented in national legislation with effect from 1 August 2001. Commission Regulation (EC) No 1234/ 2003 made the EU-wide feed controls (with a few exceptions) permanent as from 1 September 2003.

However, the Commission has identified two criteria for reviewing the controls on fishmeal. Firstly, there needs to be reasonable evidence available that the implementation of the feed ban is satisfactory in all Member States; secondly, the Commission is requiring laboratories carrying out the official method of analysis to be able to detect down to 0.1% MMBM in feed containing fishmeal, and has carried out a ring-trial of the recently improved microscopy analysis test method to see if this standard has been achieved.

For feed ban monitoring purposes the UK official laboratory has already been using the updated method, and participated in the EU-wide ring-trial. Defra report that although the ring-trial results showed an improvement, the data suggested that reliable detection of 0.1% of MMBM in the presence of fishmeal is still a challenging task for many laboratories across the EU. A summary of the ring-trial analysis reported by the Commission's Joint Research Centre included the suggestion that training at laboratories may be the most likely way to achieve a further improvement of the results.

Defra comment that while further tests, which will even detect soft tissues as well as differentiate the species of origin, are under development for future validation, official EU laboratories are reliant upon microscopy. It is unclear how the recent ring-trial results may affect the thinking of the Commission and Member States on whether or not to lift the current ban on fishmeal in ruminant feeds. At a Transmissable Spongiform Encephalopathies (TSE) working group meeting in Brussels in June 2004, an informal discussion took place following an early report on the ring-trial. Of those Member States who expressed a view, six were still against lifting the ban.

Organic Feedingstuffs Legislation

Commission Regulation (EC) No 223/2003, on labelling requirements related to the organic production method for feedingstuffs, compound feedingstuffs and feed materials was published in February 2003. One of the requirements of this Regulation was that all equipment used in units preparing organic compound feedingstuffs should be completely separated from equipment used for non-organic compound feedingstuffs. The Regulation also provided for a transition period until 31 December 2007. During this period the same equipment can be used providing that separation in terms of time is guaranteed and suitable cleaning measures, which have been checked by the organic inspection bodies, have been carried out before commencing preparation of organic products covered by the Regulation.

Consolidation of Feedingstuffs Regulations

The Feeding Stuffs Regulations 2000 contain controls on the composition and marketing of animal feeds. They mainly implement EC derived measures including provisions on undesirable substances, feed additives and feed labelling. They have been amended seven times since they were made in September 2000. Therefore, to help readers and users of this principal legislation on animal feeds, the intention is to consolidate the various amendments into one set of Regulations. The consolidated Regulations, which will also remove provisions on feed additives which have been superseded by the new Feed Additives Regulation (1831/2003), will be the subject of a public consultation.

Forward look into 2005 and beyond

The European Commission is considering a wholesale review of the current feed labelling legislation, with the intention of producing one composite Regulation to replace a number of existing measures – Council Directive 79/373/EEC on the labelling and marketing of compound feeds, Council Directive 93/74/EC on dietetic feedingstuffs, and Council Directive 96/25/EC on the labelling of feed materials.

The Commission hosted an informal exchange of views with Member State experts on 9-10 March 2004. This disclosed a wide variety of views on associated provisions, including possible extensions to the non-exclusive list of the most common feed materials, authorisation procedures for novel feed materials, information to be declared on the statutory statement, permitted nutritional claims, and a positive list of feed materials. Further meetings are likely, but a formal proposal from the Commission is not expected before 2005.

All in all, the last few years have been extremely action-packed on the regulatory front. No doubt the year ahead will bring much of the same. Watch this space.....

Glossary

BSE   -   Bovine Spongiform Encephalopathy

EC   -   European Community

ECJ   -   European Court of Justice

EFSA -   European Food Safety Authority

EU   -   European Union

FSA   -   Food Standards Agency

GM   -   Genetically modified

GMO -   Genetically modified organism

HACCP-   Hazard Analysis and Critical Control Points

MMBM-   Mammalian meat and bone meal

MPL   -   Maximum Permitted Level

MRL   -   Maximum Residue Limit

OJ   -   Official Journal of the European Communities

SCAN -   Scientific Committee on Animal Nutrition

SI   -   Statutory Instrument

TSEs   -   Transmissible Spongiform Encephalopathies

UK   -   United Kingdom

ng WHO-PCDD/F-TEQ/kg   -   nanograms of WHO Polychlorinated Dibenzo Dioxins/Polychlorinated Dibenzo Furans Toxic Equivalents per Kilogram. A unit of measurement, developed by the World Health Organisation, to express toxicity-weighted masses of mixtures of dioxins. Each individual dioxin is assigned a Toxic Equivalency Factor (TEF), relative to the most toxic dioxin, which is then multiplied by the amount of each individual dioxin and added together to calculate the overall Toxic Equivalent (TEQ) of a substance.

References

Regulation (EC) No. 183/2005 of the European Parliament and of the Council laying down requirements for feed hygiene (OJ No L35 page 1, 8 February 2005).

The Feeding Stuffs (Safety Requirements for Feed for Food-Producing Animals) Regulations 2004 (SI 2004 No. 3254).

The Feeding Stuffs, the Feeding Stuffs (Sampling and Analysis) and the Feeding Stuffs (Enforcement) (Amendment) (England) Regulations 2004 (SI 2004 No. 1301).

Regulation (EC) No 882/2004 of the European Parliament and of the Council on official controls performed to ensure the verification of compliance with feed and food law, animal health and welfare rules (OJ No L191 page 1, 28 May 2004).

Regulation (EC) No 1829/2003 of the European Parliament and of the Council on genetically modified food and feed (OJ No L268 page 1, 18 October 2003).

Regulation (EC) No 1830/2003 of the European Parliament and of the Council concerning the traceability and labelling of genetically modified organisms and the traceability of food and feed products produced from genetically modified organisms and amending Directive 2001/18/EC (OJ No L268 page 24, 18 October 2003).

Regulation (EC) No 1831/2003 of the European Parliament and of the Council on additives for use in animal nutrition (OJ No L268 page 29, 18 October 2003).

The Feeding Stuffs, the Feeding Stuffs (Sampling and Analysis) and the Feeding Stuffs (Enforcement) (Amendment) (England) Regulations 2003 (SI 2003 No. 1503).

Commission Regulation (EC) No 1334/2003 amending the conditions for authorisation of a number of additives in feedingstuffs belonging to the group of trace elements (OJ No L187 page 11, 26 July 2003).

Commission Regulation (EC) No 1234/2003 amending Annexes I, IV and XI to Regulation (EC) No 999/2001 of the European Parliament and of the Council and Regulation (EC) No 1326/2001 as regards transmissible spongiform encephalopathies and animal feeding (OJ No L173 page 6, 11 July 2003).

Commission Regulation (EC) No 223/2003 on labelling requirements related to the organic production method for feedingstuffs, compound feedingstuffs and feed materials and amending Council Regulation (EEC) No 2092/91 (OJ No L031 page 3, 6 February 2003).

Commission Directive 2003/126/EC on the analytical method for the determination of constituents of animal origin for the official control of feedingstuffs (OJ No L339 page 78, 24 December 2003).

Commission Directive 2003/100/EC amending Annex I to Directive 2002/32/EC of the European Parliament and of the Council on undesirable substances in animal feed (OJ No L285 page 33, 1 November 2003).

Commission Directive 2003/57/EC amending Directive 2002/32/EC of the European Parliament and of the Council on undesirable substances in animal feed (OJ No L151 page 38, 19 June 2003).

Regulation (EC) No 1774/2002 of the European Parliament and of the Council laying down health rules concerning animal by-products not intended for human consumption (OJ NO L273 page 1, 10 October 2002).

Directive 2002/32/EC of the European Parliament and of the Council on undesirable substances in animal feed (OJ No L140 page10, 30 May 2002).

Directive 2002/2/EC of the European Parliament and of the Council amending Council Directive 79/373/EEC on the circulation of compound feedingstuffs and repealing Commission Directive 91/357/EEC (OJ No L 63 page 23, 6 March 2002).

Regulation (EC) No 178/2002 of the European Parliament and of the Council laying down the general principles and requirements of food law, establishing the European Food Safety Authority and laying down procedures in matters of food safety (OJ No L31 page 1, 1 February 2002).

Commission Regulation (EC) No 1326/2001 laying down transitional measures to permit the changeover to the Regulation of the European Parliament and of the Council (EC) No 999/2001 laying down rules for the prevention, control and eradication of certain transmissible spongiform encephalopathies, and amending Annexes VII and XI to that Regulation (OJ No L177 page 60, 30 June 2001).

Regulation (EC) No 999/2001 of the European Parliament and of the Council laying down rules for the prevention, control and eradication of certain transmissible spongiform encephalopathies (OJ No L147 page 1, 31 May 2001).

Council Directive 2001/102/EC amending Directive 1999/29/EC on the undesirable substances and products in animal nutrition (OJ No L6 page 45, 10 January 2002).

The Feeding Stuffs Regulations 2000 (SI 2000 No 2481).

The Feedingstuffs (Sampling and Analysis) Regulations 1999 (SI 1999 No 1663).

Council Directive 1999/29/EC on the undesirable substances and products in animal nutrition (OJ No L115 page 32, 4 May 1999).

Commission Directive 98/88/EC establishing guidelines for the microscopic identification and estimation of constituents of animal origin for the official control of feedingstuffs (OJ No L318 page 45, 27 November 1998).

Regulation (EC) No 258/97 of the European Parliament and of the Council concerning novel foods and novel food ingredients (OJ No L043 page 1, 14 February 1997).

Council Directive 96/51/EC amending Directive 70/524/EEC (OJ No L235 page 39, 17 September 1996).

Council Directive 96/25/EC on the circulation of feed materials, amending Directives 70/524/EEC, 74/63/EEC, 82/471/EEC and 93/74/EEC and repealing Directive 77/101/EEC (OJ No L125 page 35, 23 May 1996).

Council Directive 93/74/EEC on feedingstuffs intended for particular nutritional purposes (OJ No L237 page 23, 22 September 1993).

Commission Directive 91/357/EEC laying down the categories of ingredients which may be used for the purposes of labelling compound feedingstuffs for animals other than pet animals (OJ No L193 page 34, 17 July 1991).

Council Directive 79/373/EEC on the marketing of compound feedingstuffs (OJ No L086 page 30, 6 April 1979).

Council Directive 70/524/EEC concerning additives in feedingstuffs (OJ No L270 page1, 14 December 1970).

7

STRATEGIES AND METHODS TO DETECT AND QUANTIFY MAMMALIAN TISSUES IN FEEDSTUFFS: A SUMMARY OF THE EU STRATFEED PROJECT (G6RD-2000-CT-00414)

I.MURRAY[1], P. DARDENNE[2], V. BAETEN[2] AND A. GARRIDO VARO[3]
[1] *Scottish Agricultural College, Aberdeen, AB21 9YA, UK;* [2] *Walloon Agricultural Research Centre, CRA-W, Gembloux, Belgium;* [3] *University of Cordoba, Spain*

Introduction

The UK BSE/vCJD and FMD epidemics were catastrophic failures in animal feed hygiene that caused great harm to the livestock industry and great loss of consumer confidence at home and abroad (Phillips *et al*, 2000, Pennington, 2003, Ridley and Baker 1998, Rhodes, 1998, North, 2001). As a result we can expect greater surveillance and traceability to be imposed on the feed industry in future (ACAF, 2003, 2004). A characteristic of EU legislation is that it is imposed with the expectation that appropriate technical methods will be devised in due course to meet the needs of compliance with the law. Resulting from the ban on the use of meat and bone meal (MBM) in animal feeds (94/381/EC; 2000/766/EC), the EU implemented the three year STRATFEED project to research "Strategies and methods to detect and quantify mammalian tissues in feedingstuffs". The study involved ten European partners who collaborated in developing and testing four different methods to detect MBM in feeds:

1. Optical Microscopy OM (the official method)
2. Near Infrared Spectroscopy, NIRS
3. Near Infrared Micro-spectroscopy, NIRM (and NIR Camera)
4. Molecular biology methods: Polymerase Chain reaction, PCR

This paper summarises the outcome of the STRATFEED project which was publicised at its recent International Symposium: "Food and feed safety in the context of prion diseases." Held in Namur, Belgium 16-18 June 2004 hosted by CRA-W, JRC-IRMM, AFSCA and Agrobiopole. Full details of the symposium, including presentations, can be found on the STRATFEED website (Vermuelen *et al*, 2003): http://stratfeed.cra.wallonie.be/

Suffice to say that no single method fulfils all the ideals of universal applicability, reliability, throughput, limit of detection (LOD) and, not least, cost. However the methods do show complementarity in forming the basis for a control strategy.

Background

It is not certain what initiated the BSE outbreak in the UK cattle herd (Phillips *et al*, 2000, Horn 2001, Asante *et al,*2002). The introduction of milk quotas in 1984 might have led to dairy farmers attempting to maximise milk yield in the previous year by increasing protein in the diet, and MBM was a cheap protein source at that time. Changes in the rendering process occurred at this time and might have contributed to transmission of infectivity. Infectivity is believed to be caused by conformational changes in prion proteins from normal alpha helix (PrPC) to beta pleated sheets (PrPSc) which recruit normal proteins to join their cause as rogue prions (Prusner, 1991, 1995, 1997).

Because of the long incubation time of TSEs and the unrecognized nature of the disease, it was some considerable time before the outbreak was realized, by which time a large cohort of animals were incubating BSE and contributing to its propagation from recycling carcass slaughter wastes as MBM. In spite of the established wisdom of taboos forbidding cannibalism and the research work of Carlton Gadjusek (1996) on cannibalism causing the TSE Kuru in the Fore Indians of Papua New Guinea, cannibalism was being imposed on dairy cattle by such practices. When the first BSE cases became recognized in 1986, the disease was seen to be similar to scrapie in sheep. Scrapie had been known in sheep since the eighteenth century and there was no evidence to suggest it could cause disease in humans. Perhaps for these reasons action to contain the outbreak was initially not rigorous enough.

Only when very young people were contracting a new form of CJD (variant CJD) did concern grow and rigorous control was imposed to remove MBM and specified risk material from the food chain and to exclude cattle over 30 months of age from the human diet. The decrease in incidence of BSE (Figure 1) is proof both that the measures were effective and that MBM was the most likely source of the outbreak. Due to the long incubation period, infected animals and feeds spread the disease to other European countries. (Figure 2) To date just under 150 cases of variant CJD have been found in the human population in the UK (NCJDSU, 2004). The UK and EU authorities have taken action to ensure eradication of BSE. In particular the EU instigated the STRATFEED research project to devise and improve methods to detect mammalian tissue in feeds.

Although the BSE epidemic might be under control and soon eradicated, intra-species recycling (cannibalism) must be prevented in future. The present

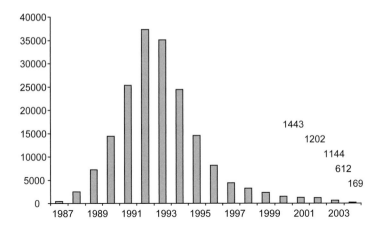

Figure 7.1 Reported BSE cases in the UK (www.oie.int/eng/info/en_esbru.htm).

Figure 7.2 Reported BSE cases in 2003 (www.oie.int/eng/info/en_esbru.htm).

rejection, within the EU, of some 14 M tonnes per year of mammalian slaughter wastes containing 55% protein (Wolf,1982) cannot be acceptable on environmental grounds. However, any subsequent reintroduction of species specific rendering and recycling will demand harmonization of high hygiene standards, underpinned by robust regulatory measures. A monograph on the risk analysis of prion diseases in animals has recently been published by OIE (Lasmézas and Adams, 2003) in which the chapter by Gizzi *et al* concerns test methods.

Methods

OPTICAL MICROSCOPY OM

Optical microscopy of feeds remains the officially recognized method for detection of animal tissues in feeds (Directive 88/1998/EC). However, its use depends particularly on the detection of dense bone tissue which may be assigned on morphological criteria (Haversian canals and lacunae) to particular animal species by an experienced microscopist. Success of the method arises from taking a large test portion (= 10 g) and using tetrachloroethylene (s.g.1.62) to form a sediment containing bone (s.g.1.67-2.0) which can be examined under low and high magnification (Figure 7.3). If bone is present, assignment to species needs skill and experience. Stratfeed microscopists developed a micrograph gallery (Figure 7.4) and the ARIES (Animal Remains Identification & Evaluation System) decision support system to assist in training and harmonization of the Stratfeed protocol for OM (Directive 126/2003/EC). A limit of detection (LOD) of 0.1% MBM in feeds can be achieved. Initially the presence of fishmeal (FM) containing fish bones confounded detection, but recent validation has shown this problem to be substantially improved (van Raamsdonk and van der Voet, 2003). Poultry long bones are still difficult to differentiate from mammal bone. Definitive species identification is still needed in suspect specimens for which taxon specific molecular biology methods (PCR, discussed later) are the most appropriate. Nevertheless OM is likely to remain important as a key forensic test.

POLYMERASE CHAIN REACTION PCR

Molecular biology methods hold the promise of unequivocal species identification of tissue as well as offering a low limit of detection (LOD 0.05%) even with multiple species present. The development of real-time polymerase chain reaction (RT-PCR) has transformed forensic science. It is important to choose a ruminant DNA target that is abundant in tissues; that is the DNA target has a large copy number. Mitochondrial DNA is most appropriate in this context because mitochondria are abundant in most tissues.

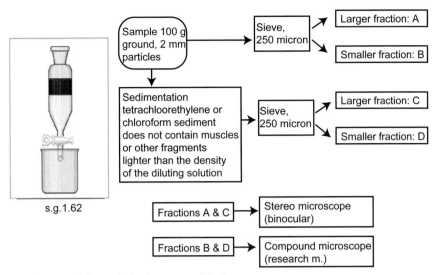

Figure 7.3 Protocol for optical microscopy of feed.

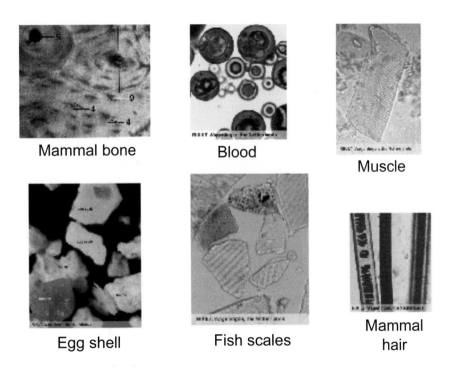

Figure 7.4 Optical Microscopy of feeds. Picture gallery from ARIES decision support system (van Raamsdonk, RIKILT, NL).

A mitochondrial DNA segment (271base pairs) was first used in 1998 to design a PCR test to detect ruminant tissue in feed at 0.125% MBM (Tartaglia *et al* 1998). However, a 24 lab ring trial (JRC, 1999) showed the method was not fit for purpose, having over 20% false positives and over 20% false negatives. Failure was ascribed to DNA extraction and clean-up, or to polymerase inhibition or thermal degradation of DNA by rendering.

Evidence for thermal fragmentation of DNA showed that shorter targets (~60bp) endured higher rendering temperatures than longer targets. PCR using RFLP (Restriction Fragment Length Polymorphism) amplifies a fragment common to several species using universal primers then applies several restriction endonucleases to detect species specific internal differences. This method allowed detection of cattle, buffalo, sheep, goat, horse, pig, chicken and turkey in feeds with 0.5% processed animal protein present in a feed (Bellagamba *et al, 2001*) Because high temperature rendering is mandatory (= 133°C), short DNA targets (~60bp) must be chosen that are inappropriate for traditional agarose gel electrophoresis PCR. For this reason RT-PCR becomes the only appropriate method. Such methods use fluorescent probes in thermocyclers that are complex and elegant. (Foy and Parkes, 2001)

PCR involves melting double stranded DNA at 90°C to form two separate complementary strands (Figure 7.5). This is followed by annealing of forward and reverse primers onto each strand at 37°C followed by chain elongation at 75°C to form two pairs of DNA molecules that subsequently repeat the cycle. Each successive thermal cycle doubles the number of copies, leading to exponential amplification shown by a fluorescent signal. Primers are chosen to bracket the target DNA region characteristic of the bovine species to be detected.

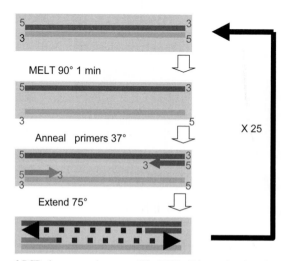

Figure 7.5 Outline of PCR thermo cycle to amplify DNA. Primers bracket the target.

The advent of real-time PCR allows visual progress of amplification (Figure 7.6) with each successive thermal cycle. If bovine DNA is present in the test portion taken for analysis (~100 mg), this becomes amplified after several cycles to produce an exponential appearance of the PCR product shown on the screen as a steeply rising curve. If the exponential rise occurs after, say, 15 cycles then the sample contains a considerable amount of the bovine target DNA. If the exponential rise occurs after, say, 35 cycles then the sample contains very small amounts of bovine DNA. If no exponential rise is observed at all then the specimen is free from any bovine target DNA. RT-PCR thus gives a semi-quantitative result as well as a qualitative 'Yes/No' detection. Thermal damage to DNA can reduce the signal, however, confounding true quantitative detection.

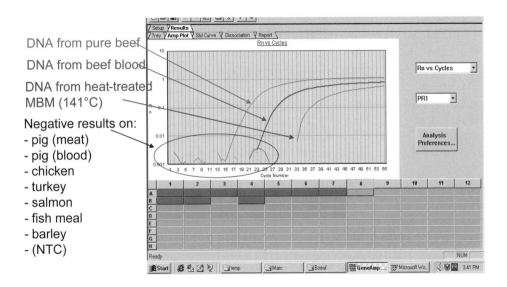

Figure 7.6 Real-time PCR screen showing exponential rise after 15 cycles for DNA from pure beef, 23 cycles for beef blood and 33 cycles from bovine MBM 141°C.

Disadvantages of PCR include the small test portion taken (100mg), which risks sampling error. Furthermore, milk, blood or rendered fat can yield traces of ruminant DNA that are detectable. The very high sensitivity of PCR risks accidental cross-contamination requiring dedicated labs as well as the high cost of a real-time thermocycler, reagents and skilled staff. In time PCR might well become the 'gold standard' for unequivocal species detection, but its use might be restricted to well resourced regulatory labs having skilled staff.

NEAR INFRARED SPECTROSCOPY NIRS

Near Infrared Spectroscopy is already well established for routine QC in feed mills. Instant analysis and reporting coupled to high throughput make NIRS attractive for screening feed for MBM adulteration or contamination. However work by the Stratfeed group showed that the limit of detection was typically 1 to 3 % MBM in a wide range of feeds (Figure 7.7). Surprisingly the spectra of FM and MBM are quite different in spite of both being high protein meals. Detection of MBM in FM proved successful at 3, 6 and 9% MBM (Murray *et al*, 2001) largely because of the more restricted range in the FM matrix. In contrast, the vast spectral range of cereal, pulses, oilseed, by-products and forage that make up plant-based feeds make for difficulties in finding robust calibration models for MBM in all types of feed matrices. Even so, if NIRS is used in feed mills for screening raw material on intake, it would be possible to establish outlier detection that would flag a material that was abnormal and raised suspicion. For example, a set of 100 FM specimens drawn from a consignment and deliberately contaminated with 8% MBM (the lowest level likely to arise in a blatant fraud) could be easily detected from another unadulterated set of 100 specimens (Figure 7.8)(Gjervik, 2001). NIRS could offer the first line of defence of the food chain in feed mills.

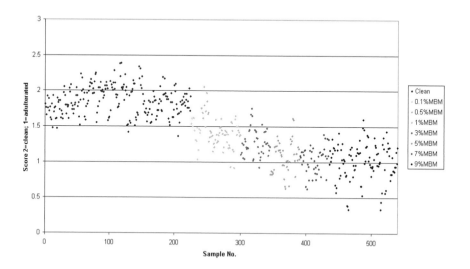

Figure 7.7 NIRS Discriminant for MBM in 540 feeds at levels from 0 to 9 %. NIRS is reliable at levels over 1 to 3% MBM. Pass/Fail threshold set at a 1.5 % MBM.

NEAR INFRARED MICRO-SPECTROSCOPY NIRM

NIR micro-spectroscopy combines the advantages of microscopy and spectroscopy so that the spectra of individual particles can be acquired to assist in their identification (Piraux and Dardenne, 1999). Comparison of spectra of particles with a data base of animal by-product meals permits

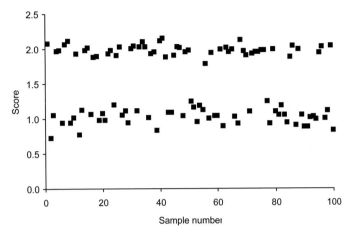

Figure 7.8 Discriminant to detect 8% MBM in fishmeal (1=FM+8%MBM; 2=FM) (H.Gjervik, 2001).

assignment of each particle to animal or vegetable categories using discriminant analysis (Figure 7.9). Since one particle of MBM will display the spectrum of 100% MBM, the NIR microscope has distinct advantages over traditional *macro*scopic NIRS. The spectra can identify and quantify the proportions of ingredients in a feed, based on counting particles. The main disadvantage, as with optical microscopy, is the need to examine large numbers of particles in order to sample a statistically significant proportion of the feed. This is time consuming and costly unless the process can be automated.

In particle counting methods, the risk of sampling error leading to false negative results can be estimated from the binomial distribution. If 1.8% false negatives are acceptable, then 400 particles must be examined to detect MBM at the 1% inclusion level, while 4000 particles must be examined to achieve the same performance at the 0.1% MBM inclusion level. NIRM can be applied to the sediment fraction from optical microscopy (Baeten *et al*, 2004) or, alternatively, a suspect particle can be picked from the microscope stage and referred to PCR to confirm animal species.

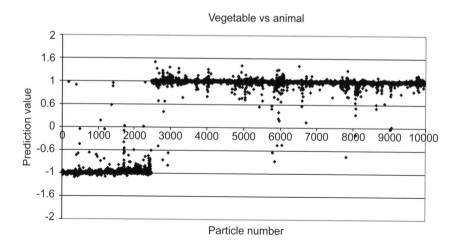

Figure 7.9 NIR Microscopy of 10,000 feed particles using Artificial Neural Networks (ANN) to discriminate animal and vegetable derived particles (CRA-W).

NIR CAMERA

The NIR Camera is a new instrument that automates the process of acquiring spectra of many hundreds or thousands of individual particles. It thus answers the main laborious disadvantage of NIRM noted above. A small test portion of feed particles is scattered over a 4 cm² area consisting of 240 x 320 = 76800 pixels. The area is scanned at 4nm intervals over the region 900-1700 nm to generate 200 data points per particle giving a hyper-spectrum of the area scanned. Typically 15 million absorbance values are acquired in 5 minutes per sample (Figure 7.10). The spectrum of each particle among 500 particles is thus acquired within 5 minutes. False colours are used to denote animal and vegetable particles and to give a particle count for suspect particles of animal origin (Figure 7.10 inset). Suspect particles can then be isolated and referred to PCR for unequivocal species identification. The high cost of the instrument means that few labs can afford the capital cost so that its use is restricted to well resourced regulatory labs. The NIR Camera is the only instrument likely to be capable of checking the proportions of ingredients in a compounded feed. Any suspect particle can be picked off the stage and referred to PCR for species identification.

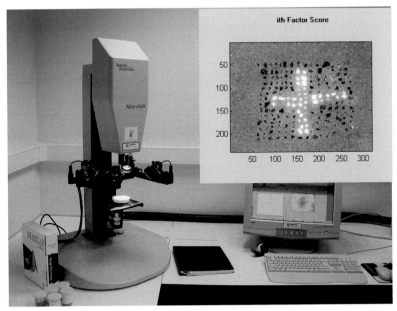

Figure 7.10 NIR Camera showing an inset of animal derived particles forming a white cross in a background of plant derived particles. (CRA-W, Gembloux, BE).

ENZYME-LINKED IMMUNO-SORBENT ASSAY ELISA

Enzyme-Linked Immuno-Sorbent Assay techniques detect bovine-specific muscle proteins such as troponin-I by single step, lateral flow immuno-chromatographic binding assay in the form of a dip stick (Figure 7.11). One such test is marketed as Agri-Screen®. The stick has two reporting lines; one for a positive detection of bovine muscle protein and a check line to ensure the device is functional. ELISA techniques became discredited when the increased official rendering temperature for slaughter waste (133°C; 3 bar; 20 min) led to denaturation of the target protein. However, since then progress has been made in finding targets that are more heat stable. Although the reporting line is diminished by increasing rendering temperature, a positive test is still reported, even at a rendering temperature of 141°C. Samples of feeds are boiled to provide the test portion applied to the test stick. ELISA techniques were not examined in the Stratfeed project. The test is simple to perform and relatively inexpensive as a screening method capable of detecting bovine MBM in feeds or ingredients.

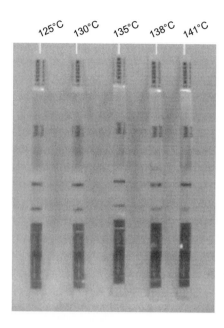

Figure 7.11 Agri-Screen® test strips with 2 zone lines. Lower line detects ruminant muscle protein, upper line checks the strip is working. The bovine signal is attenuated by increasing rendering temperature, but still detects at 141° C.

Conclusions

Table 7.1 gives a detailed summary of characteristics of the methods examined. No single method answers all the ideals for a test capable of mass screening of the vast quantities of feed and feed ingredients traded globally. Methods can be considered to have two purposes: mass screening for suspect specimens, and corroborative forensic tests conducted on suspect specimens for confirmation. A two-stage process permits greater scope for testing a larger proportion of feed samples, so that more costly tests can be used more productively.

In a deregulated context, it would be the responsibility of feed manufacturers to source materials from reputable sources. As part of GMP they would be expected to establish a product library file of NIR spectra; a QC timeline for a particular product that could be used to detect incoming specimens that were detected as abnormal outliers that warrant further scrutiny. This would benefit responsible manufacturers, their clients and consumers, whilst lifting the burden on regulators allowing them to concentrate effort on referrals that really warrant inspection. Definitive tests are likely to remain optical microscopy coupled to PCR for species confirmation. PCR and the

Table 7.1 EU STRATFEED PROJECT: Summary of features of methods to detect and quantify mammalian tissues in feedstuffs.

Features	OM	PCR	NIRS	NIRM	NIR camera	ELISA
Samples per day	10-20	5-10	100-200	3-5	?	100-200
Time per sample	1h	2days	10min	2h	1h	30min
Sample mass	1-10g	0.1-1g	5-100g	0.2-10g	0.2-10g	10g
Reagent required	Yes	Yes	No	Yes	Yes	No
Expertise required	Yes	Yes	No	No	No	No
Limit Of Detection %	0.1%	0.5%	3-5%	0.1%	<0.1%	1%
False negatives	<5%	<5%	>5%	<5%	<5%	<5%
False positives	<1%	<1%	<5%	<1%	<1%	<1%
Repeatability	High	Medium	Medium	High	High	High
Transferability	High	Medium	High	High	?	High
Contamination risk	Low	High	Low	Low	Low	Low
Matrix dependent	No	Yes	Yes	No	No	No
Interference						
Milk	No	Yes	No	No	No	No
Blood	No	Yes	?	No	No	No
Fat	No	Yes	No	No	No	Yes
Heat treatment temperature	No	No	No	No	No	No
Particle size	No	No	No	Yes	No	No
Species identified						
Terrestrial animal vs fish	Yes	Yes	Yes	Yes	Yes	Yes
Mammal/Ruminant vs poultry	No	Yes	No	No	No	Yes
Mammal vs pig	No	Yes	No	No	No	Yes

Table 7.1 Contd.

Features	OM	PCR	NIRS	NIRM	NIR camera	ELISA
Miscellaneous						
Analysis of intact feed	Yes	Yes	Yes	Yes	Yes	Yes
Analysis of sediment	Yes	Yes	No	Yes	Yes	No
Validated method	Yes	No	No	No	No	No
Existing facilities	Yes	Yes	Yes	No	No	Yes
Forensic value	Yes	Yes	No	?	?	No
Screening method	No	No	Yes	No	Yes	Yes
Quantitative method	No	No	Yes	Yes	Yes	No
Cost of instrument	60 000€	70 000€	60 000€	80 000€	250 000€	0
Cost per test	50-100€	150-200€	2-5€	50-100€	50-100€	12-15€

OM Optical microscopy
PCR Polymerase chain reaction
NIRS Near-infrared spectroscopy
NIRM Near-infrared microscopy
NIR Camera Near-infrared camera
ELISA Enzyme-Linked Immuno-Sorbent Assay

NIR Camera could play a larger part in routine inspection if their costs can be reduced and instruments made more widely available. Finally, it is important to appreciate how the inevitability of statistical sampling error grows with the setting of a diminishing limit of detection and with smaller test portions. Thus setting unrealistically low detection limits offers a false sense of security that is unhelpful in practice.

For an informative summary of the Stratfeed project, visit the website: http://stratfeed.cra.wallonie.be/.

Acknowledgements

The authors acknowledge the efforts of the following members of the Stratfeed research group: R.M. Paterson, SAC, UK; A.M. Michotte-Renier, P. Vermeulen, M. Dubois, G. Berben and O. Fumière, CRA-W, Belgium; L.W.D. van Raamsdonk, V. Pinckaers, R. Frankhuizen and J. de Jong, RIKILT, The Netherlands; J. Zegers, Nutreco, The Netherlands; J. Vancutsem and K. Haustraete, ROLT, Belgium; J. Bosch, S. Termes and A. Puigdomenech, LAGC, Spain; L. Perez, UCO, Spain; D. Portetelle, FUSAGx, Belgium, G. Vaccari and G. Brambilla, ISS, Italy and C. von Holst, JRC, Belgium.

SAC receives support from the Scottish Executive Environment and Rural Affairs Department.

References

ACAF Advisory Committee on Animal feedstuffs, *Review of Animal Feed Labeling*, Food Standards Agency, London.

ACAF (2003) Advisory Committee on Animal feedstuffs, *ACAF review of on-farm feeding practices: recommendations on identifying hazards and minimising risks*. http://www.foodstandards.gov.uk/ multimedia/ pdfs/farm.pdf

ACAF (2004) Advisory Committee on Animal feedstuffs, *Report on Feed Law Enforcement* - First Draft ACAF/04/15, (7/7/2004)

Asante, E.A., Linehan J.M., Desbruslais, M., Joiner, S., Gowland, I., Wood, A.L., Welch, J., Hill, A.F., Lloyd, S.E., Wadsworth, J.D. and Collinge, J. (2002) BSE prions propagate as either variant CJD-like or sporadic CJD-like prion strains in transgenic mice expressing human prion protein. *EMBO J* **21**, 6358-6366.

Baeten, V. and Dardenne, P. (2001). The contribution of near infrared spectroscopy to the fight against the mad cow epidemic. *NIR news*, **12**

(6), 12-13.

Baeten, V., Michotte-Renier, A., Sinnaeve, G. and Dardenne, P. (2001). Analysis of feedingstuffs by near-infrared microscopy (NIRM): detection and quantification of meat and bone meal (MBM). In: *Proc. sixth International symposium on food authenticity and safety*, 28–30 November 2000, Nantes, p. 1-11.

Baeten, V., Michotte-Renier, A., Sinnaeve, G., Garrido Varo, A. and Dardenne, P. (2004). Analysis of the sediment fraction of feed by Near-Infrared Microscopy (NIRM). In: *Proc. of the 11th International Conference on Near-Infrared Spectroscopy*, Cordoba, Spain, 6-11 April 2003. (in press)

Bellagamba, F., Moretti, V.M., Comincini, S. and Valfre, F. (2001) Identification of species in animal feedstuffs by PCR-RFLP analysis of mitochondrial DNA. *J. agric. Food Chem.*, **49**, 3775-3781.

Foy, C.A. and Parkes, H.C. (2001) Emerging Homogeneous DNA-based technologies in the clinical laboratory. *Clinical Chemistry* **47**, 990-1000 *http://www.clinchem.org/cgi/content/full/47/6/990*

Gadjusek, C. D. (1996) Kuru: from the New Guinea field journals 1957-1962, *Grand Street*, **15**, 6-33.

Gizzi, G., Baeten, V., Berben, G.,van Raamsdonk, L.W.D. and von Holst, C. (in press). Intercomparison study for the determination of processed animal proteins including meat and bone meal in animal feed. *J. A.O.A.C.* (in press)

Gizzi, G., van Raamsdonk, L.W.D., Baeten, V., Murray, I., Berben, G., Brambilla, G. & von Holst, C. (2003). An overview of tests for animal tissues in animal feeds used in the public health response against BSE. In: Risk analysis of BSE and TSEs: update on BSE and use of alternatives to MBM as protein supplements. *Rev. Sci.Tech.Off. Int. Epiz.* **22**(1), 311-331.

Gjervik, H. (2001) *Detection of adulteration of a fishmeal consignment with meat and bone meal by Near Infrared Spectroscopy*, MSc thesis, LTMD, SAC / University of Aberdeen.

Horn, G. (2001) *Review of the origin of BSE*. Department for Environment, Food and Rural Affairs (DEFRA), London.

Lasmézas, C.I. and Adams D.B. (eds)(2003) Risk analysis of prion diseases in animals OIE: World Organisation for Animal Health *Scientific and Technical Review* **22**(1), April 2003.

Murray, I., Aucott, L.S. & Pike, I.H.(2001) Use of discriminant analysis on visible and near infrared reflectance spectra to detect adulteration of fish meal with meat and bone meal, *J. Near Infrared Spectrosc.* **9**, 297-311.

NCJDSU (2004) National Creutzfeldt-Jakob Disease Surveillance Unit, http://www.cjd.ed.ac.uk/

North, R.A.E. (2001) *The death of British Agriculture: the wanton destruction of a key industry.* Duckworth, London.

Piraux, F. and Dardenne, P. (1999) Feed authentication by near-infrared microscopy. In: *Proc. of the 9th International Conference on Near-Infrared Spectroscopy* (Davies, A.M.C. and Giangiacomo, R., eds), Verona, Italy, June 1999.

Pennington, T.H. (2003) *When Food Kills: BSE, E. coli and Disaster Science,* Oxford UP, Oxford.

Phillips, N., Bridgeman, J. and Ferguson-Smith, M. (2000) *The BSE Inquiry: the report,* Vols 1-16. Vol 2: Science. HMSO, London.

Phillips, N., Bridgeman, J. and Ferguson-Smith, M. (2000) *The BSE Inquiry: the report,* Vols 1-16. Vol 10:Economic impact. HMSO, London.

Prusner, S.B. (1991) Molecular biology of prion diseases, *Science* **252**,1515-1521

Prusner, S.B. (1995) Prion diseases, *Scientific American,* **272**(1), 48-56

Prusner, S.B. (1997) Prion diseases & the BSE crisis. *Science,* **278**, 245-251

Raamsdonk L.W.D. van and van der Voet, H. (2003). *A ring trial for the detection of animal tissues in feeds in the presence of fish meal.* Report 2003.012, RIKILT, Wageningen.

Rhodes, R. (1998) *Deadly Feasts: science and the hunt for answers in the CJD crisis,* Touchstone Books, London.

Ridley, R.M. and Baker, H.F. (1998) *Fatal Protein: the story of CJD, BSE and other prion diseases,* Oxford University Press, Oxford.

Tartaglia, M. Saulle, E., Pestalozza, S., Morelli, L., Antonucci, G. and Battaglia, P.A. (1998) Detection of bovine mitochondrial DNA in ruminant feeds: a molecular approach to test for the presence of bovine-derived materials. *J. Food Protec.,* **61**(5), 513-518.

Wolff, I.A.(1982) CRC *Handbook of Processing and Utilization in Agriculture, Vol1 Animal Products,* CRC Press, Boca Raton, FL, USA

Vermeulen, P., Baeten, V., Dardenne, P., van Raamsdonk, L.W.D., Oger, R., Monjoie, A.S. and Martinez, M. (2003). Development of a website and an information system for an EU R&D project: the example of the STRATFEED project. *Biotechnol. Agron. Soc. Environ.* **7**, 161-169.

Appendix

European Commission Regulations relating to BSE & processed animal protein.

1994/381/EC   27/6/1994   *Off. J. Eur. Communities* **L 172**, 7 Jul, 23-24
1994/382/EC   27/6/1994   *Off. J. Eur. Communities* **L 172**, 7 Jul, 25-28
1996/449/EC   18/7/1996   *Off. J. Eur. Communities* **L 18**4, 24 Jul, 43-46
1997/534/EC   30/7/1997   *Off. J. Eur. Communities* **L 216**, 8 Aug, 95-98
1998/88/EC   13/11/1998   *Off. J. Eur. Communities* **L 318**, 27 Nov, 45-50
1999/534/EC   19/7/1999   *Off. J. Eur. Communities* **L 204**, 4 Aug, 37-42
2000/418/EC   29/6/2000   *Off. J. Eur. Communities* **L 158**, 30 Jun, 76-82
2000/776/EC   4/12/2000   *Off. J. Eur. Communities* **L 306**, 7 Dec, 32-33
2001/2/EC   27/12/2000   *Off. J. Eur. Communities* **L 1**, 4 Jan, 21-22
2001/165/EC   27/2/2001   *Off. J. Eur. Communities* **L 58**, 28 Feb, 43-44
2001/233/EC   14/3/2001   *Off. J. Eur. Communities* **L 84**, 23 Mar, 59-61
2002/1774/EC   3/10/2002   *Off. J. Eur. Communities* **L 273**, 10 Oct, 1-95

8

MOLECULAR TECHNIQUES FOR ANALYSING FEED RAW MATERIALS

R.G BARDSLEY AND G.A TUCKER
University of Nottingham, School of Biosciences.

Introduction

Recent food scares have led to increased pressure from consumer groups for more information about food composition and for improved labelling. Outbreaks of food poisoning caused by microbial contamination, BSE caused by prion proteins in animal offals, and allergies to plant proteins, are serious human health considerations. There is also considerable consumer resistance to the use of genetically modified organisms (GMOs) in food production. Other concerns include pork and beef in ethnic foods or milk products in vegetarian meals. At another level, some consumers are prepared to pay a premium for foods based on a specific plant variety, animal breed or geographical origin, based on perceptions of food quality. Many similar concerns are now being experienced by the feedstuffs industries supplying materials for livestock species or companion animals. These concerns relate mostly to GMOs and animal offals, but there is also evidence that different varieties of cereals, for example, may have variable nutritional value because of amino acid composition, digestibility, micronutrients or anti-nutritional factors.

There are thus several reasons for biological materials to require authentication. These include detection of genetically modified organisms (GMOs), identification of species or breed, determination of type of tissue and geographical origin. Some of these requirements would be applicable to the feed industry. Under current regulations regarding labelling, the presence of GMO material above 1%, or even lower, must be declared and thus detection of GMO is important within the feed industry. Also, labelling regulations require an accurate determination of species, for example the types of meat or fish-meal included in the feed. Authentication down to breed level might not be significant within the animal feed industry, but is becoming increasingly significant to the consumer within the context of premium human foods. The detection of tissue type, for example the presence of brain or spinal cord material in feedstuffs, is clearly of great concern. Currently, determination of geographical origin of raw materials is perhaps primarily of concern to the human food industry.

The food analyst is, therefore, called upon to devise technologies for food authentication at several levels of sophistication. In some cases, it is sufficient to report whether traces of unwanted materials are present, but in other situations a quantitative assessment is required to check that permitted levels of 'adulterants' have not been exceeded. Historically, food authentication has relied mostly on protein identification and many precise immunochemical assays have been in operation for a number of years. Diagnostic techniques could potentially also be based on analysis of RNA or metabolites, but the use of DNA is rapidly becoming the method of choice for many applications. The use of DNA, as opposed to other markers, has several advantages. Firstly, there is potential for a high level of discrimination. In the case of forensic "genetic fingerprinting" this can be down to the level of the individual. Secondly, relatively intact DNA, suitable for analytical purposes, can often be extracted from processed materials in which protein or RNA has been degraded beyond the point of usefulness. Thirdly, differences in DNA are stable and reproducible through many generations and are not influenced by environmental conditions.

There are of course limitations to the use of DNA for some aspects of authentication. For example, it is currently very difficult to develop DNA-based methods for reliable determination of tissue type, since all cells in an animal contain the same chromosomal DNA. However, as shall be seen later, this might be a possibility in the near future. Similarly, it is unlikely that DNA-based methods could be used alone to determine geographical origin of a material; metabolite analysis is probably going to be the method of choice for this type of application. Development of DNA-based methods for food authentication has largely coincided with the emergence of the Polymerase Chain Reaction (PCR). PCR has the ability to amplify minute quantities of DNA to detectable levels and has spawned a multitude of sophisticated technologies, driven mainly by developments in clinical diagnostics. This chapter will attempt an overview of the principles and practice of selected DNA-based diagnostic procedures that are finding their way into routine food and feedstuffs testing. These techniques are based on the ability to detect, identify and if possible quantify, differences in DNA sequences between closely-related biological materials. In order to understand their application, it is first necessary to consider the basis for this variation in DNA sequence.

DNA structure and variability

All living organisms, with the exception of some specific viruses, contain DNA. This DNA has the same basic chemical structure in all cells, consisting of long chains of nucleotides linked by phosphodiester bonds. There are four types of nucleotide found in DNA, which are differentiated by the presence of one of the four bases adenine (A), thymine (T), guanine (G) or cytosine (C). In most organisms, and certainly in all eucaryotes, DNA is double stranded, that is it consists of two polynucleotide strands running antiparallel to each other and held together by specific base pairs of A-T and G-C. The amount of DNA within a cell varies with organism. Thus a simple bacterium like *E.coli* has a single circular DNA molecule consisting of about 4,000,000 nucleotides and the bulk of this is thought to represent "active genes", that is DNA containing information for the synthesis of RNA and protein. In comparison, Humans have two copies of each of 23 chromosomes representing a separate linear DNA of varying length. The total number of nucleotides in the Human genome is thought to be in the region of 3×10^9. However, only a small proportion of this DNA is thought to represent "genes", which in Humans number around 30,000, the remainder being of unknown function. DNA from other mammals and plants is very similar to that from Humans. Within the actual genes there is often a high degree of conservation of DNA sequence between organisms, whilst within the remaining non-coding DNA there can be a high degree of variation. It is the variability in both coding and non-coding DNA sequences that can be exploited for authentication purposes.

All eucaryotic genes seem to have the same basic structure in that they have a promoter region concerned with regulation of expression, a structural region containing coding information for mRNA, and finally a terminator region. The structural region can be further differentiated into exons- regions that contain code that is eventually found in the RNA- and introns that contain non-coding information. In general the DNA sequence of exons tends to be conserved between species, whilst that of introns may be more variable. Moreover, the intron size and number for any given gene may vary from species to species.

The regions between genes, called intragenic spacer DNA, can also be highly variable in both sequence and size. One aspect of this variation that has been commonly exploited for authentication purposes is in the so-called micro-satellite regions, also sometimes referred to as simple sequence repeats (SSRs). These regions contain sequences of between 1 and 13 bases, but most commonly 1-4 bases, that are repeated in tandem over a stretch of about 150 bases or less. An example of such an SSR would be

ATGATGATGATGATGATG. During evolution the length of these repeats has tended to become variable among species, among cultivars and breeds and even among individuals.

Another important source of variation is the occurrence of single nucleotide polymorphisms (SNPs). A SNP may occur practically anywhere within the genome either within a coding region of a gene or within the intragenic spaces. As the name implies the variation exists as a base substitution within a single nucleotide. Even between closely-related organisms, such as the various cultivars of rice, it has been estimated that SNPs may occur as frequently as 1 nucleotide in every 300 within the genome. Detection and analysis of SNPs is rapidly becoming the method of choice for authentication down to the breed or cultivar level of discrimination.

It is thus apparent that there is a range of sequence differences between organisms that may be exploited for authentication purposes. These range from relatively large sequences, such as introns or micro-satellite regions that may differ in length between species and breeds, down to differences at the single base level, SNPs.

Sampling and DNA extraction

The first stage in any DNA-based authentication is obviously to obtain a representative sample of the material to be tested. How this may be achieved is outside the scope of this review. However, suffice it to say that this must be carried out in such a way that statistically the chance of detecting any adulteration is very high. This can be quite challenging as often the material to be tested is of a much greater volume compared to the samples being taken and it cannot be assumed that any adulterant present is at a homogenous level across the whole batch. Thus contamination of a sample with genetically modified material (GMO) may occur if the container is not completely emptied between batches and in this case contamination is likely to be unevenly distributed within the batch being tested. Similarly, contamination may arise from pollen drift in the field affecting a small but localised fraction of the crop and this inhomogeneity may be reflected in the final harvested material.

The next stage of the analysis requires extraction of the DNA. Quantitative isolation of good quality DNA is still surprisingly difficult, especially from seeds and seed oils, milk, fruit and highly-processed or cooked materials. In many cases, extracted DNA is fragmented into segments of about 100 base pairs. This can also be important in feed-stocks, especially in the case of silage. Hupfer *et al.* (1999) demonstrated that genetically modified insect-resistant maize could be detected up to seven months after ensilage, but only by use of PCR to amplify a short - 211 bases- target sequence. For this reason, many diagnostic tests in the food industry are designed to target short regions of DNA and to exploit sequence variability to identify the biological origin of the materials. There are, however, several different methods for extraction of DNA and it is normally the case that sufficient DNA can be extracted, even from intransigent materials, in a form suitable for further analysis. The key assumption to be tested is that extraction efficiency from both raw material and any adulterant(s) is identical. Several companies supply kits for optimal extraction of DNA from a wide range of feedstuffs and in many cases these are provided as part of a kit for detection of adulterants such as GMOs.

Methodologies

Many methods for authentication previously relied on restriction enzymes which cut DNA at sequence-specific sites (Restriction Fragment Length Polymorphism, RFLP). The variable number of such sites in the same region of DNA from closely-related microbial, animal or plant species will generate characteristic patterns after gel electrophoresis. This has been used for identification of fish species (Chow and Inogue, 1993), and for discrimination of turkey from chicken (Hopwood *et al.,*1999). However, this method requires a lot of good quality high molecular weight DNA and it can even yield too much information about individuals within a group rather than the group identity. The development of PCR and related technologies has greatly extended the kinds of tests that can be carried out and their sensitivity. Even low quality fragmented DNA can be used to extract information about species or variety. Indeed, PCR is probably the major technology currently employed in the area of food authentication. There are several reviews on the use of DNA-based techniques in identifying food components (Lockley and Bardsley, 2000; Marmiroli *et al.,* 2003; Lenstra, 2003).

POLYMERASE CHAIN REACTION (PCR)

The technique of PCR was first described in the mid 1980s (Saiki *et al.,* 1988). Since then, this technology has been adapted to form the basis of a very wide range of molecular techniques. The theory behind PCR is relatively simple and will be described here and in Figure 8.1, for what might be termed a targeted reaction. In this instance there is a region of DNA for which the DNA sequence from two short flanking regions is known. Two primers (short oligonucleotides) are synthesised such that they will hybridise one to each of these flanking regions. The target DNA is heated to "melt" its double stranded structure into two single strands. On subsequent cooling the primers hybridise to their respective flanking region such that one primer binds to each of the two separate strands. These primers then serve to direct the synthesis of a new second strand of DNA, by the use of DNA polymerase, and using the original DNA target as the template. The overall result is that from one original target DNA there are now two identical copies. This process is repeated over and over again with the result that with each cycle there is another doubling of the number of copies of the DNA target. Thus after 30 such cycles there will be 2^{30} copies for every single original target. This equates to an amplification of more than a thousand million. The PCR product can then easily be visualised using gel electrophoresis.

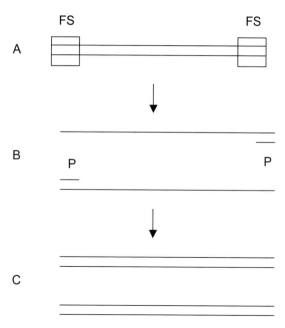

Figure 8.1 Basic PCR.

A: The flanking sequences (FS) of the target DNA are known. The double stranded DNA is heated and splits into two single strands.

B: Oligonucleotide primers (P) are hybridised to the two flanking regions.

C: DNA polymerase is used to extend the primers copying the template to form two double stranded copies of the target.

This is repeated over several cycles resulting in amplification of the target sequence. For 30 cycles the final amplification would be 2^{30} or 1,073,741,824 copies.

PCR can be specific to a particular target. The specificity lies in the sequences of the two primers used for the process. These are often around 15-30 bases in length and can be designed to be "target sequence–specific", that is they will result in amplification of a single well-defined target. However, this specificity can be heavily reliant on the temperatures used to anneal primers to their target, and mis-priming has been shown to amplify non-specific targets on many occasions. It is thus prudent when using PCR for authentication purposes to include further tests to validate the nature of the PCR product.

A range of such techniques are available. The first is the use of restriction endonucleases. These are enzymes that cut DNA at precise sequences, e.g. the enzyme *Eco*R1 cuts at the six base sequence GAATTC, and *Bam*H1 at GGATCC. If the PCR product is known to contain the recognition site for a particular restriction enzyme then it can be treated with that endonuclease and the products analysed by gel electrophoresis. The authentic product will generate a characteristic restriction pattern of bands on the gel (Mackie *et al.,* 1999; Wolf *et al.,* 1999).

Another method is that of nested PCR (Zimmermann *et al.,* 1994). In this case the original PCR product is itself used as the template for a second PCR, but in this case the primers are designed for known regions just inside the original primer regions. In this instance if the PCR product is correct then the second PCR will generate a slightly smaller PCR product. It would be extremely unlikely that a non-specific PCR product would contain either a restriction site in the correct place or regions for the second set of primers to enable either of these tests to generate the correct products, thus they are fairly fool-proof. However, the most definitive method to validate the PCR product must be to actually sequence it completely and show that it does indeed correspond to the intended target (Bartlett and Davidson, 1991; Forrest and Carnegie, 1994).

METHODS FOR QUANTITIVE PCR

Whilst straightforward PCR based techniques are effective at detecting adulteration of feedstuff, due to its exponential character PCR does not easily lend itself to quantification. There are, however, several adaptations to the basic PCR technique that enhance its ability to be quantitative. Of these, Real Time PCR is becoming the method of choice (Marmiroli *et al.,* 2003).

Real Time PCR

There are several types of Real-Time PCR, but all are usually based on measurement of fluorescence. This could either be by a relatively simple measurement of fluorescence from dyes such as SYBR-green as they intercalate between double stranded DNA. As the PCR proceeds, the amount of double stranded product increases and this can be monitored directly in Real Time using a specially constructed fluorimeter attached to a thermal cycler. However, a more accurate method employs the use of a third oligonucleotide, such as a TaqMan® probe (Applied Biosystems).

The Taqman system requires three oligonucleotides, namely two conventional PCR primers as described above and a third probe designed to bind to the DNA target between the two primers (Figure 8.2). Ideally the gap between the primers and the probe should only be a few bases in length. The probe is modified such that it contains both a fluorescent dye and a quencher. Because these are held in relative close proximity on the probe, the amount of fluorescence is limited. As the PCR progresses the probe hybridises to the PCR amplicon as it appears, but then becomes degraded in subsequent cycles as the target is copied by the DNA polymerase. This effectively separates the fluorescent dye and the quencher and as a result fluorescence increases in direct proportion to the amount of PCR product being produced. A typical trace for the development of fluorescence with PCR cycle number is shown in Figure 8.3. Given the nature of the PCR, the faster the rise in fluorescence signal the greater the number of target molecules at the start of the reaction. Providing the efficiency of the PCR for the individual target is known, then this technique can be readily used quantitatively.

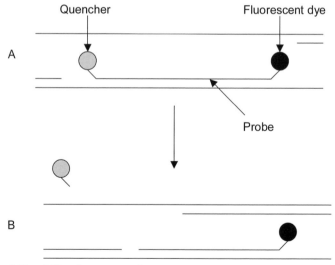

Figure 8.2 Real-Time PCR using a TaqMan® probe.

A: In this instance three oligonucleotides are required- the two PCR primers as in Figure 8.1 and a third TaqMan® probe. This probe hybridises to the target DNA between the two PCR primers. The probe carries a fluorescent dye and a quencher such that since these are in close proximity on the probe fluorescence is quenched.

B: During DNA replication the polymerase acts to break the probe thus separating the quencher from the dye and hence fluorescence increases in proportion to the number of targets copied.

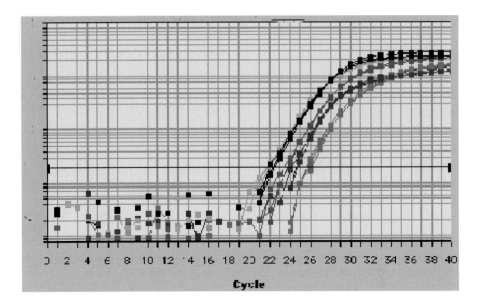

Figure 8.3 Typical Real-Time PCR trace using TaqMan® probe technology. The x-axis indicates the number of PCR cycles. Each line represents the fluorescence of an individual PCR. Thus the first and last lines to increase represent target sequences that occur in the highest and lowest copy numbers, respectively, relative to each other in the original DNA template.

Competitive PCR

Real-Time PCR, whilst perhaps the most efficient, is not the only quantitative PCR based method available. Competitive PCR is a gel-based technique that eliminates the requirement for a relatively expensive real time quantitative PCR machine. Competitive PCR relies on a PCR targeted to both an unknown target and a known internal standard. The PCR reactions are carried out as above, but are "spiked" by the addition of an artificial target that contains flanking regions for the same PCR primers but which produces an amplicon either smaller or larger than the "genuine " target. When the PCR product is run on a gel two bands are observed, one corresponding to the target and the second to the amplified internal standard. A series of PCRs is carried out each with the same level of template DNA from the sample to be tested and with an increasing known amount of the internal standard. The result is a gel as shown in Figure 8.4 (Lockley and Bardsley, unpublished). As the amount of internal standard increases there is competition between the standard and the genuine template and the intensity of the band from the PCR product for the standard increases whilst that from the target declines. At the point of equal intensity the levels of the target and internal standard are equivalent and thus the level of the target can be estimated.

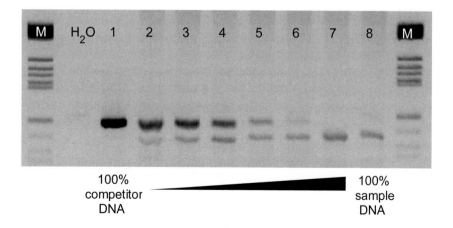

Figure 8.4 Competitive PCR to determine the quantity of chicken in a chicken:pork admixture. Lanes 1 and 8 are PCR products generated from 100% competitor and 100% chicken DNA respectively. Lanes 2 –7 contain increasing amounts (30, 20, 10, 5, 2, 1 fg) of competitor DNA. The point of equivalence in lane 5 indicated that the unknown sample contained equal copies of the chicken sequence as present in 2 fg of competitor DNA, from which the % composition of the sample could be calculated.

This technique can be very accurate and has been used for commercial quantification of target DNA. However, it is very time consuming as it requires a number of PCR and gel analyses. The development of real time RT-PCR has largely superseded competitive PCR as the technique of choice for quantification.

There still remain difficulties with both these quantitative PCRs, usually associated with the large errors inherent in an exponential reaction. There are also possible problems in the case of PCR targets originating from organelle DNA. Thus the variable mitochondrial DNA content of a sample could reflect its tissue of origin as much as the percentage adulteration of one species with another.

SNP analysis

In many cases, discrimination between closely-related species of animals or plants is based on a single nucleotide substitution or polymorphism (SNP). Specialised primers that abut the SNP can be used to determine the substitution pattern. The attraction of this kind of allele-specific 'primer extension' is that it lends itself to non gel-based detection methodologies, including mass spectrometry and microarrays ('chip' technologies) that can be automated for rapid throughput.

Once an SNP has been detected then it can be readily identified and quantified again through an extension of the PCR. In this case three primers are required, as shown in Figure 8.5. Again two "normal" PCR primers are required that can amplify the region within the template DNA corresponding to the SNP to be analysed/quantified. The third primer is designed to hybridise to the PCR product such that the final nucleotide at the 3' end of the primer is immediately adjacent to the site of the SNP. This primer is then extended by a single nucleotide by adding a complementary dideoxynucleotide (ddN). Thus, if the next base on the target is an A, the primer will be extended by the addition of ddT. However, this is a SNP and so the base on the target may be, for example, a G and in this instance the primer will be extended by the addition of ddC. The dideoxynucleotide is also known as a chain terminator since it does not allow any further nucleotide additions to the primer. The product thus consists of single base extensions of the original primer that could contain any of the four dideoxynucleotides, dependent on the nucleotide(s) present within the DNA used as the template.

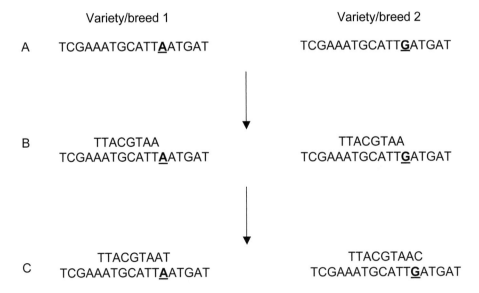

Figure 8.5 Single nucleotide polymorphisms and their detection.

A: A known polymorphism exists within a DNA target as exemplified here where the SNP can be either A or G depending on the variety or breed.

B: The target region of the DNA is amplified by conventional PCR and a primer designed to anneal directly adjacent to the site of the SNP is hybridised to single stranded DNA.

C: The primer is extended by a single nucleotide using chain terminating dideoxynucleotides. The nucleotide added being dictated by the base at the SNP in each case.

The extended primers are then ready for analysis (see Figure 8.6).

The next stage of the analysis is to identify and differentiate between these extended primers. One common method is to employ mass spectroscopy to differentiate on the basis of mass. The extended primers are subject to MALDI-ToF mass spectrometry where the primers are placed within a matrix, excited by laser to generate ionised particles and their time of flight measured. Since the four bases A,T,G and C are all slightly different in mass, the difference in ToF between the unextended and extended primers will be characteristic of the base added (Figure 8.6).

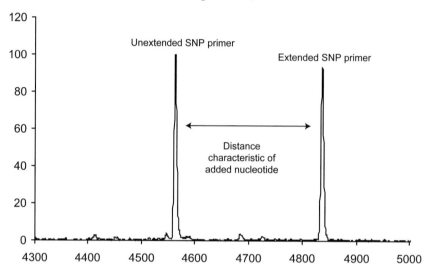

Figure 8.6 Single nucleotide polymorphism (SNP)analysis using MALDI-ToF.

The mixture of extended and unextended primers from a SNP reaction (as depicted in Figure 8.5) are analysed by MALDI-ToF. The peak position is dependent on primer size. Different bases have different molecular weights A=297Da, T=288Da, G=313Da and C=273Da. Thus the difference between the peaks is characteristic of the added base and thus identifies the SNP(s) present.

An alternative approach would be to label each of the four dideoxynucleotides-ddT, ddA, ddG and ddC used in the primer extension reaction with a different fluorescent dye, exactly as occurs in a DNA sequencing reaction. Since each ddN is labelled with a different "coloured" dye, the extended primers can be readily differentiated using a DNA sequencer. This method is often termed single base extension or SBE.

In either method of detection – MALDI-ToF or SBE – the resultant traces can be readily used for quantification. Indeed, both have been shown to be accurate down to at least the 20% level (Shifman *et al.,* 2002) and will probably work at greater sensitivities. In addition, both methods readily lend themselves to multiplex assays where several SNPs can be detected and quantified in a single reaction.

Detection of genetically modified organisms (GMO)

In the case of GMO, current practical concern within the feed industry lies with the detection of genetically modified plant material. In terms of legislation there are basically three types of genetically modified organisms- those approved and authorised to be grown within the EU, those approved for use in food and feedstuffs but not for cultivation within the EU and the third category which is the unapproved GMOs. Six plants with potential for use in animal feeds have been granted part C approval for commercialisation within the EU under the "direct release directive". These include three herbicide-tolerant and two insect-resistant maize varieties, a herbicide-tolerant soya and a herbicide-tolerant oilseed rape. However, only three of these have been licensed for cultivation within the EU (FSA, 2004). There are however, several other examples of genetically modified plants being grown commercially around the world and some of these could have potential use in feedstuffs. In theory all three categories of GMO may be encountered within animal feed. In terms of worldwide cultivation, 76 million hectares of soya bean were grown in 2002, of which 41.4 million hectares or 54% was genetically modified (FSA, 2004). In the case of maize, of 140 million hectares grown, 15.5 million or 15% was genetically modified (FSA, 2004). This makes these two of the major GMO crops currently cultivated. This, along with current EU legislation relating to the labelling and permitted levels of GMOs in animal feed, has resulted in the need to detect, identify and quantify the GMO content of animal feed raw materials.

There have been several reviews of methodologies for detection of GMO in food (Meyer, 1999; Holst-Jensen, 2003). Any detection method for GMO material must rely on the ability to monitor one or more products of the transformation process within the plant tissue. This could be the final protein, the RNA or the transgene (tDNA) itself. Protein and RNA are both present at much higher amounts than the tDNA and as such are more easily detected using direct methods. However, levels of protein and RNA are variable in relation to sample size and as such are not readily amenable to methods for quantification of GMO content. They are also both more labile than DNA and tend not to survive any silage processing. In contrast, DNA content is fixed within a sample making quantification much easier and, although DNA is degraded during processing, such as ensilaging, it is not fully destroyed and short fragments (around 200 base pairs) persist which can be employed in subsequent detection systems (Hupfer *et al.*, 1999). The problem of low levels of DNA can be overcome by the use of PCR-based technologies to amplify sequences within the tDNA. These sequences or targets can be quite diverse, as can the technology used for their detection.

Before considering the methodologies used to detect and quantify GMOs, it is useful to first describe the general nature of tDNA. A typical transformation construct is depicted in Figure 8.7. This consists of several elements; firstly promoter(s), regions of DNA able to direct the expression of the transgene within the plant tissue; secondly, a coding region which contains the genetic information used for the synthesis of RNA and eventually the protein, and lastly a terminator region. In several constructs a second "gene" is also included that is used as a selectable marker to detect the genetically modified plant. A common marker is the NptII gene which encodes for resistance to the antibiotic kanamycin. Other common markers are genes encoding for herbicide resistance. Table 8.1 lists some of the elements commonly used in genetically modified crops. Several of these elements have been used in a number of separate genetic modification events. Thus the cauliflower mosaic 35S (CaMV35S) and the nos promoters along with the nos- terminator have been used in several cases. The structural gene obviously varies between GMO since it is this that confers the trait of interest. However, even in this case there may be several different GMOs available which employ the same structural gene; for example bt11 and bt176 maize are two separate GMOs, but both have the cry1 structural gene that encodes for the bacterial toxin that confers insect resistance onto these crops. These are considered as distinct varieties and the genetic modifications are referred to as separate "events".

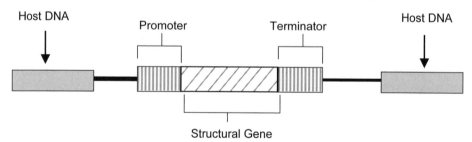

Figure 8.7 Stylised typical structure of a DNA construct (tDNA) used to transform a host organism.

Table 8.1 Common elements used in genetic modification.

Promoter	Terminator	Structural gene	Selectable marker
CaMV35S	nos	Cry1 (Insect-resistance)	nptII
nos		EPSPS (herbicide-tolerance)	bar

There are basically three levels of increasing sophistication for screening raw material for the presence of GMO. The first of these is a general screen designed to detect whether or not any form of GMO is present. The second is to identify the actual variety(ies) or "events" of GMO present to check that they comply with those allowed under EU regulations. The third is to quantify the level of GMO to comply with labelling regulations. Given the nature of tDNA, a range of targets can be employed for general detection of GMO. A general screen may target PCR primers to an individual element within the tDNA, such as the CaMV35S promoter or nos-terminator (Lin *et al.*, 2000), since these are used in a range of tDNAs and are common to many GMOs. This approach only requires a standard PCR and positive results can be readily detected using gel electrophoresis as depicted in Figure 8.8. The method is also very sensitive and can easily detect levels of GMO at around 0.1% (Figure 8.8). However, there are some disadvantages, most notably that these elements are naturally occurring and as such can be found associated with non-GMO crops leading to generation of false positives.

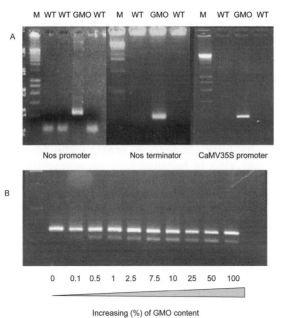

Figure 8.8 Detection of GMO using PCR targeted to common elements.

A: Detection of genetically modified tomato using PCR primers targeted at the nos or CaMV35S promoters or the nos terminator. M= Markers, WT= wild type tomato, GMO= transgenic tomatoes.

B: Sensitivity of PCR detection method. Wild type and genetically modified tomatoes were mixed and then DNA extracted and analysed for GMO content. 0=no GMO, 100= 100% GMO.

Because these elements are often common to several different tDNAs, their detection is not sufficient to provide a definitive method to identify the source of GMO.

This has been partially overcome by employing PCR primers that are specifically directed to a structural gene like cry1 in the case of bt maize or soya. However, again the same structural gene may be employed in several different GMOs and so this approach is not able to specifically identify the GMO event. Another approach is to use PCR primers that span two elements of tDNA; for instance, one primer might be targeted at the CaMV35S promoter and the second to the cry1 coding region. This makes the resultant amplicon more tDNA specific but, with the increasing number of GMOs entering the market, it may not be "event" specific and as such is unable to unequivocally identify the GMO contaminant. The most recent approach to provide "event" specific identification of the GMO is to utilise PCR primers targeted to the insertion junction; one primer is specific to a region within the tDNA, the other to a region of the host genome immediately adjacent to the site of insertion. Since it is extremely unlikely that the tDNA will insert in the same place in separate transformation events, this approach is highly specific. One disadvantage is that the site of insertion of the tDNA must be mapped. With current advances in molecular biology this is becoming easier. With all of these approaches a straight PCR followed by gel electrophoresis of the product can be used for detection of GMO contamination. However, it is often useful to confirm the identity of the product either by restriction digest, nested PCR or even sequencing.

The final level of sophistication is the quantification of GMO content. It can be difficult to quantify the level of GMO using an element such as the CaMV35S promoter. This is because it is becoming increasingly common for breeders to cross transgenic lines, for instance herbicide resistant and insect resistant maize. This so- called "stacking" of transgenes can confound any attempt to quantify the level of GMO where both tDNAs employ a common element. Thus a 1% level of a GMO with two "stacked" tDNAs each controlled by the CaMV35S would return a 2% contamination signal with any PCR method directed at the promoter. The most reliable current technique is Real-Time PCR, and this method has become the current method of choice. In order to quantify accurately there is a need not only for tDNA specific primers but also for species-specific targets with which to compare. These include endogenous genes such as invertase and lectin. Methods for accurate quantification of several event specific GMOs have been described and many of these are commercially available. It is often necessary to quantify levels of several different GMOs within the same raw material and to this effect multiplex PCR is most useful. Again there are several published methods, as exemplified by that of Huang and Pan (2004) who have described a multiplex Real-Time PCR for simultaneous detection of insect-resistant maize MON810 and herbicide-tolerant maize NK603.

Testing of GMOs has become a major industry and several companies either offer a validation service or sell kits for detection and quantification of a range of GMOs. Nearly all of these methods are for general screening based on either detection of elements such as the CaMV35S promoter or for quantification using event specific Real-Time PCR.

Recently the use of Labchip technology has been reported for quantitative multiplex detection of Roundup Ready soya beans (Burns *et al.*, 2003). This technology could replace Real-Time PCR as it requires less expensive equipment, is cheaper to run and can be quantitative. Another potential technology for multiplex detection of GMOs is microarrays. In this technology oligonucleotides representing all known GMOs are plated onto a microarray. This microarray is then challenged with fluorescently-labelled DNA from the raw material to be tested. Any hybridisation of this DNA to the oligonucleotides on the microarray would indicate the presence of GMO material. This technology can also serve to identify the event(s) and can be used to quantify levels of specific GMOs. A recent report (Germini *et al.*, 2004) has described the use of microarray technology for detection of Roundup Ready soya. In this instance they utilised peptide nucleic acid (PNA) microarrays. In PNA the sugar-phosphate backbone of "normal" oligonucleotides is replaced by a pseudopeptide chain of N-aminoethylglycine monomers. This increases the affinity of PNA oligonucleotides for the corresponding DNA in the target, improving the efficiency of the microarray.

This technology may also be adapted in the future to meet perhaps the most challenging problem in GMO detection, namely the identification of non-approved GMO varieties. All current detection methodologies rely on release of information by the producers of the GMO. There is a real possibility that unscrupulous producers may use undeclared GMO material for which there is no detection assay. The use of random sequence microarray technology currently employed for genome sequencing could be adapted to monitor the presence of any new genetic material within a plant species and thus detect non-approved GMO

Determination of species, breed and tissue of origin

The practice of including animal by-products in livestock, poultry and companion animal diets has clearly had to be reviewed in the light of recent experiences with the spongiform encephalopathies, and it is possible that speciation testing will become more widespread in the animal feedstuffs industry. Authentication of meat and dairy products destined for human foods has concentrated on development of tests that discriminate between the major livestock species, cattle, pig, sheep and goat, as well as turkey and chicken. Differentiation between seafood species has also been well studied, a well-known problem being substitution of scampi by cheaper alternatives such as monkfish. Fish speciation is also important from the point of view of conservation of fish stocks and protection of endangered species, as well as to prevent fraudulent substitutions. While PCR-based speciation tests are now available and can detect one species in the presence of another at a level of approximately 0.1-1%, the discrimination between breeds within an animal species is much more complex and is probably not of too much concern to feed manufacturers at present. What is likely to be of concern, however, is the detection by DNA-based techniques of undesirable tissues such as spinal cord, although this is also technically difficult. In this section some of the principles behind animal speciation will be described, finishing with an outline of current research directed towards breed and tissue testing.

LIVESTOCK SPECIATION BASED ON DNA HYBRIDISATION

Early methods for meat speciation were based on the fact that DNA extracted from fresh or cooked animal tissues of all major livestock species could be immobilised as discrete spots or bands on nitrocellulose-based membranes. DNA would then be extracted from a series of known species and isotopically tagged with ^{32}P-labelled nucleotides, thereby constituting a probe. It was shown by a number of groups that if cross reactivity could be kept to a minimum, labelled DNA probes would hybridise only to their own species, therefore providing a method for identification of unknown materials (Chikuni *et al.*, 1990; Ebbehøj and Thomsen, 1991). It was generally believed that the origin of species-specific hybridisation was to a large extent based on microsatellite regions in the DNA. By working with proven species-specific microsatellites it was possible to improve selectivity of probes to the point where even closely-related species such as sheep and goat could be readily discriminated (Winterø *et al.*, 1990).

The variable length of microsatellite repeats can also be reflected in the size of DNA fragments produced by digestion of DNA by restriction endonucleases. DNA that has been cut at specific points by these enzymes generates fragments that can be separated by agarose gel electrophoresis and subsequently Southern blotted onto nitrocellulose-based membranes. The use of species-specific ^{32}P labelled probes can sometimes reveal characteristic patterns, as described by Blackett and Keim (1992) for a number of deer species. The presence or absence of conserved restriction sites in related DNA sequences from different species will also contribute towards variable fragment sizes after restriction enzyme digestion. Fairbrother *et al.* (1998a) exploited the fact that members of the actin multigene family all have highly conserved protein coding regions or exons. Following DNA digestion, electrophoresis and Southern blotting, an actin-related cDNA probe revealed fragmentation patterns that were characteristic of a number of livestock and fish species (Figure 8.9).

In general, authentication methods based on DNA hybridisation require relatively large quantities of good quality DNA and lack sensitivity. They tend to be unsuitable for quantitative analysis, especially when DNA from mixed meat products is present.

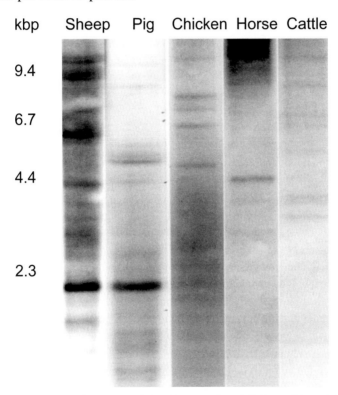

Figure 8.9 Species identification by Restriction Fragment Length Polymorphism: hybridisation to an actin cDNA probe (Fairbrother *et al.,* 1998a).

LIVESTOCK SPECIATION BASED ON PCR

PCR is designed primarily to amplify discrete DNA regions based on prior knowledge of the target sequence. However, in some cases a series of short non-specific primers has been used that bind to uncharacterised DNA at unknown sites and amplify multiple segments of DNA whose origins are completely unknown. After agarose gel electrophoresis, species-specific 'fingerprints' have been described (Figure 8.10) that are analogous to the restriction fragment-derived patterns described above (Saez *et al.*, 2004). Although this Rapid Amplification of Polymorphic DNA method (RAPD) has the capacity to work with small amounts of material for which no sequence data need be known in advance, the patterns are generally too complex for analysis of mixtures of animal materials and do not lend themselves to quantification.

Markers

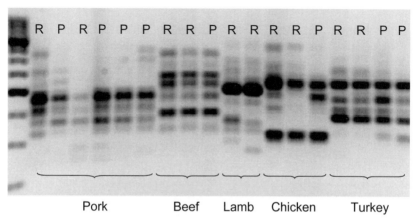

Pork   Beef   Lamb   Chicken   Turkey

Figure 8.10 RAPD analysis of livestock species in raw (R) and processed (P) meats (Saez *et al.*, 2004).

Numerous PCR methods have been devised that target known sequences in different species (for review see Lockley and Bardsley, 2000a). Generic primer pairs that amplified across a specific exon in members of the actin gene families in a number of species were found useful for generation of species-specific fingerprints (Figure 8.11; Fairbrother *et al.*, 1998b). Many other groups have amplified part of the cytochrome b gene sequence in mitochondrial DNA, partly because such sequences are present in higher copy number than those from nuclear gene targets, which facilitates product visualisation on a gel. Matsunaga *et al.* (1999) used one generic forward primer that annealed to the common SIM sequence in all livestock species in combination with a series of species-specific reverse primers. The characteristic sizes of the PCR amplicon was diagnostic of the species (Figure 8.12), even in DNA extracted from cooked meats or meat mixtures. A similar approach was used to discriminate chicken and turkey DNA (Lockley and Bardsley, 2002) and has also been extended to Real Time quantitative PCR (Sawyer *et al.*, 2003).

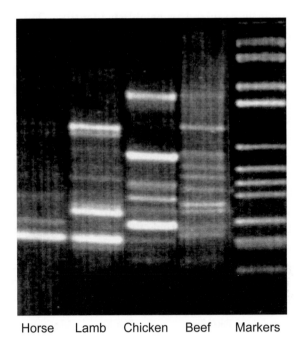

Horse   Lamb   Chicken   Beef   Markers

Figure 8.11 Species specific fingerprints generated by PCR amplification of multiple actin genes (Fairbrother *et al.,* 1998b).

Markers Goat Chicken Cattle Sheep Pig Horse

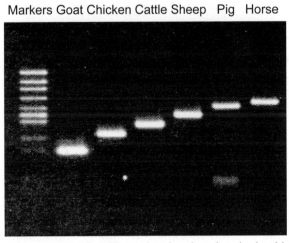

Figure 8.12 Livestock species-specific PCR reactions based on the mitochondrial cytochrome b gene (Matsunaga *et al.,* 1999)

The use of generic primer pairs that cross species is a popular approach in the speciation of meat and fish, but in many cases this leads to the appearance of amplicons of identical size but different sequence. In this case, some secondary analysis is required to establish identity. This can include full sequencing (Bartlett and Davidson, 1991), restriction enzyme digestion (Mackie *et al.*, 1999; Hopwood *et al.,* 1999; Fernandez *et al.,* 2000) or Single Strand Conformational Polymorphism analysis (SSCP: Mackie *et al.,* 1999).

In some cases, the difference between two amplicons derived from two different species is limited to just one or two Single Nucleotide Polymorphisms (SNPs). If SNPs cannot be detected by restriction enzymes, some form of allele-specific PCR can be performed. This relies on the fact that the 3' terminal one or two nucleotides of a primer must generally be an absolute match to the target sequence, whereas a certain degree of mismatching can be tolerated elsewhere in the primer. To distinguish two closely-related tuna species, for example, two forward primers of almost identical sequence but different lengths were designed such that they annealed to the same target sequence but had alternative 3' termini to discriminate the two possible versions of a species-specific SNP (Lockley and Bardsley, 2000b). The SNP, and therefore the species, was subsequently diagnosed by virtue of amplicon size on high resolution agarose gels (Figure 8.13).

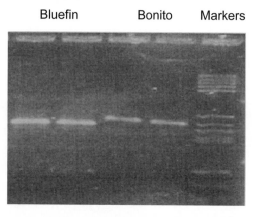

Figure 8.13 Identification of Tuna species by allele-specific PCR (Lockley and Bardsley, unpublished)

The importance of the 3' terminal nucleotide of a primer in permitting or preventing extension in order to diagnose an SNP is rapidly becoming established as the method of choice for many applications, as described elsewhere in this chapter. Indeed, it is no longer necessary to generate a whole PCR amplicon but instead to use DNA polymerase to extend a primer by one nucleotide only; subsequently, the identity of the incorporated nucleotide and therefore the genotype can potentially be assessed by mass spectroscopy (MALDI TOF) or capillary gel electrophoresis on a DNA sequencer. Examples of this were described earlier for rice varieties, although there have been few published reports of the use of primer extension for meat and fish speciation.

LIVESTOCK BREED IDENTIFICATION

Breeds are distinguished by minor variations in sequences which in some instances can have obvious effects on phenotype. As discussed earlier, variability can take the form of changes in protein coding regions, microsatellite repeats and SNPs. It is unlikely that one SNP or microsatellite will completely change a phenotype and establish a different breed, although the RyR1 mutation ('halothane') in pigs is a well-known example of a single SNP with a powerful effect on lean growth and it has a particular association with the Pietran breed (Fujii *et al.,* 1990). Verkaar *et al.* (2002) described PCR reactions followed by restriction enzyme digestion (PCR RFLP) that discriminate bovine species such as buffalo, yak and zebu cattle from taurine breeds normally used in western agriculture. However, to discriminate more closely-related western breeds it is necessary to analyse variability at multiple sites in the genome. Currently this requires identification of microsatellite length and/or SNP genotype in a large population of each breed, followed by statistical analysis to assign the probability that an individual exhibits a distribution characteristic of a particular breed. An example of this approach is that of Paszek *et al.* (1998) who examined the variable length of a microsatellite on Chromosome 6 in five pig breeds. Although one microsatellite allele was apparently unique to the Meishan breed, in general the assignation of breed to an unknown sample required statistical analysis. A theoretical approach that estimated the number of markers or SNPs necessary to achieve a 95% accurate breed identification suggested that up to 10 microsatellites or 50 SNPs would be necessary if these were randomly distributed throughout the genome (Blott *et al.,* 1999). However, provisional studies from our laboratories indicate that the distribution of three SNPs in a 600 bp region of the porcine α skeletal actin can discriminate Large White and Duroc from the Yorkshire breed with approximately 80% reliability (McAnulty, 2002). Similarly a panel of 11 SNPs analysed by the MALDI TOF approach has recently been used to identify varieties of Basmati rice (Graham *et al.,* unpublished data). Although a reliable analysis of breeds can be achieved, given sufficient resources and animal populations of authentic pedigrees, simple tests that unequivocally identify all breeds are not yet available. Recently, however, a commercial Aberdeen Angus–specific test for use in the food industry and for selective breeding has been advertised (http://www.viagen.com).

METHODOLOGIES FOR QUANTITATIVE DETERMINATION OF MEAT MIXTURES

Most of the species-specific PCR tests described above have been based on mitochondrial DNA and these are now being adapted for the Real Time PCR techniques described earlier for the detection of GMOs. Quantitative species-specific tests for analysis of beef, pork, lamb, chicken and turkey DNA based on the mitochondrial cytochrome b gene have recently been adapted for the TaqMan® system (Dooley *et al.*, 2004; Sawyer *et al.*, 2003). In meat admixtures, Dooley *et al.* (2004) found that the limit of detection for any one species in the presence of any other was 0.5%. Quantitative competitive PCR (QC PCR) is an alternative to real time PCR for quantification of meat mixtures and the principles behind this method were described earlier in this chapter. Wolf and Lüthy (2001) described a QC PCR method based on the porcine growth hormone gene that would allow measurement of pork in meat admixtures. Although the sensitivity was similar to that of real time methods, and did not require expensive instrumentation and materials, the main drawback is that a series of PCR reactions and a gel electrophoresis step need to be carried out for each sample analysed. This is in contrast to real time methods where many samples can often be analysed simultaneously.

One of the remaining questions for quantitative analysis of DNA from meat admixtures is whether the natural variability in mitochondrial DNA copy number between the same tissue from different species could lead to a biased assessment compared to measurement made on nuclear gene targets. The error could be compounded if mixtures of dissimilar tissues from different species were present in an admixture. In this context, recent work from this laboratory using quantitative Real Time PCR based on nuclear gene (α-skeletal actin) and mitochondrial gene (cytochrome b) sequences revealed that the ratios of mitochondrial to nuclear targets in porcine skeletal muscle, heart, liver and kidney were 26:1, 111:1, 150:1 and 171:1 respectively (Jeyapalan, 2004). Few data are available on whether mitochondrial DNA or nuclear DNA are preferentially extracted from fresh or processed animal tissues.

DETERMINATION OF TISSUE TYPE

Although species-specific PCR reactions are now well-established, it has proved more difficult to devise tissue-specific tests, since the sequence of DNA in all cells of an organism is necessarily identical. Tissue-specific tests would clearly be advantageous in order to exclude brain and spinal cord from feedstuffs or to monitor the presence of bone-derived DNA as an indicator of mechanically recovered materials.

Clearly what would be ideal is a test that detects tissue-specific gene products rather than the gene itself, although mRNA is too unstable and does not survive unless tissue is immediately frozen post-mortem and stored at -80°C. However, in order to regulate the expression of genes in different tissues, many gene promoter sequences are believed to be methylated on cytosine residues, which reduces the transcriptional efficiency of the promoter in tissues where the gene is not to be expressed (Jones *et al.,* 1981). Attempts are accordingly underway to devise PCR-based procedures that can differentiate between methylated and unmethylated promoter sequences and perhaps therefore the tissues of origin. For example, identification of a region of DNA that is demethylated only in cells of the spinal cord could lead to a specific test for this kind of contamination. To detect methylated sequences in a particular gene sequence first requires treatment of isolated DNA with sodium bisulphite, which converts all cytosines to uracil unless they are methylated, as for example in a quiescent gene (Clark *et al*., 1994). Since deoxyuracil is read as thymidine in the PCR reaction, the bisulphite reaction essentially creates novel sequences for which appropriate gene-specific primers can be designed. Any C residues remaining after bisulphite reaction and PCR must have been methylated, and thus the methylation status of a particular gene sequence can be assessed. Using this general approach, there have been recent indications that demethylated sequences in the bovine and ovine GFAP and porcine connexin genes may be useful for tissue-specific diagnostic testing (www.food.gov.uk/multimedia/pdfs/q01q02programme review.PDF)

Conclusions

In the past decade, DNA based methods for authentication of human food have steadily increased in versatility, sophistication and reliability. Unlike most applications in clinical diagnostics, tests for use with food DNA usually need an additional quantitative dimension so that upper limits of contamination can be monitored. Similar considerations will apply to deployment of PCR–based tests for animal feedstuffs. In contrast to clinical diagnostics, where a few microlitres of blood or saliva are usually all that is required, considerable attention needs to be paid to reliable methods for DNA extraction when testing bulk food or feedstuffs of uncertain homogeneity. Although a PCR test can readily be used with 10-100 nanograms of DNA, it will not be easy to ensure that DNA obtained from many tons of a feedstuff is truly representative of the sources of its ingredients.

References

Bartlett S.E. and Davidson W.S. (1991) FINS (Forensically Informative Nucleotide Sequencing): A procedure for identifying the animal origin of biological specimens. *Biotechniques* **12**, 408-411.

Blackett R.S. and Keim P. (1992) Big game species identification by deoxyribonucleic acid (DNA) probes. *Journal of Forensic Sciences* **37**, 590-596.

Blott S.C., Williams J.L. and Haley C.S. (1999) Discriminating among cattle breeds using genetic markers. *Heredity* **82**, 613-619.

Burns M., Shanahan, B.M., Valdiva H. and Harris N. (2003) Quantitiative event-specific multiplex PCR detection of roundup ready soya using LabChip technology. *European Food Research and Technology* **216**, 428-433.

Chikuni K., Ozutsumi K., Koishikawa T. and Kato S. (1990) Species identification of cooked meats by DNA hybridization assay. *Meat Science* **27**, 119-128.

Chow S. and Inogue S. (1993) Intra –and inter-specific Restriction Fragment Length Polymorphism in mitochondrial genes of *Thunnus* tuna species. *Bulletin of the National Institute of Far Seas Fisheries* **30**, 207-224.

Clark S.J., Harrison J., Paul C.L. and Frommer M. (1994). High sensitivity mapping of methylated cytosines. *Nucleic Acid Research* **22**, 2990-2997.

Dooley J.J., Paine K.L., Garrett S.D. and Brown H.M. (2004) Detection of meat species using TaqMan real time PCR assays. *Meat Science* **68**, 431-438.

Ebbehøj K.F. and Thomsen P.D.(1991) Differentiation of closely related species by DNA hybridisation. *Meat Science* **30**, 359-366.

FSA (2004) Food Standards Agency WEB site http://www.food.gov.uk/ gmfoods/gm-animal.

Fairbrother K.S., Hopwood A.J., Lockley A.K.and Bardsley R.G. (1998a) Meat speciation by restriction fragment length polymorphism analysis using an a-actin cDNA probe. *Meat Science* **50**, 105-114.

Fairbrother K.S., Hopwood A.J., Lockley A.K.and Bardsley R.G. (1998b) The actin multigene family and livestock speciation using the Polymerase Chain Reaction. *Animal Biotechnology* **9**, 89-100.

Fernández A., García T, Asensio L., Rodríguez M.A., González I., Céspedes A., Hernández P.E. and Martín R. (2000) Identification of the clam species *Ruditapes decussatas* (Grooved carpet shell, *Venerupis pullastra* (Pullet carpet shell) and *Ruditapes philippinarum* (Japanese carpet shell) by PCR-RFLP. *Journal of Agricultural and Food Chemistry* **48**, 3336-3341.

Forrest A.R.R. and Carnegie P.R. (1994) Identification of gourmet meat using FINS (Forensically Informative Nucleotide Sequencing*). Biotechniques* **17**, 24-26.

Fujii J.,Otsu K., Zorzato F., de Leon S., KhannaV.K., Weiler J.E., O'Brien P.J. and MacLennan D.H. (1991). Identification of a mutation in porcine ryanodine receptor associated with malignant hyperthermia. *Science* **253**, 448-451.

Germini A., Mezzelani A., Lesignoli F., Corradini R., Marchelli R., Bordoni R., Conslandi C and De Bellis G (2004). Detection of genetically modified soybean using peptide nucleic acids (PNAs) and microarray technology. *Agricultural and Food Chemistry* **52**, 4535-4540.

Holst-Jenson A. (2003) Advanced DNA-based detection techniques for genetically modified food. In *Food Authenticity and Traceability.* Lees M. (Ed). Woodhead Publishing Ltd pp 575-594.

Hopwood A.J., Fairbrother K.S., Lockley A.K. and Bardsley R.G. (1999). An actin gene–related Polymerase Chain Reaction (PCR) test for identification of chicken in meat mixtures. *Meat Science* **53**, 227-231.

Huang H-Y. and Pan T-M. (2004) Detection of genetically modified maize MON810 and NK603 by multiplex and Real-Time polymerase chain reaction methods. *Agricultural and Food Chemistry* **52**, 3264-3268.

Hupfer C., Mayer J., Hotzel H., Sachse K. and Engel K.H. (1999) The effect of ensiling on PCR-based detection of genetically modified Bt maize. *European Food Research and Technology* **20**, 301-304.

Jeyapalan J.N. (2004) *Comparison of nuclear and mitochondrial DNA markers for meat authentication*. PhD Thesis, University of Nottingham.

Jones R.E., DeFeo D. and Piatigorsky J. (1981) Transcription and site-specific hypomethylation of the d-crystallin genes in the embryonic chicken lens. *Journal of Biological Chemistry* **256**, 8172-8176.

Lenstra J.A. (2003) DNA methods for identifying plant and animal species in food. In *Food Authenticity and Traceability.* Lees M. (Ed). Woodhead Publishing Ltd pp 34-53.

Lin HY., Chiueh L.C. and Shih D.Y.C. (2000) Detection of genetically modified soybeans and maize by the polymerase chain reaction method. *Journal of Food and Drug Analysis* **6**, 200-207.

Lockley A.K. and Bardsley R.G. (2000a) DNA-based methods for food authentication. *Trends in Food Science and Technology* **11**, 67-77.

Lockley A.K. and Bardsley R.G. (200b) Novel method for the discrimination of tuna (*Thunnus thynnus*) and bonito (*Sarda sarda*) DNA. *Journal of Agricultural and Food Chemistry* **48**, 4463-4468.

Lockley A.K. and Bardsley R.G. (2002) Intron variability in an actin gene can be used to discriminate between chicken and turkey DNA. *Meat*

Science **61**, 163-168.

Mackie I.M., Pryde S.E., Gonzales-Sotelo C., Medina I., Peréz-Martin R., Quintero J., Rey-Mendez M. and Rehbein H. (1999). Challenges in the identification of species of canned fish. *Trends in Food Science and Technology* **10**, 9-14.

Marmiroli N., Peano C.and Maestri E. (2003) Advanced PCR techniques in identifying food components. In *Food Authenticity and Traceability*. Lees M. (Ed). Woodhead Publishing Ltd pp 3-33.

Matsunaga T., Chikuni K., Tanabe R., Muroya S., Shibata K., Yamada J and Shinmura Y. (1999) A quick and simple method for the identification of meat speciesand meat products by PCR assay. *Meat Science* **51**, 143-148.

McAnulty C.M. (2002) *Actin gene variability in different porcine breeds*. PhD Thesis University of Nottingham, UK.

Meyer R. (1999) Development and application of DNA analytical methods for the detection of GMOs in food. *Food Control* **10**, 391-399.

Paszek A.A., Flickinger G.H., Fontanesi L., Rohrer G.A., Alexander L., Beattie C.W. and Schook L.B. (1998) Livestock variation of linked microsatellite markers in diverse swine breeds. *Animal Biotechnology* **9**, 55-66.

Saez R., Sanz Y. and Toldrá F. (2004) PCR-based techniques for rapid detection of animal species in meat products. *Meat Science* **66**, 659-665.

Saiki R.K., Gelfand D.H., Stoffel S., Scharf S.J., Higuchi R., Horn G.T., Mullis K.B. and Ehrlich H.A. (1988). Primer–directed enzymatic amplification of DNA with thermostable DNA-polymerase. *Science* **239**, 487-491.

Sawyer J., Wood C., Shanahan D., Gout S. and McDowell D (2003) Real-Time PCR for quantitative meat species testing. *Food Control* **14**, 579-583.

Shifman S., Pisante-Shalom A., Yakir B. and Darvasi A. (2002).Quantitative technologies for allele frequency estimation of SNPs in DNA pools. *Molecular and Cellular Probes* **16,** 429-434.

Verkaar E.L.C., Nijman I.J., Boutaga K. and Lenstra J.A. (2002) Differentiation of cattle species in beef by PCR-RFLP of mitochondrial and satellite DNA. *Meat Science* **60**, 365-369.

Winterø A.K., Thomsen P.D. and Davis W. (1990) A comparison of DNA-hybridization, immunodiffusion, countercurrent immunoelectrophoresis and iso-electric focussing for detecting the admixture of pork to beef. *Meat Science* **27**,75-85.

Wolf C., Rentsch J. and Lüthy J. (1999). PCR-RFLP analysis of mitochondrial DNA: a reliable method for species identification. *Journal of Agricultural and Food Chemistry* **47**, 1350-1355.

Wolf C. and Lüthy J. (2001). Quantittive competitive PCR for quantification of porcine DNA. *Meat Science* **57**, 161-168.

Zimmermann K., Pischinger K. and Mannhalter J.W. (1994). Nested primer PCR detection limits of HIV-1 in the background of increasing numbers of lysed cells. *Biotechniques* **17**, 18-20

9

CHANGES IN THE ASIAN LIVESTOCK INDUSTRIES: IMPLICATIONS FOR EUROPEAN PRODUCERS AND THE FEED INDUSTRY

DR PAUL MEGGISON

Aust-Asia Business Solutions, Wagga Wagga, NSW, Australia

Introduction

The world's population will continue to grow at a steady, but slightly declining, rate of between 0.5 and 1.3% for the next 20 to 50 years (US Census Bureau, 2000). By far the majority of this growth will occur in the Eastern Hemisphere; Asia, the Pacific Rim and the Indian Sub-continent (Figure 9.1).

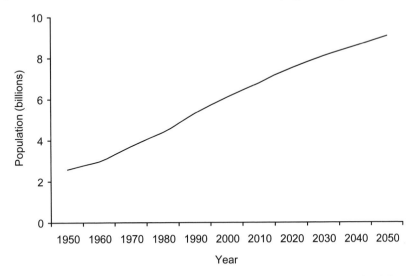

Figure 9.1 World Population – 1950-2050. (Source – U.S. Census Bureau, International Data Base 5-10-00).

Population alone cannot stimulate human food consumption. Africa is a clear example of this. However, a combination of economic growth and sound work ethics, offering elevated disposable income amongst the masses, together with population growth, has led to massive developments in the production and consumption of livestock products in the Far East region in recent years. This will continue to grow at an increasing rate for years to come providing interesting commercial opportunities for feed and feed related companies.

Analysing world population developments and world wide compound feed production, there appears to be a declining trend, since 1995, in the amount of feed produced per capita of population. Further, according to the International Feed Industry Federation (IFIF), world compound feed production peaked at 605million MT in 1997 (see Table 9.1) and has since fallen by an average of over 1% per annum. This being a real trend, and in consideration of the foregoing comments regarding compound feed developments in the Asian region, one begins to believe that there is a real movement in the focus of feed production from 'west' to 'east'. Is this a sustainable trend and is there a reaction in the 'west' away from meat and other animal products?

Table 9.1 World compound feed production.

Year	Tonnes (million)	Per capita consumption (kg)
1975	290	71
1980	370	82
1985	440	90
1990	537	101
1995	590	105
1997	605	104
1999	586	98
2001	591	96

Source: International Feed Industry Federation (IFIF)

This chapater will focus much on China, since it is China that will have the greatest impact on livestock and feed production in the region. China already produces more compound feed than the rest of Asia together. However, due to similar cultural origins and developments in dietary demands, feed production in other Asian countries will be along the same lines as China but on different scales and at different times, according to size of population and status of economic development, respectively. Foreign investment policy and practice are also not too dissimilar from country to country within the region.

Creating a demand for livestock production

The livestock industry in Asia is developing in a way and at a rate that has not been seen before at global level. Only in South America has a similar change been experienced but not on the same scale. Compound feed production growth has been forecast to exceed 10% per annum for the foreseeable future. The growth is primarily consumer-based precipitated by a number of interrelated factors:-

•   Population growth - In terms of absolute numbers, despite the one-child-one family policy, China still leads the Far East (Table 9.2). In addition, countries like Vietnam, Philippines and Indonesia all have high forecast population development rates for the coming years. And if the countries of the sub-continent (India, Pakistan and Bangladesh) are included, there will be a colossal increase in the number of mouths to feed in the next 20 – 25 years in the Eastern Hemisphere.

Table 9.2 Forecast population development to 2050 in the most populated countries (million).

	2000	*2050*	*Population change*	*Change (%)*
China	1,261	1,470	209	17
India	1,014	1,620	606	60
USA	276	404	128	47
Indonesia	225	338	113	50
Brazil	173	207	34	20
Russia	146	118	-28	-19
Pakistan	142	268	126	89
Bangladesh	129	205	76	59
Japan	127	101	-25	-20
Nigeria	123	304	180	146
Mexico	100	153	53	53
Philippines	81	154	73	90
Vietnam	79	119	40	51
Egypt	68	113	45	65

Source: International Feed Industry Federation (IFIF)

•   Economic Development – The Far East, especially China, Thailand, Vietnam and India have been the focus of foreign investment over recent years. With investment comes GDP and internal wealth and, particularly, disposable income for the people. This is especially relevant in Asia where spending priorities are quite different from those in the 'west'. Food and dining out have a high priority amongst ordinary people, as do cosmetics, followed by cars but ownership of a 'flash' family home is well down the priority list. These traits are common to most Asian countries and behave as strong key indicators to development in the region.

• Changes in Eating Habits – Japan, South Korea, Hong Kong and Thailand - have come through similar economic development pathways and all are now classified as Newly Industrialised Countries. As more disposable income becomes available there are several common cultural trends. Firstly, the basic diet of poor people generally consisted of rice or noodles, a small amount of vegetable and an even smaller amount of meat or fish. This kind of diet is still staple in most emerging countries today and often only varies according to flavour or type and amount of spices used. However, Asians love their food and as people become wealthier so an increased amount of income is spent on food; and food of high quality. At this point, rice ceases to be interesting and diets are often dominated by various types of meats, offal, eggs and fish products. Asian people also like to entertain for dinner and to 'eat out'. This also comes with wealth and so does a desire to consume more and more high quality protein foods.

It is not difficult to see why the demand for livestock, and therefore animal and aqua feed, has increased in recent years and how the forecasts for the future are very positive.

Demand and potential for compound feed in Asia

Many people have attempted to quantify the demand for animal and aquatic feeds in Asia and there are as many forecasts as there are reports. The main reasons for the variability are:-

• Rapid developments in compound feed production from 'home mix'

• Interpretation of the use of protein concentrates

• Backyard farming is still the norm rather than the exception

• Method of calculation – entrails, offal and adipose tissue, "the fifth quarter" are almost totally consumed as a general practice. Little is wasted.

- Reliability of census information in countries of industry and geographical diversity.

- Non-conventional farming and feeding – domestic by-products and swill feeding is common practice in most Asian countries

- Variable statistics (see Table 9.5).

Looking at the pig industry in China as an example, the China Statistics Yearbook, 2003 states that there are just over 450million pigs in China at any one time and that around 600million are slaughtered annually. The Yearbook also states that the Chinese consumed 33.8kg per capita of pig meat in 2002; a surprising figure when compared with that of western developed countries. These data are supported by Pig International, July 2003 where it was reported that swine meat consumption in China was 44.5million MT. Assuming that each pig weighs around 75kg at slaughter then, if these pigs consumed compound feed and grew at a feed conversion rate of, say 4:1, pigs in China alone would consume more than 160million MT of feed each year. In contrast, International Feed Industry Federation (IFIF), 2001 states that total compound feed production in China is still less than 60million MT! Asian Agribusiness Research Pte Ltd, have indicated that in 2002, 82million MT of feed were produced in Chinese feed mills of which 28million MT were pig feed.

If these figures are even broadly true, even in consideration of concentrate feeding, then the amount of 'home mix' and by-product feeding may be grossly underestimated representing a huge potential for the future of compound feeds.

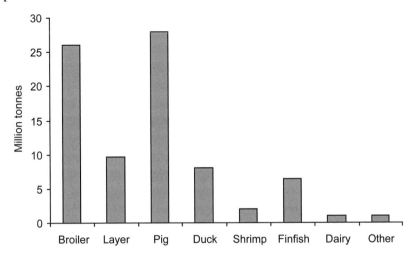

Figure 9.2 Compound feed production in China by species - 2002. (Source – Asian Agribusiness Research Pte Ltd).

Table 9.3a Pig farm size analysis in China.

Size of farm (No. slaughtered / annum)	Number of farms	Yearly slaughtered ('000)
1–9	99,894,369	338,588
10–49	4,438,302	105,344
50–99	790,307	53,637
100–499	212,909	51,651
500–2,999	27,495	29,363
3,000–9,999	3,242	16,432
10,000–49,999	862	12,839
>50,000	28	2,058
Total	105,367,514	609,914

Source: China Statistics Yearbook, 2003

Table 9.3b Broiler farm size analysis in China.

Size of farm (No. slaughtered / annum)	Number of farms	Yearly slaughtered ('000)
1–99	51,203,222	864,223
100–1,999	1,284,361	776,501
2,000–9,999	327,836	1,659,800
10,000–49,999	62,956	1,176,358
50,000–99,999	4,232	270,403
100,000–499,999	1,317	267,642
500,000–999,999	110	77,799
>1,000,000	81	153,003
Total	52,884,115	5,245,728

Source: China Statistics Yearbook, 2003

Tables 9.3a and 9.3b give some indication of why the actual and theoretical figures are so different. According to the Chinese Statistics Year Book, 2003 there are 105.4million registered pig farms in China. Some 95% of these slaughter less than 9 pigs per year. In other words, of the 610million pigs slaughtered each year, 338million (55%) come from farms sending fewer than 9 pigs per year for slaughter.

So far only pigs have been discussed; however, China is a big grower of broilers, layers and ducks (see Table 9.3b).

In a similar, but not quite so spectacular way as pigs, 15% of broilers slaughtered are produced on farms raising less than 99 birds per year and 85% come from farms slaughtering less than 50,000 per year. There is a similar pattern for layers and for ducks.

Undoubtedly, the industry will change. The smaller farms will get bigger and there will be fewer and fewer farm units. This is a natural progression of economic and social development and one experienced already in most advances economies. The burning question is when and what impact can the Asian livestock business have on Europe?

Feed production data for the rest of Asia is equally open to conjecture for similar reasons to those for China.

Table 9.4 Current annual compound and total feed production – 2003 and market potential for developing Asian countries.

Country	Population (million)	Main livestock produced	Compound and concentrate (million MT / anum)	Est. total feed consumption (million MT)	Feed produced per capita (kg)	Market 'ceiling' production (million MT)
Developing						
China	1,261.0	Po / P	82.0	190?	65	275
Thailand	66.0	Po / P	10.2	12.3	154	19
Indonesia	225.0	Po	7.0	15.0	32	49
Philippines	81.6	P	5.6	11.0	69	18
Vietnam	81.7	P	3.8	12?	47	18
Indo China	70.6	P	0.4	N/A	75	15
Malaysia	24.2	Po	4.8	9.0	198	6
Sub total	1,810.1		113.8			400
Developed						
Japan	127.0	All	23.5	N/A	185	25
Korea	46.9	All	15.6	18.5	371	16
Taiwan	23.6	Po / P	7.5	N/A	317	7.5
Sub total	197.5		46.6			48.5
Total	2,007.6		160.4			448.5
USA	276		142		514	145

Source – Asian Agribusiness Research Pte Ltd
Key: Po – Poultry, P – Pig

Table 9.4 offers estimates of compound feed production in other Asian countries and speculates about the market 'ceiling' potential. Market 'ceiling' refers to the amount of feed potentially produced when a market is fully developed. Japan, USA and Korea would be examples of this. Japan, and Korea produce 23.5million MT and 15.6million MT of compound feed each year, respectively, and since their economies are almost fully developed and per capita GDPs relative to the rest of the developed world are comparable, it is unlikely that there will be much increase in feed volume in the foreseeable future and therefore they can be said to have reached market "ceiling".

Clearly, despite the already sizeable compound feed market in Asia there is much more to come as national economies develop.

Implications for European producers

THREATS

Logic would indicate that, as internal demand for livestock products increases amongst Asian countries in line with socio-economic growth, threats of cheap exports would decline. This may be the case in the long term but at present logic is not followed. Both China and Thailand are big exporters of broilers; Thailand directly into Europe and China, competing with potential exports from Europe into Japan and Korea.

Table 9.5 clearly shows that 'face value' logic does not always prevail. Regardless of the source of data it would appear that China has a balance of production and consumption of broilers but it also imports and exports between half a million and a million tonnes per year (5 - 8% of production), depending on the source of information.

Table 9.5 Poultry meat supply and consumption in China in 2002 – data from two different sources.

000's MT	1999 (WP, 2004)	1999 (CSYB, 2002)	Difference (%)	2002 (WP, 2004)	2002 (CSYB, 2002)	Difference (%)
Production	8,550	11,150	23%	9,558	12,545	24%
Imports	591	946	38%	435	950	54%
Exports	375	404	7%	438	530	17%
Consumption	8,766	11,692	25%	9,555	12,965	26%

Source: World Poultry, 2004 (WP,2004) and China Statistics Yearbook, 2002 (CSYB,2002)

In relation to exports from Asia to Europe on a large scale, only broilers, beef and some aquaculture products are significant. For many reasons, especially those of a phyto-sanitary nature, it is unlikely that pigs or pig meat and eggs will arrive on European shores, although from an economic perspective it would appear to be quite feasible.

First and foremost it should be noted that most Asian countries are net importers of feed grains. These are bought, in general, unsubsidised, at world market prices and used in association with locally grown/produced raw materials. The quality of feeds manufactured in modern feed mills is generally good with a high technical input often of western influence. Many Asian students attend our universities and learn from our hard-earned experience.

The key drivers towards low cost exports are low production costs (especially labour), efficient growth performance, low cost processing and finally, but nonetheless significant, lower profit aspirations compared to their European counterparts.

State of the art growing units, high quality feeds and 'high-tech' processing plants maintain food chain integrity and, as much as European competitors would like to 'fault' Asian imports, hygiene has few weaknesses.

The question must be asked however, is the control of the sources of meat imports the same as that control is by our own standards. It is not sufficient just to set standards of production and sanitation and to pay the occasional visit to processing plants. There must be full confidence in feed safety, OH&S, animal welfare and raising conditions as well as suitable processing sanitation. Food chain integrity must be guaranteed. The use and abuse of in-feed and supplementary medications is still a big issue, especially now that EU is moving to a medication-free nutritional environment.

For the future, competition is likely to become weaker rather than stronger. It is unlikely that newly developing economies such as Philippines, Indonesia and Vietnam will focus on exports to Europe. There may still be some movement within the region but local requirements will begin to take priority as demand increases, not forgetting that with economic development come salary increases and higher profit aspirations thereby reducing the incentive to export. But, as mentioned above, logic does not always prevail in Asia! Clearly, guarantees on bio-security standards and food chain responsibility must be an integral part of exports.

OPPORTUNITIES

Compound feed production

Cross-cultural investments in basic industries in Asia are fraught with danger. Some have succeeded quite well but others have failed miserably. Intra-Asian investments have clearly been more successful, but even these are not without issues. The fundamental problem is that livestock farming and its associated industries are basic to the very culture of Asian people. The presence of a 'foreigner' can be seen to be a threat, even alien to this culture

It is therefore concluded that the potential for making profits by European feed companies in the Asian stock feed arena is limited. However, where the correct approach and flexibility to adaptation are implemented, success is quite possible. Cultural understanding, local knowledge and communication and language proficiency all play crucial roles in success and if not correctly addressed they remain firm barriers to entry. Many companies from many countries have tried and almost as many have failed. The list of failures is long and alarming.

There is now sufficient experience to be able to classify the type of company that has a high chance of success. Clearly, of those that *have* been successful, by far the majority come from similar cultural origins, ie. Asian companies, or have longstanding local experience. Although languages are widely different in most Asian countries, Asians are able to learn other Asian languages faster than their European counterparts.

Another key to success is boldness. Many unsuccessful companies have "had a try" without real financial and personnel commitment. A crucial error made by many 'new entrants' is to try to run an Asian operation from European headquarters. Certainly, policy can be set at head office but key decision making must be a matter for local management.

The most significant of the successful companies is Thai-based feed giant, Charoen Pokaphand (CP) who, during a period of little over 10 years, established over 100 foreign invested ventures with feed manufacturing capacity in excess of 12million MT producing over 8million MT in 2001, in China, Vietnam, and Indonesia.

Success for European companies over a similar period has been, at best, poor but there are success stories. Gold Coin, a division of FE Zuellig, Switzerland is present in Thailand, China, Malaysia, Indonesia, Vietnam and is run from its Asian base in Singapore. Gold Coin has had market presence for many years by having the vision to enter when risks were high. Through its personnel policy and local management concept, Gold Coin is accepted as a local company by the Asian feed industries. Another example of some success is Associated British Nutrition (ABN) from the UK who invested in China in the early 1990s. Another successful company has been Cargill from USA.

Investment in the compound feed industries in the mature markets such as Japan, Korea, Taiwan and Thailand would not be easy. Competition is high and cultural barriers have already left a history of carnage. In the intermediately developed markets such as China, Malaysia and Vietnam, opportunities do still exist but it must be understood that since these markets were 'opened up' in the 1980s and early 1990s, most of the rich opportunities have already been taken. The Philippines, Myanmar and Indonesia represent interesting options. Their feed industries are just starting to move. Politics, albeit somewhat turbulent, offer some hope for stability and they are on a similar course of economic development as their Asian predecessors.

Like any investment, choosing the correct vehicle for entry is absolutely critical. Chemistry between partners must fit to achieve longevity. Partner research and selection is most important. In many cases, financial rewards from operations alone would not be sufficient to satisfy most European Shareholders. Synergies need to be sought. Technology transfer is a crucial 'attractant' to Asian partners as is a strong reputation in the home market. And, remember in Asia, although money counts for a lot it does not mean everything and as one learned Chinese colleague once told me as I was pressing for a joint venture decision, "We have been waiting 5000 years for this, we can continue to wait"!

The use of cross-cultural communication mediators who really know the markets, the industry, the culture and the personalities is a valuable way to become familiar with potential partners at a distance that is close enough to make investment decisions.

Investments, in general, can be classified into Wholly Owned Foreign Enterprises or Joint Invested Ventures. Each has its own merits. Strong but fair partners, on both sides, are necessary for success and longevity in joint ventures, whereas a sound local knowledge is paramount for WOFE's where success can be at the hands of fickle governments.

Other feed-related production

Opportunities, however, do still exist for foreign co-operations in the production and sale of specialty feeds, feed additives and advanced technology products, especially where some know-how can be transferred and trade names are well known. The companion animal market is still quite small in Asia compared with Europe but the rate of growth is alarmingly high. Asians in general, are fond of pets and pet food manufacture offers interesting Joint Venture opportunities with good returns.

Similarly, products such as Specialty Pig Starter Feeds, and milk-based products also provide interesting opportunities, especially where technical developments from foreign research can accompany the deal. Asia is one of the leading regions in the world for aquaculture and aquatic feed today is a significant proportion of total feed production (see Figure 9.2). The main target species are shrimp and fresh water and marine warm-water fish. Technology in this sector of the industry has largely been developed locally and therefore foreign partners are not seen to be able to add value to what are highly profitable businesses. Several European companies have ventured into this sector. Few have succeeded. Most had been hurt.

The potential for commodity supply

From the foregoing, it can be seen that meeting the raw materials demand for feed production in the future will require vast quantities of feed ingredients and grains. Most Asian countries have a high population density with preciously small amounts of land available for crop production apart from cash crops and rice. Others are not fertile or have major climatic restrictions to support high yields of feed grains. It is therefore likely that, for a long time into the future, the Asian feed industry will be a net importer of vast quantities of feed ingredients and grain.

It should also be noted that the Asian eating habit leaves little by-product from slaughter for feeding to animals. Most offal is consumed in one form or another and indeed often brings premium prices from consumers! Spent hens are consumed almost as a delicacy. All kind and sizes of fish are eaten in various forms. There is no fat left for rendering.

Consequently, locally-produced meat meals, fish meals and tallows are rarely available in Asia. This is also notable for the pet food industry as a restriction of raw material supply.

In conclusion, Asia is rapidly becoming the leading feed producing region in the world.

Expansion is fuelled by economic and population growth together with changing eating habits. Participation in this business in one form or another by members of the European feed industry carries risks and a high potential for failure. Cost of entry is high and controversy will never be far away. With the correct approach, success is possible and returns can be high.

10

THE ROLE OF THE LEPTIN AXIS IN MODULATING ENERGY PARTITIONING AND NUTRIENT UTILIZATION IN LIVESTOCK SPECIES.

RA HILL
University of Idaho

Introduction.

Leptin a hormone first discovered in 1994, was initially thought to be synthesized only by adipocytes (Zhang *et al.*, 1994); its actions being mediated via the central nervous system, directing satiety and energy expenditure (Pelleymounter *et al.*, 1995; Campfield *et al.*, 1995). As well as its centrally mediated actions, leptin has many actions which are directly manifested in the periphery (reviewed in Margetic *et al.* (2002b). These peripheral actions include modulation of energy expenditure and repartitioning of energy substrate utilization favoring fatty acid oxidation over glucose. Thus, manipulation of the leptin axis has potential in animal production in the context of improving production efficiency and in modulating fat/lean composition. To enable meaningful manipulations at the animal level, either through nutritional/management strategies or through genetic selection or advanced gene modification technologies, we need a greater understanding of the regulation of energy partitioning modulated via the leptin axis. Unfortunately, very few studies have been conducted into the energy partitioning activity of the leptin axis in livestock species. Much of what we understand about the leptin axis has been demonstrated in experimental animals, mostly rodents and in humans. Nonetheless, the studies which have been conducted in livestock species, or *in vitro* models derived from livestock species, have shed light upon leptin actions and it may be stated generally that there is good agreement between the models in helping us understand the mechanisms of regulation of the leptin axis in all animals.

Leptin interacts directly with other hormones, modulating nutrient transfer and partitioning. These include insulin (Muoio *et al.*, 1997; Muoio *et al.*, 1999; Ceddia *et al.*, 1999b; Liu *et al.*, 1997b; Nowak *et al.*, 1998), glucagon

(Zhao *et al*., 2000), glucocorticoids (Slieker *et al*., 1996; Wabitsch *et al*., 1996; Tataranni *et al*., 1996; Dagogo-Jack *et al*., 1997; Miell *et al*., 1996), growth hormone (Considine, 1997; Nyomba *et al*., 1999; Carro *et al*., 1997; Carro *et al*., 1999; Tannenbaum *et al*., 1998; Dyer *et al*., 1997; Roh *et al*., 1998; Williams *et al*., 1999; Barb *et al*., 1998; Chen *et al*., 1998; Houseknecht *et al*., 2000; Hardie *et al*., 1996) insulin-like growth factor-1(Boni-Schnetzler *et al*., 1996; Boni-Schnetzler *et al*., 1999) other members of the cytokine superfamily (Janik *et al*., 1997; Zumbach *et al*., 1997; Matarese, 2000; Mantzoros *et al*., 1997a; Grunfeld *et al*., 1996; Langhans & Hrupka, 1999; Mantzoros & Moschos, 1998), and the thyroid axis (Fain *et al*., 1997; Yoshida *et al*., 1997; Mantzoros *et al*., 1997b; Sesmilo *et al*., 1998).

In the living animal, muscle and adipose tissue exist not only as separate tissues, but also in intimate contact, a well known example of which is marbling in beef. Thus, there is potential for endocrine signaling, paracrine signaling, and direct cell to cell contact interactions. As a major source of leptin is adipose tissue, there is great scope to investigate each of these levels of interaction and perhaps to intervene at fat depot-specific or individual muscle-specific mechanisms to produce desired qualities in the final meat product. Factors which add to the complexity of these interactions include modulation by both systemic and locally produced, specific, leptin binding proteins (LBP), cell membrane-associated and matrix-associated factors, and their influence on regulation of hormone-receptor signaling, receptor number, and intracellular events downstream from receptor activation, including second messenger signaling, and metabolic response within the cell or tissue.

Research into the leptin axis in livestock species has been active. The development of more reliable leptin assays for ruminants (Delavaud *et al*., 2000; Ehrhardt *et al*., 2000; Kauter *et al*., 2000) has been an important step in improving our understanding. At least one commercial leptin assay has now been shown to be inadequate and even misleading when used to assay ruminant leptin (Delavaud *et al*., 2000; Ehrhardt *et al*., 2000).

Leptin and food intake

Leptin appears to have a role in the central nervous system (CNS) in regulating food intake in several species including rodents (Pelleymounter *et al*., 1995), macaques (Tang-Christensens *et al*., 1999), and sheep (Henry *et al*., 1999; Clarke *et al*., 2001; Morrison *et al*., 2001), but when intake in restricted in sheep, centrally infused leptin has little effect (Figure 10.1). Although peripherally dosed leptin appears to similarly affect food intake in rodents (Levin *et al*., 1996), peripherally dosed leptin in sheep has little effect

(Morrison *et al.*, 2002), (Figure 10.2). Furthermore, as outlined below, later pregnancy features increased circulating leptin in rodents (Chehab *et al.*, 1997), humans (Masuzaki *et al.*, 1997) and ruminants (Block *et al.*, 2001; Ashworth *et al.*, 2000; Ehrhardt *et al.*, 2001; Bispham *et al.*, 2003) but is a physiological state in which food intake is increased rather than diminished.

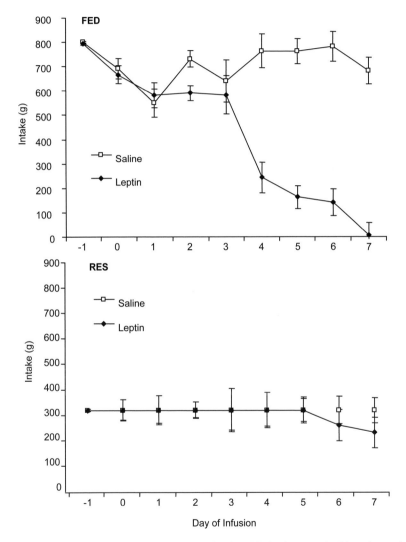

Figure 10.1 Feed intake in fed (FED) and diet-restricted (RES) lambs treated with an increasing dose of leptin or saline ICV. Leptin decreased feed intake in fed lambs on days 4 to 7 compared to that in saline infused controls (P < 0.001), but had no effect on feed intake in diet-restricted lambs (P > 0.25). (After Morrison et al., 2001. Reproduced by permission of the Society for Endocrinology.)

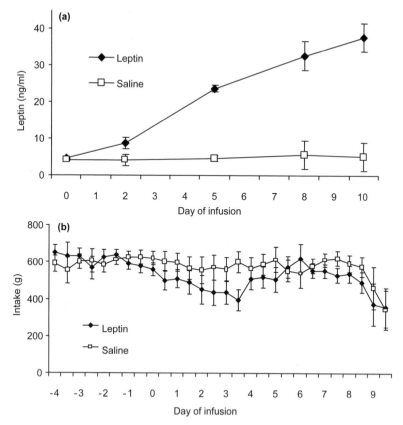

Figure 10.2 (a) Serum concentrations of leptin in lambs receiving a linearly increasing intravenous infusion of leptin or saline. Lambs infused with leptin had significantly greater serum concentrations of leptin by day 2 of infusion (P = 0.05). By day 10, leptin-infused lambs had serum concentrations of leptin approximately 9-fold greater than their saline-treated counterparts (P < 0.001). (b) Twice-daily feed intake values for leptin and saline-infused lambs. A significant day x treatment interaction was detected (P = 0.027), however an analysis of the amount of food consumed by saline vs. leptin-treated lambs on congruent days revealed a significant difference only on day 3.5 (P = 0.01). (Reprinted from Morrison et al., 2002 with permission from Elsevier.)

Leptin and body condition

Leptin is related to live weight and level of body fat in both sheep and cattle (Ehrhardt *et al.*, 2000; Leon *et al.*, 2004; McFadin *et al.*, 2002; Thomas *et al.*, 2001; Blache *et al.*, 2000; Delavaud *et al.*, 2000; Morrison *et al.*, 2001; Minton *et al.*, 1998; Tokuda & Yano, 2001). Table 10.1 summarizes these relationships.

Table 10.1 Correlations between leptin and animal or carcass parameters.

Animal model	Treatment	Body condition	Body fat %	Live weight	Backfat : live weight	Reference
Ovariectomized ewe	Diet ±	0.72	0.68	0.47		(Delavaud *et al.*, 2000)
Dairy calves	Diet ±		0.83			(Ehrhardt *et al.*, 2000)
Dairy cows (late lactation)		0.37				(Ehrhardt *et al.*, 2000)
Pregnant ewes	Diet (x period)	0.56	0.73	0.54		(Thomas *et al.*, 2001)
Postparturent ewe		0.14		0.12		(McFadin *et al.*, 2002)
ewe	Diet ±				0.3	(Blache *et al.*, 2000)
wether					0.31	
Cross-bred heifer	Restricted nutrition	0.47				(León *et al.*, 2004)
	Weight gain	0.83				

The table shows that under conditions in which animals are growing, and there is little confounding influence of other hormone axes (ovariectomized ewes, dairy calves), leptin is highly correlated to body condition (fat%) and live weight. The study of Thomas *et al.* (2001) also shows strong correlations, but it must be stated that this model included a dietary treatment in which young pregnant ewes were offered an extreme, high energy diet, resulting in animals with 20-40% carcass fat at slaughter, and is atypical when compared to other studies of pregnant animals. Thus, leptin appears to have potential as a physiological indicator of body condition. The problem of confounding influences may be seen in the table as in late lactating dairy cows (Ehrhardt *et al.*, 2000) (presumably, the animals were in the second trimester of pregnancy), the correlation between leptin and body condition appears to be diminished, while in post-partum ewes (McFadin *et al.*, 2002) there is only a weak relationship between leptin and body condition. Thus, as outlined above, the relationship between leptin and body condition in breeding females during the postpartum period appears to be affected by many other physiological interactions.

Leptin and reproduction/lactation

The role of leptin in reproduction including during pregnancy and lactation has been examined in pigs (Estienne *et al.*, 2000), sheep (Newby *et al.*, 2001; Forhead *et al.*, 2002; Thomas *et al.*, 2001; Ehrhardt *et al.*, 2001; Laud *et al.*, 1999; McFadin *et al.*, 2002; Yuen *et al.*, 2004; Bispham *et al.*, 2003; Yuen *et al.*, 2003; Symonds *et al.*, 2003; Muhlhausler *et al.*, 2003; Yuen *et al.*, 2002b; Mostyn *et al.*, 2002; Yuen *et al.*, 2002a; Yuen *et al.*, 1999) and in dairy (Block *et al.*, 2001; Kadokawa *et al.*, 2000) and beef (Amstalden *et al.*, 2002; Amstalden *et al.*, 2000; Garcia *et al.*, 2002; Williams *et al.*, 2002) cows and heifers. The investigation of leptin actions in beef animals has mainly focused upon the pubertal heifer and ovariectomized mature cows. An excellent review of this work summarised this group's findings to date (Williams *et al.*, 2002).

Early studies of the role of leptin in reproduction suggested that leptin produced in the placenta is likely to affect maternal, placental and foetal function through both autocrine and paracrine mechanisms. It is possible that placental leptin may have physiological effects on the placenta including angiogenesis, growth and immunomodulation (Ashworth *et al.*, 2000). Leptin may also be involved in regulation of foetal and uterine metabolism.

In the model which first facilitated the discovery of leptin, the *ob/ob* mouse (leptin deficient), both males and females are sterile (Bray & York, 1979). Replacing leptin using peripheral injections of recombinant leptin activates the reproductive axis and restores fertility in both sexes (Barash *et al.*, 1996; Chehab *et al.*, 1996). Thus, leptin has an essential, permissive role in the endocrinological processes which control fertility. If leptin is truly an indicator of body condition, the notion that leptin is a permissive factor allowing development through puberty and facilitating reproduction is consistent with the evolution of control of reproduction. Reproduction, particularly for the female, is an energetically demanding process and once committed to the gravid state, the female physiology invests heavily in all ways (including on energetic terms) in the successful outcome of the pregnancy, and subsequent lactation. Thus, in its permissive role, leptin is signaling to the maternal metabolism that there is sufficient energy in store to commit to an energetically demanding process, with good probability of a successful outcome.

LEPTIN AND PREGNANCY

During the second and third trimesters, serum leptin is elevated in mice (Chehab *et al.*, 1997) and in humans (Masuzaki *et al.*, 1997). Interestingly, in humans the actions of fetal- and placental-derived leptin appear to be directed at the maternal metabolism mobilizing maternal energy/nutrient storage for the benefit of the fetus (Linnemann *et al.*, 2000). At present there appear to be number of possible explanations for the increase in leptin levels in pregnancy: increased production by maternal fat; increased expression by the placenta; and increased levels of binding protein(s) (also of placental origin in the case of the mouse (Gavrilova *et al.*, 1997)), and in the maternal circulation (Hoggard *et al.*, 1998). However, hyperleptinaemia during pregnancy is not associated with decreased food intake or a decline in metabolic efficiency as might be expected given one of the roles of leptin is as a satiety factor. Explanations for this may be a possible pregnancy-induced state of 'leptin resistance', or a change in leptin bioavailability (Barb, 1999; Holness *et al.*, 1999). The increase in leptin concentration during pregnancy appears to be counter-intuitive as this is a period of increased nutritional demands, and not one in which the actions of a satiety factor are expected to increase (Henson & Castracane, 2000). Others have hypothesised that the soluble form of the leptin receptor mediates leptin actions during pregnancy as the circulating concentrations of this protein are increased especially in the mouse, (Mounzih *et al.*, 1998; Gavrilova *et al.*, 1997) although much less so in humans (Lewandowski *et al.*, 1999). Given the range of interactions between leptin and other hormone axes, and the complex pattern of central and peripheral pathways invoked, if leptin does in fact have a role as a satiety factor, this role is centrally mediated and may not be significantly modulated by the soluble leptin receptor. Interactions such as those between leptin and insulin, which modulate oxidation of free fatty acids (FFA) and lipogenesis as part of the process which mobilizes energy reserves during pregnancy, are peripherally modulated and more influenced by the balance of energy reserve status and the demands of the foetus. Studies of the interaction of leptin binding protein status with plane of nutrition and energy reserve status during pregnancy will be required to confirm or refute this.

As in humans and rodents, circulating leptin concentrations in ruminants are considerably elevated in pregnancy (Block *et al.*, 2001; Ashworth *et al.*, 2000; Ehrhardt *et al.*, 2001; Bispham *et al.*, 2003) (Table 10.2). In ewes the increase in plasma leptin is reported to be approximately a doubling from around 5 to 10 ng/ml between pre-breeding and mid-pregnancy values (Ehrhardt *et al.*, 2001), while a report from the same laboratory shows a similar order of increase in pregnant dairy cows, from approximately 3 to 6 ng/ml (Block *et al.*, 2001). The increase in ewes is observed as early as day 28 of pregnancy with no change in body weight or body condition (Figure 10.3) (Bispham *et al.*, 2003). In no case is there any indication that elevated leptin reduces appetite or food intake. On the contrary, this is a period of increased food intake. These data suggest that leptin is up-regulated as a response to early pregnancy in ruminants as well as other species, implicating the placenta as a source of leptin or LBP as observed in rodents and humans. Thus, it may be that concomitant increases in leptin binding proteins, as demonstrated in humans and mice may modulate leptin actions in the periphery and prevent the CNS effect of leptin (appetite suppression) from being manifested. When late pregnant ewes previously fed to maintenance, plus energy required to maintain a growing foetus, were subsequently offered a high quality diet *ad libitum*, in comparison to ewes that continued to receive the control diet, plasma leptin was significantly elevated (Muhlhausler *et al.*, 2003; Muhlhausler *et al.*, 2002) while foetal leptin remained unchanged (Figure 10.4). It appears that the maternal physiology was driving this increase in leptin in response to increased nutrients rather than a pregnancy-related regulatory mechanism.

Table 10.2 Feed intake, body weight, plasma metabolites, insulin and leptin in ewes measured across physiological state. (Modified after Ehrhardt *et al.*, 2001 with permission from Elsevier.).

	Physiological state[1]				
	Pre-breeding	*Mid pregnancy*	*Late pregnancy*	*Lactation*	*SE*
Feed Intake (g/d)[2]	856±21	894±28	1275±50	2850±155	-
Body weight (Kg)	50.3[a]	52.6[a]	62.2[b]	49.5[a]	1.0
Glucose (mg/dL)	65	67	60	68	2.4
NEFA (μM)[3]	344	300	387	359	46
Insulin (ng/mL)	0.22	0.29	0.30	0.32	0.04
Leptin (ng/mL)	5.3[a]	9.5[b]	7.9[ab]	6.0[a]	

[1]Ewes (n=8) were studied 20-40 d before breeding, during mid pregnancy (d 50-60 PC), during late pregnancy (d 125-135 PC) and during early lactation (d 15-22 post partum).
Means with different superscripts (a,b) were different at $P<0.05$ by Scheffé's test.
[2]Means±SE are shown. Statistical analysis was not performed because feed was offered according to requirements during pre-breeding and pregnancy and offered *ad libitum* during lactation.
[3]NEFA, non-esterified fatty acids.

In a study of human pregnancy, the relative contributions of fetal-derived leptin and maternal leptin to maternal total leptin has been described by Linnemann *et al.* (2000) using an *in vitro* perfusion technique. This study suggested that only 1-2% of placental leptin enters the fetal circulation, while

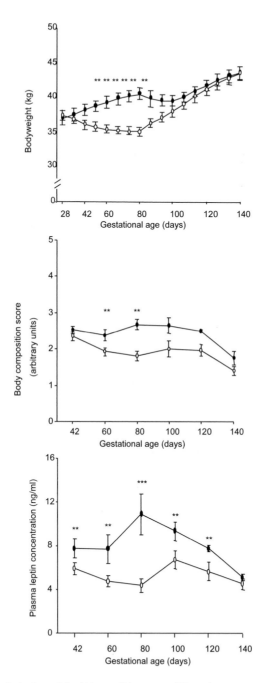

Figure 10.3 Changes in body weight (A), condition score (B), and mean maternal plasma leptin (C) for ewes nutrient restricted between 28 and 80 days gestation and then fed to appetite for the remainder of gestation (open symbols) or fed to appetite throughout pregnancy (closed symbols). (After Bispham *et al.*, 2003. Reproduced with permission from The Endocrine Society, Copyright 2003.)

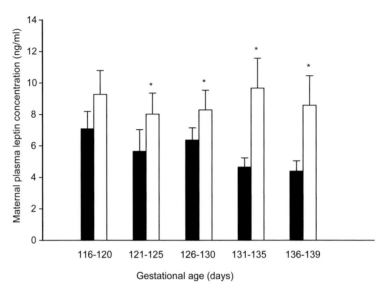

Figure 10.4 Maternal plasma leptin concentrations in control (closed histograms, n = 5) and well-fed (open histograms, n = 7) ewes between 116 and 139 days gestation. Asterisks denote a significant effect of increased maternal nutrient intake compared with control group (P < 0.05). (Reprinted from Muhlhausler et al., 2002 with permission from the Society for the Study of Reproduction.)

approximately 98% enters the maternal circulation. Although not yet determined, the interaction of leptin with insulin may also be important in pregnancy. Leptin appears to act as a permissive factor with respect to its effects on insulin-mediated fuel storage in muscle and in the liver and on insulin-mediated fuel utilization by muscle (reviewed Margetic *et al.* (2002b)). Thus, one of the actions of fetal- and placental-derived leptin may be to act on maternal liver and muscle, changing the dynamics of fuel utilization. Particularly in the case where maternal nutrition may become limiting, and thus, maternal leptin reduced; fetal- and placental-derived leptin would act on the maternal metabolism to ensure that fuel is preferentially utilized by the fetus, rather than rebuilding maternal stores. This hypothesis is consistent with the observation that leptin concentration is higher in arterial than in venous cord blood, and suggests that both fetal- and placental-derived leptin targets maternal metabolism. An increase in plasma leptin in breeding cows and ewes is also consistent with this hypothesis, as in ruminants glucose is the main energy source available to the fetus. Thus, leptin's actions repartitioning energy substrate utilization towards fatty acids in the maternal metabolism, would preserve glucose for utilization by the fetus (Figure 10.5). As demonstrated by Ehrhardt *et al.* (2001) in the ewe, peripheral insulin resistance increases through pregnancy (Figure 10.6), concomitant with increasing maternal plasma leptin. The antagonism between insulin and leptin actions on partitioning of energy substrate utilization demonstrated in other tissues, notably muscle, is also consistent with this glucose-sparing effect.

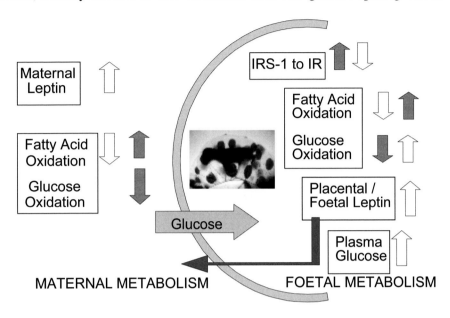

Figure 10.5 Hypothetical leptin-insulin interaction in modulation of energy substrate partitioning between the dam and foetus. Leptin mobilizes fatty acids in the dam sparing glucose for use by the foetus. As pregnancy advances the dam becomes more 'insulin resistant,' increased leptin antagonizing the anabolic effects of insulin. Leptin binding proteins may also play a role by increasing the total leptin pool. Insulin effects open arrows, leptin effects hatched arrows. (Photograph reproduced with permission from Senger, 2003).

LEPTIN IN THE FOETUS AND NEONATE

Apparently leptin mRNA is detectable in the foetal lamb from about day 75 of gestation (Yuen *et al.*, 1999; Devaskar *et al.*, 2002), although there appear to be no investigations of leptin at earlier gestational ages. Leptin in the foetus may be expressed at a higher level per adipocyte in mid-gestation compared to later gestation (Devaskar *et al.*, 2002). Leptin is increased in the latter part of gestation, as may be expected when foetal adipose tissue in lambs begins to accumulate more rapidly (Muhlhausler *et al.*, 2003; Yuen *et al.*, 1999; Muhlhausler *et al.*, 2002). It has been suggested that although placental and foetal leptin contribute to the foetal leptin pool in humans, that adipose tissue may be the major source of ovine foetal leptin (Devaskar *et al.*, 2002). This may be so, however it is still unclear what the role of placental leptin binding protein may be in the sheep and whether it contributes to

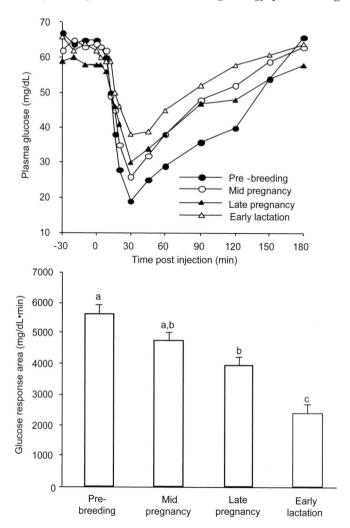

Figure 10.6 Changes in plasma glucose in ewes in response to a single injection of insulin. Left Panel. Temporal pattern of plasma glucose in a representative ewe before breeding, during mid-pregnancy, during late pregnancy and during early lactation. Right Panel. Mean ± SE of plasma glucose response area to a single insulin injection in ewes (n = 8) studied across physiological state. Means with different letters are different (P < 0.05). (Reprinted from Ehrhardt et al., 2001 with permission from Elsevier.)

maintaining the leptin pool in the ovine foetus. It appears that foetal plasma leptin at term, is elevated in comparison to newborn plasma leptin (Forhead *et al.*, 2002) (Table 10.3).

Table 10.3 Mean (± SEM) plasma leptin concentration in intact sheep fetuses near term, lambs over the first five weeks of life and adult non-pregnant sheep. Within each column, values with different letters are significantly different from each other ($p < 0.05$). (After Forhead *et al.* (2002), Copyright 2002, The Endocrine Society.)

Animals	Number	Gestational or postnatal age (days)	Plasma leptin ($pg\ ml^{-1}$)
Intact untreated fetuses	10	136-140	350 ± 26[ab]
Neonatal lambs	4	0-1	211 ± 36[b]
	8	7	635 ± 149[a]
	7	14	670 ± 143[a]
	8	35	612 ± 126[a]
Adult non-pregnant sheep	7	Adult	1318 ± 254[c]

In the neonatal lamb, it appears that leptin may have an important role in regulating uncoupling protein-1 in brown adipose tissue (Mostyn *et al.*, 2004; Mostyn *et al.*, 2002). The role of leptin in the context of nutritional manipulation of the foetus and neonate is an active area of research in which the lamb has served as a useful model for clinical interpretation. Much of this work was recently reviewed (Symonds *et al.*, 2003; Mostyn *et al.*, 2001; Symonds *et al.*, 2001).

LEPTIN – POSTPARTUM

Parturition is an event which is characterized by a rapid change in physiology. The dam moves from a condition of positive energy balance to a negative energy balance, most exaggerated in dairy cows, and there are significant and rapid hormonal changes during this period.

In cattle there appears to be a clear link between pulsatile secretion of luteinizing hormone (LH) and circulating leptin concentration. Amstalden *et al.* (2002) have shown that intracerebro-ventricular (ICV) infusion of leptin led to hypersecretion of LH. Similarly, Henry *et al.* (2001) have shown that leptin can stimulate LH secretion in underfed, ovarectomized ewes. Furthermore, in both cattle (Kadokawa *et al.*, 2000) and rodents (Brogan *et al.*, 1999), it appears that serum leptin is depressed during lactation, although the current evidence indicates that this is not so in sheep (McFadin *et al.*, 2002). It appears that the postpartum depression in leptin is coincident with suppression of LH secretion, which also occurs during the first 60 to 80 days of lactation in cattle. Thus, plasma leptin in the early post-partum period appears to be one of many physiological regulators which is in a state of flux.

Leptin stability – temporal issues

The effect of fasting on serum leptin concentrations in cattle appears to show little acute response. Serum leptin remained unchanged until 48 hours after commencement of the fasting period (Amstalden *et al.*, 2000). Given the dynamics of the rumen, this is not surprising and suggests that leptin may not have a role in acute regulation of satiety as suggested for monogastric species. This notion is supported by the study of Morrison *et al.* (2001), as ICV infusion of an increasing dose of leptin in ewes had no effect on feed intake until day 4 (Figure 10.1). Plasma leptin appears to be stable on a diurnal basis, one study in ewes (Tokuda *et al.*, 2000) showing evidence of irregular leptin pulses of only a small magnitude.

At least one study has addressed the issue of leptin variation due to photoperiod in sheep (Marie *et al.*, 2001). The findings of this study were inconclusive and suggest that the increased appetite of sheep coincident with photoperiod, increases fat mass and thus circulating leptin.

Role of leptin receptor isoforms in bovine leptin biology

Recently, there has been a great improvement in our understanding of the distribution of bovine leptin receptor isoforms, in mammary tissue (Sayed-Ahmed *et al.*, 2003; Silva *et al.*, 2002; Smith & Sheffield, 2002), in different adipose tissue depots (Ren *et al.*, 2002), muscle (Chelikani *et al.*, 2003; Silva *et al.*, 2002) and other tissues (Chelikani *et al.*, 2003; Yanagihara *et al.*, 2000; Parhami *et al.*, 2001; Silva *et al.*, 2002). Interestingly, the long form of the leptin receptor (ObR$_L$, equivalent to the ObRb of rodents) appears to be strongly expressed in liver and in the fat depots although less so in muscle (Chelikani *et al.*, 2003; Silva *et al.*, 2002), and unlike in rodents, the short form of the receptor (ObR$_S$, equivalent to rodent ObRa), appears to be restricted to only a few tissues, pituitary, liver, spleen, adrenal cortex and brain stem. Thus the isoform with greatest intracellular signaling capacity, ObR$_L$, appears to be more abundant. These data provide further evidence supporting the importance of the leptin axis in energy substrate utilization in ruminants. Despite the strong evidence for these plasma membrane-bound forms of the leptin receptor, we still have little knowledge of circulating forms of the leptin receptor or other leptin binding protein moieties.

Modulation of peripheral interactions of leptin by specific leptin binding proteins (LBP)

Leptin interactions may be modulated by both systemic and cell-associated binding proteins. Furthermore, leptin is unusual in that it is hydrophobic (Margetic *et al.*, 2002b). This is an important characteristic of leptin, as it indicates that it has potential to bind non-specifically to hydrophobic sites, such as those on serum albumin, which is known to be a non-specific carrier for a range of hydrophobic molecules. This characteristic may also be important in the interaction of leptin with its specific binding proteins and its receptor.

A range of circulating LBP have been identified from humans and rodents (Sinha *et al.*, 1996; Houseknecht *et al.*, 1996; Diamond *et al.*, 1997; Hill *et al.*, 1998; Birkenmeier *et al.*, 1998; Patel *et al.*, 1999) including a soluble form of the leptin receptor (sObR) (Liu *et al.*, 1997a). In the following discussion, where the specific soluble form of the leptin receptor is identified, the abbreviation sObR will be used and in cases where a less specific term is required LBP will be used. In addition to sObR, there appears to be a range of cell-associated leptin binding moieties which have no resemblance to the ObR (Corp *et al.*, 1998; Golden *et al.*, 1997; Stephens *et al.*, 1995). In both humans and rodents, leptin circulates in both the free and bound forms. The sObR has been shown to bind leptin at a 1:1 ratio in humans (Lewandowski *et al.*, 1999). Recent development of assays for hu-sObR (Wu *et al.*, 2002; Chan *et al.*, 2002; Kratzsch *et al.*, 2002a; Kratzsch *et al.*, 2002b) and mouse sObR (Lammert *et al.*, 2002) have facilitated rapid progress in these studies. The major LBP in human circulation was found to be sObR that occurs in two different isoforms (Kratzsch *et al.*, 2002b). Consistent with the increased concentration of leptin during late pregnancy in all species studied to date (including ruminants), sObR at least in humans (Lewandowski *et al.*, 1999) and in mice (Lammert *et al.*, 2002) has been shown to be elevated, 25% (3.5 nM) and 290-fold (63 nM) respectively. In an *in vitro* study, Maamra *et al.* (2001) reported that sObR may be obtained via the cleavage of membrane bound receptors ObRa and ObRb. sObR was up-regulated by PKC and sulfhydryl group activation. Metaloprotease inhibitors could inhibit the production of sObR, indicating that the enzyme responsible for receptor cleavage may belong to the metalloprotease family. Apparently nothing is known about enzymatic generation of sObR in any livestock species.

Pharmacokinetics studies have provided a hint of the importance of circulating LBP in leptin biology. At least two studies in normal rats (Hill *et al.*, 1998; Zeng *et al.*, 1997) and one study in ewes (Hunt *et al.*, 2003) has demonstrated that the loss of leptin from the circulation is best described by a two-pool model, consistent with the occurrence of leptin in two forms: free, and leptin bound to other proteins. The biological forms which constitute the bound pool of leptin include the hormone bound to tissue receptors, to non-specific sites in the tissues, and to a carrier molecule(s) in plasma (Hill *et al.*, 1998; Margetic *et al.*, 2002a). The rapid removal of the free form of leptin from plasma (half-life, 3.4 min) is consistent with data for other peptide hormones which occur in the free form such as insulin (half-life, ≈ 4 min (Gray *et al.*, 1985)) or insulin-like growth factor-1 (IGF-1) in which the free form is cleared rapidly (half-life, 12 to 15 min (Bastian *et al.*, 1993; Hill *et al.*, 1997)). Furthermore, clearance of the bound form of leptin which is retained in the plasma for a much longer period (half-life, 71 min (Hill *et al.*, 1998)) is consistent with the values obtained for IGF-1, which is known to also occur bound to one of several specific binding proteins (half-life, 148 to 264 min; Bastian *et al.*, 1993; Hill *et al.*, 1997) suggesting that a specific binding molecule for leptin may also be important. It is likely that leptin association with carrier molecules which appear to be different across the species investigated to date (Houseknecht *et al.*, 1996; Sinha *et al.*, 1996), is an important factor affecting leptin clearance. Thus, it will be important to determine the characteristics of the leptin binding proteins unique to ruminants, and their potential to modify leptin actions in the tissues. Recently, Sandowski *et al.* (2002) have cloned, expressed and purified large amounts of a soluble, recombinant protein which corresponds to the leptin binding domain of the hu-ObR, which will be an invaluable tool in characterizing leptin interactions with the sObR and other LBPs.

There is little published data describing LBP in ruminants. In preliminary studies (Hill *et al.*, 2002), we have characterized the binding of leptin in bovine serum by incubating ^{125}I-labelled leptin with bovine serum samples over a range of times and resolved the bound peaks using Sephadex S300 column chromatography (Figure 10.7, Table 10.4).

Of these proteins, it is possible that some represent the various forms of the sObR. The literature suggests that sObR may be present in multiple isoforms, or proteolytically degraded forms which retain leptin binding. In humans, leptin binding molecules include moieties with molecular masses of 66, 68, 75, 80, 97, 100, 116, 130 and 280 kDa (Sinha *et al.*, 1996), 85, 176, and 240 kDa (Houseknecht *et al.*, 1996), 85 kDa (Quinton *et al.*, 1999), 450 kDa (Diamond *et al.*, 1997), and 180 and 380 kDa (van Dielen *et al.*, 2002). The evidence for the affinity of leptin binding to these binding proteins is scant. Nor is the abundance of these proteins well described.

Variation in body fat not explained by variation in leptin

Studies in cattle (Delavaud *et al.*, 2002), horses (Cartmill *et al.*, 2003) and sheep (Kauter *et al.*, 2003) have demonstrated that there is variation in body condition which cannot be correlated with plasma leptin. Perhaps this is not

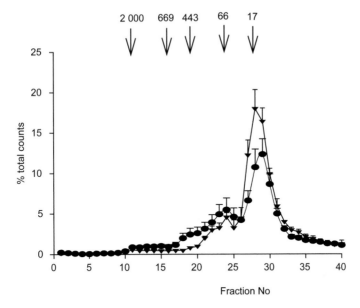

Figure 10.7 Chromatographic resolution of bovine serum following incubation with ^{125}I-leptin *in vitro*. Bovine serum was incubated with the tracer for 12 hr at 4°C in the presence (▼) or absence (•) of excess unlabelled leptin (50 µg) and chromatographed using the same column. The plots show the recovery of radioactivity in 1 ml fractions, as a percentage of total radioactivity eluted. The positions of molecular weight markers (kDa) are indicated. Each data point represents the mean ± SEM for three runs.

Table 10.4 Percent of total radioactivity in areas under the peaks from Sephacryl S-300 chromatography of bovine serum incubated *in vitro* with radio-labelled leptin for different times at 4°C, in the absence or presence of unlabelled leptin (50 µg). Values are means ± SEM (n= 3) (Hill *et al.*, 2002).

Incubation period (hr)	Peak I (~ 670 kDa)	Peak II (~ 220- kDa)	Peak III (~ 66 kDa)	Free leptin (17 kDa)
0	-	-	-	93.88 ± 1.92
2	1.86 ± 0.08	5.58 ± 0.51	11.81 ± 0.88	71.26 ± 0.87
4	3.33 ± 0.52	4.06 ± 1.21	15.27 ± 3.21	60.80 ± 2.15
8	4.45 ± 1.59	7.17 ± 1.12	15.7 ± 1.63	57.23 ± 2.19
12	6.18 ± 0.93	6.33 ± 1.39	25.72 ± 2.13	47.65 ± 3.21
+ unlabelled leptin				
12	4.51 ± 0.52	2.52 ± 0.32	4.83 ± 1.34	79.27 ± 3.91

surprising because of the complex interplay of the hormonal milieu as outlined above. Ewes which had been selected for subcutaneous fatness showed a significant decrease in plasma leptin concentrations when placed on a 60% diet restriction. Control ewes on the other hand showed no change in leptin despite losing a similar amount of weight (Kauter *et al.*, 2003). It is possible that, as in humans, as body fat percent increases to extreme, circulating sObR is reduced, and during short-term fasting, the drop in plasma leptin observed in lean individuals is buffered by the proportionally higher circulating concentrations of sObR (Landt *et al.*, 2001; Klein *et al.*, 1996). The variation observed in plasma leptin in otherwise similar animals may be due to differences in circulating sObR, and thus the ratio of total leptin to bound or free leptin may be an important variable in determining leptin actions *in vivo*.

Leptin and control of fatty acid and glucose metabolism

Much of the above discussion sets the stage to point to the role of the leptin axis in modulating physiology to favor the oxidation of fatty acids as an energy source. For animal nutritionists, the prospect of exploiting regulation of metabolic pathways to optimize animal performance, and to possess the knowledge to closely match feed requirements to life-stage in livestock, is an ideal worthy of scientific pursuit. Some of the essential information required to get to a point where nutrient requirements may be precisely defined, includes an understanding of the hormonal regulation of intracellular metabolic pathways. Much of what we presently know about leptin mediation of energy substrate utilization has been developed in rodent models. Much of this information has not yet been gathered in livestock species.

Skeletal muscle metabolism is the major pathway for FFA oxidation contributing to greater than 20% of basal metabolism (Zurlo *et al.*, 1990; Matzinger *et al.*, 2002). Furthermore, the role of FFA increases in relative importance to other energy/fuel sources as muscle activity (energy expenditure) increases above basal. In ruminants this is of the order of 60% during moderate activity (Hocquette *et al.*, 1998). This important observation appears to have been overlooked in many studies of insulin-mediated oxidation of energy sources. To recapitulate, insulin a major regulator of energy metabolism, inhibits oxidation of FFA in muscle. Leptin suppresses this insulin-mediated action (Ceddia *et al.*, 2001; Muoio *et al.*, 1997; Muoio *et al.*, 1999; Bryson *et al.*, 1999). Thus, there appears to be significant interactions between leptin and insulin in the regulation of energy substrate utilization.

The underlying mechanisms which form the pathway(s) linking leptin-insulin interactions have been partially revealed in studies of insulin resistance in humans and in rodents. Although leptin has not been widely recognized as interacting and contributing to insulin resistance, the link between fatty acid metabolism and glucose utilization in muscle has been widely studied (Matzinger *et al.*, 2002; Cortright *et al.*, 1997; Yu *et al.*, 2001; Yu *et al.*, 1997; Yamauchi *et al.*, 2001; Griffin *et al.*, 1999; Virkamaki *et al.*, 2001; Dyck *et al.*, 2001; Dyck *et al.*, 2000; Boden *et al.*, 1996; Boden *et al.*, 2001; Cha *et al.*, 2001; Winder & Holmes, 2000; Jacob *et al.*, 1999; Roden *et al.*, 1999; Roden *et al.*, 1996).

Two forms of the leptin receptor (denoted here, Ob-R_L, long isoform and Ob-R_S, short isoform) have been shown to activate intracellular second messengers (Murakami *et al.*, 1997; Bjorbaek *et al.*, 1999; Bjorbaek C, 1997), and as outlined above both isoforms have been identified in skeletal muscle in ruminants (Chelikani *et al.*, 2003; Silva *et al.*, 2002). The extracellular domains of the two isoforms are identical, the differences occurring in the intracellular domains. Ob-R_L has both Janus kinase (Jak)/ signal transducers and activators of transcription (STAT) box motifs, whereas Ob-R_S has a single putative Jak motif. Ob-R activation is thought to occur following leptin binding and receptor homo-dimerisation (White & Tartaglia, 1999). The precise mechanism of signaling of Ob-R_S is highly speculative, but appears to involve activation of Jak (Bjorbaek C, 1997; Murakami *et al.*, 1997).

We have recently observed that Jak is constituitively associated with the leptin receptor in bovine myogenic cells, and that insulin stimulation leads to depletion of Jak from the leptin receptor. This may be the result of its recruitment to the insulin signaling molecule, insulin receptor substrate-1 (IRS-1, unpublished data). Activation of the insulin receptor and downstream events have been extensively studied (reviewed by Kido *et al.*, 2001), and recently at least one mechanism of insulin-mediated FFA transport, via phoshphatidylinositol (PI)-3 kinase and CD36 has been identified (Dyck *et al.*, 2001; Luiken *et al.*, 2002). Evidence for the intersection of the leptin and insulin pathways is supported as an increase in fatty acid oxidation in muscle associated with insulin resistance appears to be linked to leptin signaling and is described in several complex intracellular pathways (Boden *et al.*, 2001; Yu *et al.*, 2001; Griffin *et al.*, 1999); with elements of the early insulin signaling cascade being affected (Griffin *et al.*, 1999); as well as inhibition of glucose transport and glucose phosphorylation (Roden *et al.*, 1996). The effects of leptin treatment on muscle may be partially attenuated by synthetic blockade of PI-3 kinase activity (Muoio *et al.*, 1999), and leptin has profound effects on tyrosine phosphorylation of IRS-1, with no associated effect on insulin receptor phosphorylation (Cohen *et al.*, 1996).

Most recently it was reported that leptin directly stimulates fatty-acid oxidation in muscle by activating the 5'-AMP-activated protein kinase, an enzyme that phoshorylates and subsequently inactivates CoA carboxylase (Minokoshi *et al.*, 2002; Minokoshi & Kahn, 2003). Thus a direct mechanism for the action of leptin on fatty acid oxidation consistent with the work of (Muoio *et al.* (1997; 1999) has now been defined. Figure 10.8 (Kokta *et al.*, 2004) shows the pathways in muscle which may be unique to leptin binding/ activation, those which may be unique to insulin binding/activation, and those which appear to be activated following binding/activation of either. It is not known whether any of the members of these cascades is limiting, or if activation of either the leptin or insulin axis causes recruitment of the entire cascade. Thus, knowledge of the level of expression together with the level of phosphorylation induced is required to aid our understanding of the leptin-insulin interaction in ruminant muscle. Note that there are alternate pathways to each of these shared mechanisms. For example, insulin may activate glucose transport via insulin receptor interaction with the adapter protein CAP and the proto-oncogene product, cbl (Baumann *et al.*, 2000); a pathway which has not been implicated in studies of leptin action in muscle.

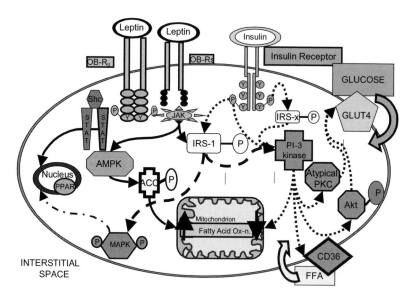

Figure 10.8 The pathways in muscle that may be unique to leptin binding/activation (solid lines), those that may be unique to insulin binding/activation (short dots), and those that appear to be activated following binding/activation of either (long dashes). Newly discovered interactions in these pathways are being reported regularly. Although the figure is not a comprehensive summary of the known pathways, it indicates the level of complexity of signaling interactions. (Reprinted from Kokta et al., 2004 with permission from Elsevier.)

Furthermore, leptin's actions are mediated via both Jak and STAT, and no intersection of activation of STAT with the insulin axis has been described. Thus, for each pathway there appear to be both independent and intersecting mechanisms. Therefore, it is possible that under appropriate physiological conditions, that each axis may function with minimal interaction or alternatively with maximal interaction.

Three separate studies have shown that leptin did not change insulin-stimulated glucose metabolism in isolated rodent muscle (Muoio *et al.*, 1997; Furnsinn *et al.*, 1998; Zierath *et al.*, 1998). In agreement with this finding Ranganathan *et al.* (1998) showed that leptin had no effect on basal and insulin-stimulated glucose transport in cultured rat and human skeletal muscle cells. In another study of normal mice (Kamohara *et al.*, 1997), infusion of murine leptin led to an increase in glucose turnover and 2-deoxyglucose uptake into skeletal muscle and BAT, increased whole-body glucose turnover and increased glucose oxidation; despite no change in plasma insulin or glucose concentrations. Furthermore, the leptin-induced increase in glucose uptake into *edl* and *soleus* muscles was attenuated by denervation. Taken together, all of these data indicate that the effects of acute leptin administration on glucose metabolism in muscle are likely to be mediated via the CNS. Unfortunately, a number of these studies did not or could not investigate any down-stream effects on glucose metabolism, such as effects on glycogenesis, or glucose oxidation as the form of tagged glucose used was non-metabolizable (2-deoxy- or 3-O-methyl-glucose (Zierath *et al.*, 1998; Ranganathan *et al.*, 1998)). In only three of these studies has the leptin-insulin interaction effects on both glucose and fatty acid oxidation been investigated. All of these studies have reported increased oxidation of fatty acids in response to leptin treatment (Bryson *et al.*, 1999; Muoio *et al.*, 1997; Muoio *et al.*, 1999).

The responses to leptin-insulin interaction and their intracellular biochemical pathways have not yet been reported in ruminant muscle. An understanding of these pathways has important implications for directing intervention strategies. Furthermore, the fundamental interaction of leptin with its receptor and down-stream activation of Jak-2 and STAT in ruminant muscle has not been reported. Leptin has been shown to activate this second messenger system in a range of cell types (Matsuoka *et al.*, 1999; Morton *et al.*, 1998; Ghilardi & Skoda, 1997; Ghilardi *et al.*, 1996), and the Jak/STAT pathway has been shown to be active in skeletal muscle (Guillet-Deniau *et al.*, 1997). However, there have been no reports of leptin-mediated Jak/STAT activation in ruminant muscle.

Conclusion

A better understanding of the actions of the leptin axis in livestock species may lead to closer targeting of dietary components to physiological demand, or to manipulate carcass composition to provide a product to precise specifications. However, our understanding is superficial at this time. It seems clear that blood leptin concentrations provide an indicator of body condition, but this can be perturbed in several physiological states including pregnancy and lactation. Ideally all mature females in production systems are either pregnant or lactating or both. Thus, simply measuring blood leptin alone, in general will tell us very little about the animal's nutritional state or its requirements.

It appears that leptin may be categorized as working at two levels. Firstly, its actions in the CNS appear to be closely related to signaling of satiety and in promoting energy expenditure. Leptin actions in the periphery appear to be equally complex, and there appear to be multiple interactions with other hormones which regulate energy metabolism and other pathways. In addition, the role of leptin binding proteins in modulating leptin actions are not even under investigation as yet in livestock species, as they are in the rodent model and in humans. Further investigation of this aspect of leptin biology is likely to greatly improve our understanding.

One aspect of leptin action in the periphery appears to be becoming better understood. Its modulation of fatty acid oxidation in peripheral tissues appears to be an action which may be exploited for the manipulation of carcass composition and in the design of diets precisely matched to the animal's needs. Unfortunately, this is not likely to be of general use until some of the complexities of the functions of the leptin axis are more clearly elucidated.

References

Amstalden, M., Garcia, M. R., Stanko, R. L., Nizielski, S. E., Morrison, C. D., Keisler, D. H. & Williams, G. L. 2002. Central infusion of recombinant ovine leptin normalizes plasma insulin and stimulates a novel hypersecretion of luteinizing hormone after short-term fasting in mature beef cows. *Biol Reprod*, **66,** 1555-61.

Amstalden, M., Garcia, M. R., Williams, S. W., Stanko, R. L., Nizielski, S. E., Morrison, C. D., Keisler, D. H. & Williams, G. L. 2000. Leptin gene expression, circulating leptin, and luteinizing hormone pulsatility are acutely responsive to short-term fasting in prepubertal heifers: relationships to circulating insulin and insulin-like growth factor I(1). *Biol Reprod*, **63,** 127-133.

Ashworth, C. J., Hoggard, N., Thomas, L., Mercer, J. G., Wallace, J. M. & Lea, R. G. 2000. Placental leptin. *Rev Reprod*, **5**, 18-24.

Barash, I., Cheung, C. C., Weigle, D. S., Ren, H., Kabigting, E. B., Kuijper, J. L., Clifton, D. K. & Steiner, R. A. 1996. Leptin is metabolic signal to the reproductive system. *Endocrinology*, **137**, 3144-3147.

Barb, C. R. 1999. The brain-pituitary-adipocyte axis: Role of leptin in modulating neuroendocrine function. *J Anim Sci*, **77**, 1249-1257.

Barb, C. R., Yan, X., Azain, M. J., Kraeling, R. R., Rampacek, G. B. & Ramsay, T. G. 1998. Recombinant porcine leptin reduces feed intake and stimulates growth hormone secretion in swine. *Domest Anim Endocrinol*, **15**, 77-86.

Bastian, S. E. P., Walton, P. E., Wallace, J. C. & Ballard, F. J. 1993. Plasma clearance and tissue distribution of labelled insulin-like growth factor-1 (IGF-1) and an analogue LR3IGF-1 in pregnant rats. *J Endocrinol*, **138**, 327-336.

Baumann, C. A., Ribon, V., Kanzaki, M., Thurmond, D. C., Mora, S., Shigematsu, S., Bickel, P. E., Pessin, J. E. & Saltiel, A. R. 2000. CAP defines a second signaling pathway required for insulin-stimulated glucose transport. *Nature*, **407**, 202-207.

Birkenmeier, G., Kampfer, I., Kratzsch, J. & Schellenberger, W. 1998. Human leptin forms complexes with a2-macroglobulin which are recognised by the a2-macroglobulin receptor/low density lipoprotein receptor-related protein. *Eur J Endocrinol*, **139**, 224-230.

Bispham, J., Gopalakrishnan, G. S., Dandrea, J., Wilson, V., Budge, H., Keisler, D. H., Broughton Pipkin, F., Stephenson, T. & Symonds, M. E. 2003. Maternal endocrine adaptation throughout pregnancy to nutritional manipulation: consequences for maternal plasma leptin and cortisol and the programming of fetal adipose tissue development. *Endocrinology*, **144**, 3575-85.

Bjorbaek, C., El-haschimi, K., Frantz, J. D. & Flier, J. S. 1999. The role of SOCS-3 in leptin signaling and leptin resistance. *J Biol Chem*, **274**, 30059-30065.

Bjorbaek C, Uotani, S., Da Silva B, Flier JS. 1997. Divergent signaling capacities of the long and short isoforms of the leptin receptor. *J Biol Chem*, **272**, 32686-32695.

Blache, D., Tellam, R. L., Chagas, L. M., Blackberry, M. A., Vercoe, P. E. & Martin, G. B. 2000. Level of nutrition affects leptin concentrations in plasma and cerebrospinal fluid in sheep. *J Endocrinol*, **165**, 625-637.

Block, S. S., Butler, W. R., Ehrhardt, R. A., Bell, A. W., Van Amburgh, M. E. & Boisclair, Y. R. 2001. Decreased concentration of plasma leptin in periparturient dairy cows is caused by negative energy balance. *J Endocrinol*, **171**, 339-348.

Boden, G., Chen, X., Mazzoli, M. & Ryan, I. 1996. Effect of fasting on serum leptin in normal human subjects. *J Clin Endocrinol Metab*, 2419-3423.

Boden, G., Lebed, B., Schatz, M., Homko, C. & Lemieux, S. 2001. Effects of acute changes of plasma free fatty acids on intramyocellular fat content and insulin resistance in healthy subjects. *Diabetes*, **50**, 1612-7.

Boni-Schnetzler, M., Gosteli-Peter, M. A., Moritz, W., Froesch, E. R. & Zapf, J. 1996. Reduced ob mRNA in hypophysectomised rats is not restored by growth hormone (GH), but further suppressed by exogenously administered insulin-like growth factor (IGF) 1. *Biochem Biophys Res Commun*, **225**, 296-301.

Boni-Schnetzler, M., Hauri, C. & Zapf, J. 1999. Leptin is suppressed during infusion of recombinant human insulin-like growth factor I (rhIGFI) in normal rats. *Diabetologia*, **42**, 160-166.

Bray, G. A. & York, D. A. 1979. Hypothalamic and genetic obesity in experimental animals: an autonomic and endocrine hypothesis. *Physiol Rev*, **59**, 719-809.

Brogan, R. S., Mitchell, S. E., Trayhurn, P. & Smith, M. S. 1999. Suppression of leptin during lactation: contribution of the suckling stimulus versus milk production. *Endocrinology*, **140**, 2621-7.

Bryson, J. M., Phuyal, J. L., Swan, V. & Caterson, I. D. 1999. Leptin has acute effects on glucose and lipid metabolism in both lean and gold thioglucose-obese mice. *Am J Physiol Endocrinol Metab*, **277**, E417-422.

Campfield, L. A., Smith, F. J., Guisez, Y., Devos, R. & Burn, P. 1995. Recombinant mouse OB protein: evidence for a peripheral signal linking adiposity and central neural networks. *Science*, **269**, 546-549.

Carro, E., Senaris, R., Considine, R. V., Casanueva, F. F. & Dieguez, C. 1997. Regulation of in vivo growth hormone secretion by leptin. *Endocrinology*, **138**, 2203-2206.

Carro, E., Senaris, R. M., Seoane, L. M., Frohman, L. A., Arimura, A., Casanueva, F. F. & Dieguez, C. 1999. Role of growth hormone (GH)-releasing hormone and somatostatin on leptin-induced GH secretion. *Neuroendocrinology*, **69**, 3-10.

Cartmill, J. A., Thompson, D. L., Jr., Storer, W. A., Gentry, L. R. & Huff, N. K. 2003. Endocrine responses in mares and geldings with high body condition scores grouped by high vs. low resting leptin concentrations. *J. Anim Sci.*, **81**, 2311-2321.

Ceddia, R. B., William, W. N., Jr. & Curi, R. 2001. The response of skeletal muscle to leptin. *Front Biosci*, **6**, D90-7.

Ceddia, R. B., William, W. N. J. & Curi, R. 1999b. Comparing effects of leptin and insulin on glucose metabolism in skeletal muscle; evidence for an effect of leptin on glucose uptake and decarboxylation. *Int J Obes*, **23**, 75-82.

Cha, B. S., Ciaraldi, T. P., Carter, L., Nikoulina, S. E., Mudaliar, S., Mukherjee, R., Paterniti, J. R., Jr. & Henry, R. R. 2001. Peroxisome proliferator-activated receptor (PPAR) gamma and retinoid X receptor (RXR) agonists have complementary effects on glucose and lipid metabolism in human skeletal muscle. *Diabetologia*, **44,** 444-52.

Chan, J. L., Bluher, S., Yiannakouris, N., Suchard, M. A., Kratzsch, J. & Mantzoros, C. S. 2002. Regulation of circulating soluble leptin receptor levels by gender, adiposity, sex steroids, and leptin: observational and interventional studies in humans. *Diabetes*, **51,** 2105-12.

Chehab, F. F., Lim, M. E. & Lu, R. H. 1996. Correction of the sterility defect in homozygous obese female mice is treatment with the human recombinant leptin. *Nat Genet*, **12,** 318-320.

Chehab, F. F., Mounzih, K., Lu, R. & Lim, M. E. 1997. Early onset of reproductive function in normal female mice treated with leptin. *Science*, **275,** 88-90.

Chelikani, P. K., Glimm, D. R. & Kennelly, J. J. 2003. Short Communication: Tissue Distribution of Leptin and Leptin Receptor mRNA in the Bovine. *J Dairy Sci*, **86,** 2369-2372.

Chen, X. L., Hausman, D. B., Dean, R. G. & Hausman, H. J. 1998. Hormonal regulation of leptin mRNA expression and preadipcyte recruitment and differentiation in porcine primary cultures and S-V cells. *Obes Res*, **6,** 164-172.

Clarke, I. J., Henry, B., Iqbal, J. & Goding, J. W. 2001. Leptin and the regulation of food intake and the neuroendocrine axis in sheep. *Clin Exp Pharmacol Physiol*, **28,** 106-7.

Cohen, B., Novick, D. & Rubinstein, M. 1996. Modulation of insulin activities by leptin. *Science*, **274,** 1185-1188.

Considine, R. V. 1997. Weight regulation, leptin growth hormone. *Horm Res*, **48,** 116-121.

Corp, E. S., Conze, D. B., Smith, F. & Campfield, L. A. 1998. Regional localisation of specific [125I] leptin binding sites in rat forebrain. *Brain Res*, **789,** 40 - 47.

Cortright, R. N., Muoio, D. M. & Dohm, G. L. 1997. Skeletal muscle lipid metabolism: a frontier for new insights into fuel homeostasis. *Nutr Biochem*, **8,** 228-245.

Dagogo-Jack, S., Selke, G., Melson, A. K. & Newcomer, J. W. 1997. Robust leptin secretory responses to dectamethasone in obese subjects. *J Clin Endocrinol Metab*, **82,** 3230-3233.

Delavaud, C., Bocquier, F., Chilliard, Y., Keisler, D. H. & Gertler, A. 2000. Plasma leptin determination in ruminants: effect of nutritional status and body fatness on plasma leptin concentration assessed by a specific RIA in sheep. *J Endocrinol*, **165,** 519-526.

Delavaud, C., Ferlay, A., Faulconnier, Y., Bocquier, F., Kann, G. & Chilliard, Y. 2002. Plasma leptin concentration in adult cattle: Effects of breed, adiposity, feeding level, and meal intake. *J Anim Sci*, **80,** 1317-1328.

Devaskar, S. U., Anthony, R. & Hay, W., Jr. 2002. Ontogeny and insulin regulation of fetal ovine white adipose tissue leptin expression. *Am J Physiol Regul Integr Comp Physiol*, **282,** R431-8.

Diamond, F. B., Eichler, D. C., Duckett, G., Jorgensen, E. V., Shulman, D. & Root, A. W. 1997. Demonstration of a leptin binding factor in human serum. *Biochem Biophys Res Commun*, **233,** 818-822.

Dyck, D. J., Steinberg, G. & Bonen, A. 2001. Insulin increases FA uptake and esterification but reduces lipid utilization in isolated contracting muscle. *Am J Physiol Endocrinol Metab*, **281,** E600-607.

Dyck, J. R., Berthiaume, L. G., Thomas, P. D., Kantor, P. F., Barr, A. J., Barr, R., Singh, D., Hopkins, T. A., Voilley, N., Prentki, M. & Lopaschuk, G. D. 2000. Characterization of rat liver malonyl-CoA decarboxylase and the study of its role in regulating fatty acid metabolism. *Biochem J*, **350 Pt 2,** 599-608.

Dyer, C. J., Simmons, J. M., Matteri, L. & Keisler, D. H. 1997. Leptin receptor mRNA is expressed in ewe anterior pituitary and adipose tissue and is differentially expressed in hypothalamic regions of well-fed and feed-restricted ewes. *Domest Anim Endocrinol*, **14,** 119-128.

Ehrhardt, R. A., Slepetis, R. M., Bell, A. W. & Boisclair, Y. R. 2001. Maternal leptin is elevated during pregnancy in sheep. *Domest Anim Endocrinol*, **21,** 85-96.

Ehrhardt, R. A., Slepetis, R. M., Siegal-Willott, J., Van Amburgh, M. E., Bell, A. W. & Boisclair, Y. R. 2000. Development of a specific radioimmunoassay to measure physiological changes of circulating leptin in cattle and sheep. *J Endocrinol*, **166,** 519-528.

Estienne, M. J., Harper, A. F., Barb, C. R. & Azain, M. J. 2000. Concentrations of leptin in serum and milk collected from lactating sows differing in body condition. *Domest Anim Endocrinol*, **19,** 275-280.

Fain, J. N., Coronel, E. C. & Beauchamp MJ, B. S. 1997. Expression of leptin and b_2-adrenergic receptors in rat adipose tissue in altered thyroid states. *Biochem J*, **322,** 145-150.

Forhead, A. J., Thomas, L., Crabtree, J., Hoggard, N., Gardner, D. S., Giussani, D. A. & Fowden, A. L. 2002. Plasma leptin concentration in fetal sheep during late gestation: ontogeny and effect of glucocorticoids. *Endocrinology*, **143,** 1166-73.

Furnsinn, C., Brunmair, B., Furtmuller, R., Roden, M., Englisch, R. & Waldhausl, W. 1998. Failure of leptin to affect basal and insulin-stimulated glucose metabolism of rat skeletal muscle in vitro. *Diabetologia*, **41,** 524-529.

Garcia, M. R., Amstalden, M., Williams, S. W., Stanko, R. L., Morrison, C. D., Keisler, D. H., Nizielski, S. E. & Williams, G. L. 2002. Serum leptin and its adipose gene expression during pubertal development, the estrous cycle, and different seasons in cattle. *J Anim Sci*, **80,** 2158-67.

Gavrilova, O., Barr, V., Marcus-Samuels, B. & Reitman, M. 1997. Hyperleptinaemia of pregnancy associated with the appearance of a circulating form of the leptin receptor. *J Biol Chem*, **272**, 30546-30551.

Ghilardi, N. & Skoda, R. C. 1997. The leptin receptor activates Janus Kinase 2 and signals for proliferation in a factor-dependent cell line. *Mol Endocrinol*, **11**, 393-399.

Ghilardi, N., Ziegler, S., Wiestner, A., Stoffel, R., Heim, M. H. & Skoda, R. C. 1996. Defective STAT signaling by the leptin receptor in diabetic mice. *Proc Natl Acad Sci USA*, **93**, 6231-6235.

Golden, P. L., Maccagnan, T. J. & Pardridge, W. M. 1997. Human blood-brain barrier leptin receptor. *J Clin Invest*, **99**, 14-18.

Gray, R. S., Cowan, P., di Mario, U., Elton, R. A., Clarke, B. F. & Duncan, L. J. P. 1985. Influence of insulin antibodies on pharmacokinetics and bioavailability of recombinant human and highly purified beef insulins in insulin dependent diabetics. *Br Med J*, **290**, 1687-1691.

Griffin, M. E., Marcucci, M. J., Cline, G. W., Bell, K., Barucci, N., Lee, D., Goodyear, L. J., Kraegen, E. W., White, M. F. & Shulman, G. I. 1999. Free fatty acid-induced insulin resistance is associated with activation of protein kinase C theta and alterations in the insulin signaling cascade. *Diabetes*, **48**, 1270-4.

Grunfeld, C., Zhao, C., Fuller, J., Pollock, A., Moser, A., Friedman, J. & Feingold, K. R. 1996. Endotoxin and cytokines induce expression of leptin, the ob gene product, in hamster. *J Clin Invest*, **97**, 2152-2157.

Guillet-Deniau, I., Burnol, A. F. & Girard, J. 1997. Identification and localization of a skeletal muscle secrotonin 5-HT2A receptor coupled to the Jak/STAT pathway. *J Biol Chem*, **272**, 14825-9.

Hardie, L. J., Guilhot, N. & Trayhurn, P. 1996. Regulation of leptin production in cultured mature white adipocytes. *Horm Metab Res*, **28**, 685-689.

Henry, B. A., Goding, J. W., Alexander, W. S., Tilbrook, A. J., Canny, B. J., Dunshea, F., Rao, A., Mansell, A. & Clarke, I. J. 1999. Central administration of leptin to ovariectomized ewes inhibits food intake without affecting the secretion of hormones from the pituitary gland: evidence for a dissociation of effects on appetite and neuroendocrine function. *Endocrinology*, **140**, 1175-1182.

Henry, B. A., Goding, J. W., Tilbrook, A. J., Dunshea, F. R. & Clarke, I. J. 2001. Intracerebroventricular infusion of leptin elevates the secretion of luteinising hormone without affecting food intake in long-term food-restricted sheep, but increases growth hormone irrespective of bodyweight. *J Endocrinol*, **168**, 67-77.

Henson, M. C. & Castracane, V. D. 2000. Leptin in pregnancy. *Biol Reprod*, **63**, 1219-1228.

Hill, R. A., Flick-Smith, F. C., Dye, S. & Pell, J. M. 1997. Actions of an IGF-1-enhancing antibody on IGF-1 pharmacokinetics and tissue distribution: increased IGF-1 bioavailability. *J Endocrinol*, **152**, 123-130.

Hill, R. A., Margetic, S. & Hughes, N. 2002. Leptin binding moieties in bovine serum. *J Anim Sci*, **80 (suppl 1)**, 195.

Hill, R. A., Margetic, S. & Pegg, G., Gazzola C. 1998. Leptin: its pharmokinetics and tissue distribution. *Int J Obes*, **22**, 765-770.

Hocquette, J. F., Ortigues-Marty, I., Pethick, D., Herpin, P. & Fernandez, X. 1998. Nutritional and hormonal regulation of energy metabolism in skeletal muscles of meat-producing animals. *Livest Prod Sci*, **56**, 115-143.

Hoggard, N., Hunter, L., Trayhurn, P., Williams, L. M. & Mercer, K. G. 1998. Leptin and reproduction. *Proc Nutr Soc*, **57**, 421-427.

Holness, M. J., Munns, M. J. & Sugden, M. C. 1999. Current concepts concerning the role of leptin in reproductive function. *Mol Cell Endocrinol*, **157**, 11-20.

Houseknecht, K. L., Mantzoros, C. S., Kuliawat, R., Hadro, E., Flier, J. S. & Kahn, B. B. 1996. Evidence for leptin binding proteins in serum of rodents and humans; modulation with obesity. *Diabetes*, **45**, 1638-1643.

Houseknecht, K. L., Portocarrero, C. P., Lamenager, R. & Spurlock, M. E. 2000. Growth hormone regulates leptin gene expresion in bovine adipose tissue: correlation with adipose IGF-1 expression. *J Endocrinol*, **164**, 51-57.

Hunt, M. L., Kokta, T. A., O'Shea, T., Hill, R. A. & McFarlane, J. R. 2003. Leptin Clearance Rates During the Early Follicular Phase are Higher Than During the Luteal Phase of the Estrus Cycle in Ewes. *Proc Endocrin Soc Aust*, **46**, 77.

Jacob, S., Hauer, B., Becker, R., Artzner, S., Gruaer, P., Loblein, K., Nielsen, M., Renn, W., Rett, K., Wahl, H.-G., Stumvoll, M. & Haring, H.-U. 1999. Lipolysis in skeletal muscle is rapidly regulated by low physiological doses of insulin. *Diabetologia*, **42**, 1171-1174.

Janik, J. E., Cutri, B. D., Considine, R. V., Rager, H. C., Powers, G. C., Alvord, W. G., Smith II, J. W., Gause, B. L. & Kopp, W. C. 1997. Interleukin 1a increases serum leptin concentrations in humans. *J Clin Endocrinol Metab*, **82**, 3084-3086.

Kadokawa, H., Blache, D., Yamada, Y. & Martin, G. B. 2000. Relationships between changes in plasma concentrations of leptin before and after parturition and the timing of first post-partum ovulation in high-producing Holstein dairy cows. *Reprod Fert Devel*, **12**, 405-11.

Kamohara, S., Burcelin, R., Halaas, J. L., Friedman, J. M. & Charron, M. J. 1997. Acute stimulation of glucose metabolism in mice by leptin treatment. *Nature*, **389**, 374-377.

Kauter, K., Ball, M. & Kearney, P. 2000. Adrenaline, insulin and glucagon do not have acute effects on plasma leptin levels in sheep: development and characterisation of an ovine leptin ELISA. *J Endocrinol*, **166,** 127-135.

Kauter, K. G., O'Shea, T. & McFarlane, J. R. 2003. Different leptin responses to fasting in merino wethers. *Proc Endocrin Soc Aust*, **46,** 138.

Kido, Y., Nakae, J. & Accili, D. 2001. Clinical review 125: The insulin receptor and its cellular targets. *J Clin Endocrinol Metab*, **86,** 972-979.

Klein, S., Coppack, S. W., Mohamed-Ali, V. & Landt, M. 1996. Adipose tissue leptin production and plasma leptin kinetics in humans. *Diabetes*, **45,** 984-7.

Kokta, T. A., Dodson, M. V., Gertler, A. & Hill, R. A. 2004. Intercellular signaling between adipose tissue and muscle tissue. *Domest Anim Endocrinol*, **27,** 303-331.

Kratzsch, J., Berthold, A., Lammert, A., Reuter, W., Keller, E. & Kiess, W. 2002a. A rapid, quantitative immunofunctional assay for measuring human leptin. *Horm Res*, **57,** 127-32.

Kratzsch, J., Lammert, A., Bottner, A., Seidel, B., Mueller, G., Thiery, J., Hebebrand, J. & Kiess, W. 2002b. Circulating soluble leptin receptor and free leptin index during childhood, puberty, and adolescence. *J Clin Endocrinol Metab*, **87,** 4587-94.

Lammert, A., Brockmann, G., Renne, U., Kiess, W., Bottner, A., Thiery, J. & Kratzsch, J. 2002. Different isoforms of the soluble leptin receptor in non-pregnant and pregnant mice. *Biochem Biophys Res Commun*, **298,** 798-804.

Landt, M., Horowitz, J. F., Coppack, S. W. & Klein, S. 2001. Effect of short-term fasting on free and bound leptin concentrations in lean and obese women. *J Clin Endocrinol Metab*, **86,** 3768-71.

Langhans, W. & Hrupka, B. 1999. Interleukins and tumor necrosis factor as inhibitors of food intake. *Neuropeptides*, **33,** 415-424.

Laud, K., Gourdou, I., Belair, L., Keisler, D. H. & Djiane, J. 1999. Detection and regulation of leptin receptor mRNA in ovine mammary epithelial cells during pregnancy and lactation. *Febs Letters*, **463,** 194-8.

Leon, H. V., Hernandez-Ceron, J., Keislert, D. H. & Gutierrez, C. G. 2004. Plasma concentrations of leptin, insulin-like growth factor-I, and insulin in relation to changes in body condition score in heifers. *J Anim Sci*, **82,** 445-51.

Levin, N., Nelson, C., Gurney, A., Vandlen, R. & de Sauvage, F. 1996. Decreased food intake does not completely account for adiposity reduction after ob protein infusion. *Proc Natl Acad Sci U S A*, **93,** 1726-30.

Lewandowski, K., Horn, R., O'Callaghan, C. J., Dunlop, D., Medley, G., O'Hare, P. & Brabant, G. 1999. Free leptin, bound leptin and soluble leptin receptor in normal and diabetic pregnancies. *J Clin Endocrinol Metab*, **84,** 300-306.

Linnemann, K., Malek, A., Sager, R., Blum, W. F., Schneider, H. & Fusch, C. 2000. Leptin production and release in the dually *in vitro* perfused human placenta. *J Clin Endocrinol Metab*, **85**, 4298-4301.

Liu, C., Liu, X.-J., Barry, G., Ling, N., Maki, R. A. & De Souza, E. 1997a. Expression and characterization of a putative high affinity human soluble leptin receptor. *Endocrinology*, **138**, 3548-3554.

Liu, Y.-L., Emilsson, V. & Cawthorne, M. A. 1997b. Leptin inhibits glycogen synthesis in the isolated soleus muscle of obese (ob/ob) mice. *FEBS Lett*, **411**, 351-355.

Luiken, J. J. F. P., Dyck, D. J., Han, X.-X., Tandon, N. N., Arumugam, Y., Glatz, J. F. C. & Bonen, A. 2002. Insulin induces the translocation of the fatty acid transporter FAT/CD36 to the plasma membrane. *Am J Physiol Endocrinol Metab*, **282**, E491-495.

Maamra, M., Bidlingmaier, M., Postel-Vinay, M. C., Wu, Z., Strasburger, C. J. & Ross, R. J. M. 2001. Generation of Human Soluble Leptin Receptor by Proteolytic Cleavage of Membrane-Anchored Receptors. *Endocrinology*, **142**, 4389-4393.

Mantzoros, C., S, & Moschos, S. J. 1998. Leptin: in search of role(s) in human physiology and pathophysiology. *Clin Endocrinol*, **49**, 551-567.

Mantzoros, C. S., Moschos, S., Avramopoulos, L., Kaklamani, V., Liolios, A., Doulgerakis, D. E., Griveas, I., Katsilambros, N. & Flier, J. S. 1997a. Leptin concentrations in relation to body mass index and the tumour necrosis factor-a system in humans. *J Clin Endocrinol Metab*, **82**, 3408-3413.

Mantzoros, C. S., Rosen, H. N., Greenspan, S. L., Flier, J. S. & Moses, A. C. 1997b. Short-term hyperthyroidism has no effect on leptin levels in man. *J Clin Endocrinol Metab*, **82**, 497-499.

Margetic, S., Gazzola, C., Pegg, G. G. & Hill, R. A. 2002a. Characterization of leptin binding in bovine kidney membranes. *Domest Anim Endocrinol*, **23**, 411-24.

Margetic, S., Gazzola, C., Pegg, G. G. & Hill, R. A. 2002b. Leptin: a review of its peripheral actions and interactions. *Int J Obes*, **26**, 1407-1433.

Marie, M., Findlay, P. A., Thomas, L. & Adam, C. L. 2001. Daily patterns of plasma leptin in sheep: effects of photoperiod and food intake. *J Endocrinol*, **170**, 277-86.

Masuzaki, H., Ogava, Y., Sagawa, N., Hosoda, K., Matsumoto, T., Mise, H., Nishimura, H., Yoshimasa, Y., Tanaka, I., Mori, T. & Nakao, K. 1997. Nonadipose tissue production of leptin; Leptin as a novel placenta-derived hormone in humans. *Nat Med*, **3**, 1029-1033.

Matarese, G. 2000. Leptin and the immune system: how nutritional status influences the immune response. *Eur Cytokine Netw*, **11**, 7-13.

Matsuoka, T., Tahara, M., Yokoi, T., Masumoto, N., Takeda, T., Yamaguchi, M.,

Tasaka, K., Kurachi, H. & Murata, Y. 1999. Tyrosine phosphoylation of STAT3 by leptin through leptin receptor in mouse metaphase 2 stage oocyte. *Biochem Biophys Res Commun*, **256,** 480-484.

Matzinger, O., Schneiter, P. & Tappy, L. 2002. Effects of fatty acids on exercise plus insulin-induced glucose utilization in trained and sedentary subjects. *Am J Physiol Endocrinol Metab*, **282,** E125-131.

McFadin, E. L., Morrison, C. D., Buff, P. R., Whitley, N. C. & Keisler, D. H. 2002. Leptin concentrations in periparturient ewes and their subsequent offspring. *J Anim Sci*, **80,** 738-43.

Miell, J. P., Englaro, P. & Blum, W. F. 1996. Dexamethasone induces an acute and sustained rise in circulating leptin levels in normal human subjects. *Horm Metab Res*, **28,** 704-707.

Minokoshi, Y. & Kahn, B. B. 2003. Role of AMP-activated protein kinase in leptin-induced fatty acid oxidation in muscle. *Biochem Soc Trans*, **31,** 196-201.

Minokoshi, Y., Kim, Y. B., Peroni, O. D., Fryer, L. G., Muller, C., Carling, D. & Kahn, B. B. 2002. Leptin stimulates fatty-acid oxidation by activating AMP-activated protein kinase. *Nature*, **415,** 339-343.

Minton, J. E., Bindel, D. J., Drouillard, J. S., Titgemeyer, E. C., Grieger, D. M. & Hill, C. M. 1998. Serum leptin is associated with carcass traits in finishing cattle. *J Anim Sci*, **76 Suppl. 1,** 231.

Morrison, C. D., Daniel, J. A., Holmberg, B. J., Djiane, J., Raver, N., Gertler, A. & Keisler, D. H. 2001. Central infusion of leptin into well-fed and undernourished ewe lambs: effects on feed intake and serum concentrations of growth hormone and luteinizing hormone. *J Endocrinol*, **168,** 317-324.

Morrison, C. D., Wood, R., McFadin, E. L., Whitley, N. C. & Keisler, D. H. 2002. Effect of intravenous infusion of recombinant ovine leptin on feed intake and serum concentrations of GH, LH, insulin, IGF-1, cortisol, and thyroxine in growing prepubertal ewe lambs. *Domest Anim Endocrinol*, **22,** 103-12.

Morton, N. M., Emilsson, V., Liu, Y.-L. & Cawthorne, M. A. 1998. Leptin action in intestinal cells. *J Biol Chem*, **273,** 26194-26201.

Mostyn, A., Bispham, J., Pearce, S., Evens, Y., Raver, N., Keisler, D. H., Webb, R., Stephenson, T. & Symonds, M. E. 2002. Differential effects of leptin on thermoregulation and uncoupling protein abundance in the neonatal lamb. *Faseb J*, **(July 18, 2002),** 10.1096/fj.02-0077fje.

Mostyn, A., Keisler, D. H., Webb, R., Stephenson, T. & Symonds, M. E. 2001. The role of leptin in the transition from fetus to neonate. *Proc Nutr Soc*, **60,** 187-194.

Mostyn, A., Pearce, S., Stephenson, T. & Symonds, M. E. 2004. Hormonal and nutritional regulation of adipose tissue mitochondrial development and function in the newborn. *Exp Clin Endocrinol Diabetes*, **112**, 2-9.

Mounzih, K., Qui, J., Ewart-Toland, A. & Chehab, F. F. 1998. Leptin is not necessary for gestation and parturition but regulates maternal nutrition via a leptin resistance state. *Endocrinology*, **139**, 5259-5262.

Muhlhausler, B. S., Roberts, C. T., McFarlane, J. R., Kauter, K. G. & McMillen, I. C. 2002. Fetal leptin is a signal of fat mass independent of maternal nutrition in ewes fed at or above maintenance energy requirements. *Biol Reprod*, **67**, 493-9.

Muhlhausler, B. S., Roberts, C. T., Yuen, B. S., Marrocco, E., Budge, H., Symonds, M. E., McFarlane, J. R., Kauter, K. G., Stagg, P., Pearse, J. K. & McMillen, I. C. 2003. Determinants of fetal leptin synthesis, fat mass, and circulating leptin concentrations in well-nourished ewes in late pregnancy. *Endocrinology*, **144**, 4947-54.

Muoio, D. M., Dohm, G. L., Fiedorek, F. T., Tapscott, E. B. & Coleman, R. A. 1997. Leptin directly alters lipid partitioning in skeletal muscle. *Diabetes*, **46**, 1360-1363.

Muoio, D. M., Dohm, G. L., Tapscott, E. B. & Coleman, R. A. 1999. Leptin opposes insulin's effects on fatty acid partitioning in muscle isolated from obese ob/ob mice. *Am J Physiol*, **276**, E913-E921.

Murakami, T., Yamashita, T., Ilida, M., Kuwajima, M. & Shima, K. 1997. A short for of the leptin receptor performs signal transduction. *Biochem Biophys Res Commun*, **231**, 26-29.

Newby, D., Gertler, A. & Vernon, R. G. 2001. Effects of recombinant ovine leptin on in vitro lipolysis and lipogenesis in subcutaneous adipose tissue from lactating and non-lactating sheep. *J Anim Sci*, **79**, 445-452.

Nowak, K., Mackowiak, P., Nogowski, L., Szkudelski, T. & Malendowicz. 1998. Acute action on insulin blood level and liver insulin receptor in the rat. *Life Sci*, **63**, 1347-1352.

Nyomba, B. L. G., Johnson, M., Berard, L. & Murphy, L. J. 1999. Relationship between serum leptin and the insulin-like growth factor-I system in humans. *Metab-Clin Exp*, **48**, 840-844.

Parhami, F., Tintut, Y., Ballard, A., Fogelman, A. M. & Demer, L. L. 2001. Leptin enhances the calcification of vascular cells: artery wall as a target of leptin. *Circ Res*, **88**, 954-60.

Patel, N., Brinkman-Van der Linden, E. C. M., Altmann, S. W., Gish, K., Balasubramanian, S., Timans, J. C., Peterson, D., Bell, M. P., Bazan, J. F., Varki, A. & Kastelein, R. A. 1999. OB-BP1/Siglec-6; a leptin- and sialic acid-binding protein of the immunoglobulin superfamily. *J Biol Chem*, **274**, 22729-22738.

Pelleymounter, M. A., Cullen, M. J., Baker, M. B., Hecht, R., Winters, D., Boone, T. & Collins, F. 1995. Effects of the obese gene product on body weight regulation in ob/ob mice. *Science*, **269**, 540-543.

Quinton, N. D., Smith, R. F., Clayton, P. E., Gill, M. S., Shalet, S., Justice, S. K., Simon, S. A., Walters, S., Postel-Vinay, M. C., Blakemore, A. I. & Ross, R. J. 1999. Leptin binding activity changes with age: the link between leptin and puberty. *J Clin Endocrinol Metab*, **84**, 2336-41.

Ranganathan, S., Ciaraldi, T. P., Henry, R. R., Mudalair, S. & Kern, P. A. 1998. Lack of effect of leptin on glucose transport, lipoprotein lipase, and insulin action in adipose and muscle cells. *Endocrinology*, **139**, 2509-2513.

Ren, M. Q., Wegner, J., Bellman, O., Brockman, G. A., Schneider, F., Teuscher, F. & Ender, K. 2002. Comparing mRNA levels of genes encoding leptin, leptin receptor, and lipoprotein lipase between dairy and beef cattle. *Domest Anim Endocrinol*, **23**, 371-381.

Roden, M., Krssak, M., Stingl, H., Gruber, S., Hofer, A., Furnsinn, C., Moser, E. & Waldhausl, W. 1999. Rapid impairment of skeletal muscle glucose transport/phosphoylation by free fatty acids in humans. *Diabetes*, **48**, 358-364.

Roden, M., Price, T. B., Perseghin, G., Petersen, K. F., Rothman, D. L., Cline, G. W. & Shulman, D. 1996. Mechanism of free fatty acid-induced insulin resistance in humans. *J Clin Invest*, **97**, 2859-2865.

Roh, S. H., Clarke, I. J., Xu, R. W., Goding, J. W., Loneragan, K. & Chen, C. 1998. The in vitro effect of leptin on basal and growth hormone-releasing secretion from the ovine pituitary gland. *Neuroendocrinology*, **68**, 361-364.

Sandowski, Y., Raver, N., Gussakovsky, E. E., Shochat, S., Dym, O., Livnah, O., Rubinstein, M., Krishna, R. & Gertler, A. 2002. Subcloning, expression, purification, and characterization of recombinant human leptin-binding domain. *J Biol Chem*, **277**, 46304-9.

Sayed-Ahmed, A., Elmorsy, S. E., Rudas, P. & Bartha, T. 2003. Partial cloning and localization of leptin and leptin receptor in the mammary gland of the Egyptian water buffalo. *Domest Anim Endocrinol*, **25**, 303-314.

Sesmilo, G., Casamitjana, R., Halperin, I., Gomis, R. & Vilardell, E. 1998. Role of thyroid hormones on serum leptin levels. *Eur J Endocrinol*, **139**, 428-430.

Silva, L. F., VandeHaar, M. J., Weber Nielsen, M. S. & Smith, G. W. 2002. Evidence for a local effect of leptin in bovine mammary gland. *J Dairy Sci*, **85**, 3277-86.

Sinha, M. K., Opentanova, I. & Ohannesian, J. P. 1996. Evidence of free and bound leptin in human circulation. Studies in lean and obese subjects and during short-term fasting. *J Clin Invest*, **98**, 1277-1282.

Slieker, L. J., Sloop, K. W., Surface, P. L., Kriauciunas, A., LaQuier, F., Manetta, J., Blue-Valleskey, J. & Stephens, T. W. 1996. Regulation of expression of ob mRNA and protein by glucocorticoids and cAMP. *J Biol Chem*, **271**, J Biol Chem.

Smith, J. L. & Sheffield, L. G. 2002. Production and regulation of leptin in bovine mammary epithelial cells. *Domest Anim Endocrinol*, **22**, 145-54.

Stephens, T. W., Basinski, M., Bristow, P., K., Blue-Valleskey, J., M., Burgett, G., Craft, L., Hale, J., Hoffmann, J., Hsiung, H., M., Kriauciunas, A., MacKeller, W., Rosteck, P., R., Schoner, B., Jr., Smith, D., Tinsley, F. C., Zhang, X. Y., & Heiman, M. 1995. The role of neuropeptide Y in the antiobesity action of the obese gene product. *Nature*, **377**, 530-532.

Symonds, M. E., Mostyn, A., Pearce, S., Budge, H. & Stephenson, T. 2003. Endocrine and nutritional regulation of fetal adipose tissue development. *J Endocrinol*, **179**, 293-9.

Symonds, M. E., Mostyn, A. & Stephenson, T. 2001. Cytokines and cytokine receptors in fetal growth and development. *Biochem Soc Trans*, **29**, 33-7.

Tang-Christensens, M., Havel, P. J., Jacobs, R. R., Larsen, P. J. & Cameron, J. L. 1999. Central administration of leptin inhibits food intake and activates the sympathetic nervous system in rhesus macaques. *J Clin Endocrinol Metab*, **84**, 711-717.

Tannenbaum, G. S., Gurd, W. & Laponte, M. 1998. Leptin is a potent stimulator of spontaneous pulsatile growth hormone (GH) secretion and the GH response to GH-releasing hormone. *Endocrinology*, **139**, 3871-3875.

Tataranni, P., Larsen, D. E., Snitker, S., Young, J. B. & Flatt, J. P. 1996. Ravussin E. Effects of glucocorticoids on energy metabolism and food intake in humans. *Am J Physiol*, **271**, E317-E325.

Thomas, L., Wallace, J. M., Aitken, R. P., Mercer, J. G., Trayhurn, P. & Hoggard, N. 2001. Circulating leptin during ovine pregnancy in relation to maternal nutrition, body composition and pregnancy outcome. *J Endocrinol*, **169**, 465-76.

Tokuda, T., Matsui, T. & Yano, H. 2000. Effects of light and food on plasma leptin concentrations in ewes. *Anim Sci*, **71**, 235-242.

Tokuda, T. & Yano, H. 2001. Blood leptin concentrations in Japanese Black cattle. *Anim Sci*, **72**, 309-313.

van Dielen, F. M., van 't Veer, C., Buurman, W. A. & Greve, J. W. 2002. Leptin and soluble leptin receptor levels in obese and weight-losing individuals. *J Clin Endocrinol Metab*, **87**, 1708-16.

Virkamaki, A., Korsheninnikova, E., Seppala-Lindroos, A., Vehkavaara, S., Goto, T., Halavaara, J., Hakkinen, A. M. & Yki-Jarvinen, H. 2001. Intramyocellular lipid is associated with resistance to in vivo insulin actions on glucose uptake, antipolysis, and early insulin signaling pathways in human skeletal muscle. *Diabetes*, **50**, 2337-43.

Wabitsch, M., Jensen, P. B., Blum, W. F., Christoffersen, C. T., Englaro, P., Heinze,

E., Rascher, W., Teller, W., Tornqvist, H. & Hauner, H. 1996. Insulin and cortisol promote leptin production in cultured human fat cells. *Diabetes*, **45**, 1435-1438.

White, D. W. & Tartaglia, L. A. 1999. Evidence for ligand-independent homo-oligomerisation of leptin receptor (OB-R) isoforms: A proposed mechanism permitting productive long-form signalling in the presence of excess short-form expression. *J Cell Biochem*, **73**, 278-288.

Williams, G. L., Amstalden, M., Garcia, M. R., Stanko, R. L., Nizielski, S. E., Morrison, C. D. & Keisler, D. H. 2002. Leptin and its role in the central regulation of reproduction in cattle. *Domest Anim Endocrinol*, **23**, 339-49.

Williams, L. M., Adam, C. L., Mercer, J. G., Moar, K. M., Slater, D., Hunter, L., Findlay, P. A. & Hoggard, N. 1999. Leptin receptor and neuropeptide Y gene expression in the sheep brain. *J Neuroendocrinol*, **11**, 165-169.

Winder, W. W. & Holmes, B. F. 2000. Insulin stimulation of glucose uptake fails to decrease palmitate oxidation in muscle if AMPK is activated. *J Appl Physiol*, **89**, 2430-7.

Wu, Z., Bidlingmaier, M., Liu, C., De Souza, E. B., Tschop, M., Morrison, K. M. & Strasburger, C. J. 2002. Quantification of the soluble leptin receptor in human blood by ligand-mediated immunofunctional assay. *J Clin Endocrinol Metab*, **87**, 2931-9.

Yamauchi, T., Kamon, J., Waki, H., Murakami, K., Motojima, K., Komeda, K., Ide, T., Kubota, N., Terauchi, Y., Tobe, K., Miki, H., Tsuchida, A., Akanuma, Y., Nagai, R., Kimura, S. & Kadowaki, T. 2001. The Mechanisms by Which Both Heterozygous Peroxisome Proliferator-activated Receptor gamma (PPARgamma) Deficiency and PPARgamma Agonist Improve Insulin Resistance. *J Biol Chem*, **276**, 41245-41254.

Yanagihara, N., Utsunomiya, K., Cheah, T. B., Hirano, H., Kajiwara, K., Hara, K., Nakamura, E., Toyohira, Y., Uezono, Y., Ueno, S. & Izumi, F. 2000. Characterization and functional role of leptin receptor in bovine adrenal medullary cells. *Biochem Pharmacol*, **59**, 1141-5.

Yoshida, T., Monkawa, T., Hayashi, M. & Saruta, T. 1997. Regulation of expression of leptin mRNA and secretion of leptin by thyroid hormone in 3T3-L1 adipocyte. *Biochem Biophys Res Commun*, 822-826.

Yu, H. Y., Inoguchi, T., Kakimoto, M., Nakashima, N., Imamura, M., Hashimoto, T., Umeda, F. & Nawata, H. 2001. Saturated non-esterified fatty acids stimulate de novo diacylglycerol synthesis and protein kinase c activity in cultured aortic smooth muscle cells. *Diabetologia*, **44**, 614-20.

Yu, W. H., Kimura, M., Walczewska, A., Karanth, S. & McCann, S. M. 1997. Role of leptin in hypothalamic-pituitary function. *Proc Natl Acad Sci USA*, **94**, 1023-1028.

Yuen, B. S., McMillen, I. C., Symonds, M. E. & Owens, P. C. 1999. Abundance of leptin mRNA in fetal adipose tissue is related to fetal body weight. *J Endocrinol*, **163**, R11-4.

Yuen, B. S., Owens, P. C., McFarlane, J. R., Symonds, M. E., Edwards, L. J., Kauter, K. G. & McMillen, I. C. 2002a. Circulating leptin concentrations are positively related to leptin messenger RNA expression in the adipose tissue of fetal sheep in the pregnant ewe fed at or below maintenance energy requirements during late gestation. *Biol Reprod*, **67**, 911-6.

Yuen, B. S., Owens, P. C., Muhlhausler, B. S., Roberts, C. T., Symonds, M. E., Keisler, D. H., McFarlane, J. R., Kauter, K. G., Evens, Y. & McMillen, I. C. 2003. Leptin alters the structural and functional characteristics of adipose tissue before birth. *Faseb J*, **17**, 1102-4.

Yuen, B. S., Owens, P. C., Symonds, M. E., Keisler, D. H., McFarlane, J. R., Kauter, K. G. & McMillen, I. C. 2004. Effects of leptin on fetal plasma adrenocorticotropic hormone and cortisol concentrations and the timing of parturition in the sheep. *Biol Reprod*, **70**, 1650-7.

Yuen, B. S. J., Owens, P. C., McFarlane, J. R., Symonds, M. E., Edwards, L. J., Kauter, K. G. & McMillen, I. C. 2002b. Circulating Leptin Concentrations Are Positively Related to Leptin Messenger RNA Expression in the Adipose Tissue of Fetal Sheep in the Pregnant Ewe Fed at or Below Maintenance Energy Requirements During Late Gestation. *Biol Reprod*, **67**, 911-916.

Zeng, J., Patterson, B., Klein, S., Martin, D. R., Dagogo-Jack, S., Kohrt, W. M., Miller, S. B. & Landt, M. 1997. Whole body leptin kinetics and renal metabolism in vivo. *Am J Physiol*, **273**, E1102-E1106.

Zhang, Y., Proenca, R., Maffei, M., Barone, M., Leopold, L. & Friedman, J. M. 1994. Positional cloning of the mouse obese gene and its human homologue. *Nature*, **372**, 425-432.

Zhao, A. Z., Shinohara, M. M., Huang, D., Schimizu, M., Eldar-Finkelman, H., Krebs, E. G., Beavo, J. A. & Bornfeldt, K. E. 2000. Leptin induces insulin-like signaling that antagonises cAMP elevation by glucagon in hepatocytes. *J Biol Chem*, **275**, 11348-11354.

Zierath, J. R., Frevert, E. U., Ryder, J. W., Berggren, P.-O. & Kahn, B. B. 1998. Evidence against a direct effect of leptin on glucose transport in skeletal muscle and adipocytes. *Diabetes*, **47**, 1-4.

Zumbach, M. S., Boehme, M. W. J., Wahl, P., Stremmel, W., Ziegler, R. & Nawroth, P. P. 1997. Tumor necrosis factor increases serum leptin levels in humans. *J Clin Endocrinol Metab*, **82**, 4080-4082.

Zurlo, F., Larson, K., Bogardus, C. & Ravussin, E. 1990. Skeletal muscle metabolism is a major determinant of resting expenditure. *J Clin Invest*, **86**, 1423-1427.

11

CARBOHYDRATE NUTRITION OF TRANSITION DAIRY COWS

T. R. OVERTON, K. L. SMITH, AND M. R. WALDRON
Department of Animal Science, Cornell University, Ithaca NY

Introduction

Nutrition and management of the transition cow continues to attract substantial research attention. Despite the tremendous quantity of research conducted on nutrition and physiology of transition cows, the transition period remains a problematic area on many commercial dairy farms, and metabolic disorders continue to occur at economically important rates (Burhans, Bell, Nadeau and Knapp, 2003). Data summarized recently by researchers at the University of Minnesota (Godden, Stewart, Fetrow, Rapnicki, Cady, Weiland, Spencer and Eicker, 2003) indicate that approximately 25% of cows that left dairy herds in Minnesota from 1996 to 2001 did so during the first 60 days in milk, with an uncertain additional proportion leaving before the end of the lactation as a consequence of problems during the transition period. The economic ramifications of culling high value cows, together with the costs associated with occurrence of various metabolic disorders in both clinical and sub-clinical form, are large. Therefore research attention will continue to focus on understanding the biology of transition cows and implementing management schemes on dairy farms to optimize production and profitability.

During the past fifteen years, a substantial amount of research has been conducted on the physiological adaptations involved in transition to lactation and on developing nutritional strategies to ensure that the process is successful (reviewed by Overton and Waldron, 2004). Recent information has caused the dairy industry to rethink the length of the dry period, which potentially has carryover effects for implementation of nutritional systems for transition cows on commercial dairy farms. This chapter will review physiological changes, nutritional management strategies, dynamics of dry matter intake, and practical considerations in transition cows.

Physiological adaptations of transition cows

For cows to successfully transition to lactation, metabolic adaptations must enable increased synthesis of glucose, mobilization of sufficient (but not excessive) body fat reserves to meet the energetic demands of lactation, and calcium mobilization to meet the increased demands for calcium. Adaptations needed to successfully adapt to increased demand for calcium by increasing calcium resorption from bone and facilitating calcium absorption from the intestine were reviewed by Goff and Horst (1997). From an energy metabolism standpoint, the need to synthesize and direct glucose to the mammary gland for synthesis of lactose represents the overriding metabolic demand during the first few weeks of lactation. The cow accomplishes this by concurrently increasing hepatic gluconeogenesis (Reynolds, Aikman, Lupoli, Humphries and Beever, 2003) and decreasing oxidation of glucose by peripheral tissues (Bennink, Mellenberger, Frobish and Bauman, 1972). Reynolds *et al.* (2003) recently reported that there was little net use or release of glucose by the gastrointestinal tract during the transition period and early lactation; therefore, the 267% increase in total glucose output by the gastrointestinal tract and liver from 9 d before expected parturition to 21 d after parturition resulted almost completely from increased hepatic gluconeogenesis (Figure 11.1). The major substrates for hepatic gluconeogenesis in ruminants are propionate from ruminal fermentation, lactate from Cori cycling, amino acids (AA) from amino acid catabolism, and glycerol released during lipolysis of adipose tissue (Seal and Reynolds, 1993). The calculated maximal contribution of propionate to net glucose release by liver ranged from approximately 50 to 60% during the transition period; that for lactate ranged from 15 to 20%; and that for glycerol ranged from 2 to 4% (Reynolds *et al.*, 2003). By difference, amino acids accounted for a minimum of approximately 20 to 30% during the transition period; the maximal contribution of alanine increased from 2.3% at 9 d prepartum to 5.5% at 11 d postpartum. These data are consistent with those of Overton, Drackley, Douglas, Emmert and Clark (1998), who reported that hepatic capacity to convert [1-^{14}C]alanine to glucose was approximately doubled on 1 d postpartum compared with 21 d prepartum. Although AA are not likely to be quantitatively important in terms of the amount of milk that the AA pool will support during early lactation, these results lend support to the use of AA as an adaptational substrate pool for glucose synthesis during the immediate postpartum period.

A third key metabolic adaptation relates to mobilization of body reserves, particularly body fat stores, in support of increased energetic demands during early lactation when there is insufficient energy intake. Mobilization of body fat occurs through release of non-esterified fatty acids (NEFA) into the

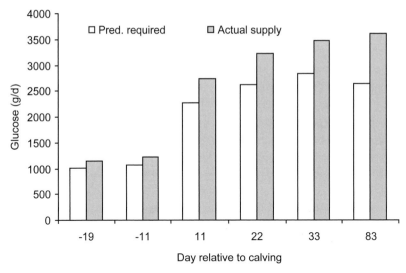

Figure 11.1 Predicted whole-body glucose requirements compared with actual supply of glucose by gut and liver during the transition period and early lactation. Data are from Reynolds *et al.* (2003). Predictions are as described by Overton (1998).

bloodstream (Figure 11.2). These NEFA are used for energy by body tissues and as precursors for synthesis of milk fat; however, available data suggest that the liver takes up NEFA in proportion to their supply (Emery, Liesman and Herdt, 1992). Unfortunately, the liver typically does not have sufficient capacity to completely dispose of NEFA through export into the blood or catabolism for energy (Figure 11.2), and thus transition cows are predisposed to accumulate triglycerides in liver tissue. The primary consequence of this triglyceride accumulation appears to be impaired liver function, including decreased capacity for ureagenesis and gluconeogenesis.

Strategies to increase energy intake of the cow during the prepartum period, in an effort to minimize mobilization of NEFA from adipose tissue combined with the need to supply substrates for oxidative metabolism and gluconeogenesis, have led to increased emphasis on the carbohydrate portion of the diet fed during the prepartum period. Mixed ruminal fermentation of carbohydrates provides primarily acetate, propionate, and butyrate – all of which are used as fuel sources by various tissues in the body and as building blocks for synthesis of milk components. Fermentation of starch in the rumen will favour production of propionate – the primary gluconeogenic precursor. Ruminal fermentation of carbohydrates in the presence of adequate amounts of ruminally available nitrogen also leads to synthesis and passage of microbial protein to the small intestine. Protein is absorbed as amino acids to support synthesis of milk and body proteins and potentially to provide an adaptational source of gluconeogenic substrate as described above.

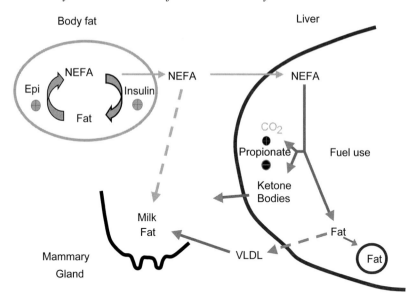

Figure 11.2 Schematic representation of metabolism of non-esterified fatty acids (NEFA) in the dairy cow (adapted from Drackley, 1999).

Nutritional strategies to support physiological adaptations

GROUPING STRATEGIES

The primary goal of nutritional management strategies of dairy cows during the transition period should be to support the metabolic adaptations described above. Industry-standard nutritional management of dairy cows during the dry period in the US consists of a two-group nutritional scheme. The NRC (2001) recommended that a diet containing approximately 1.25 Mcal/kg of NE_L (approximately 8.2 MJ/kg of ME) be fed from drying off until approximately 21 d before calving and that a diet containing 1.54 to 1.62 Mcal/kg of NE_L (approximately 10.3 MJ/kg of ME) be fed during the last three weeks preceding parturition. The primary rationale for feeding a lower energy diet during the early dry period is to minimize body condition score (BCS) gain during the dry period; furthermore, Dann, Litherland, Underwood, Bionaz and Drackley (2003) reported recently that supplying excessive energy to dairy cows during the early dry period may actually have detrimental carryover effects during the subsequent early lactation period. The nature of these carryover effects is not known. One could speculate, however, that effects could be mediated through metabolic pathways responsible for tissue responsiveness to endocrine signals during the late prepartum period.

In general, information supports feeding the higher energy close-up diet for only two to three weeks prior to parturition (Mashek and Beede, 2001; Corbett, 2002; Contreras, Ryan and Overton, 2004). Results from two of these experiments indicated farm-specific negative effects on subsequent production and health if cows were fed the higher energy diet for the entire dry period (Contreras *et al.*, 2004) or for an average of 37 d prepartum (Mashek and Beede, 2001). These responses may correspond to the negative carryover effects of overfeeding energy during the early dry period described by Dann *et al.* (2003).

Furthermore, recent results (Contreras *et al.*, 2004) support managing cows to achieve a BCS of approximately 3.0 (1 to 5 scale) at dry off rather than the traditional 3.5 to 3.75 BCS – perhaps partially due to the decreased dry matter intake (DMI) associated with higher BCS during the prepartum period (Hayirli, Grummer, Nordheim and Crump, 2002). Studies conducted with limited replication indicate increased DMI and milk yield for cows of BCS 2 to 2.5 at calving versus those with a BCS of 3.5 to 4 on a 4-point scale (Garnsworthy and Topps, 1982a,b; Treacher, Reid and Roberts, 1986; Garnsworthy and Jones, 1987). These results are also consistent with those of Domecq, Skidmore, Lloyd and Kaneene (1997), who reported that as BCS of cows at dry off increased, milk yield during the first 120 days in milk decreased; furthermore, thinner cows that gained BCS during the dry period yielded more milk during the first 120 days in milk. Collectively, results published in the scientific literature support the concept that cows of moderately lower BCS within a well-managed transition management system are more likely to have positive transition period outcomes than cows of greater BCS due to their propensity to have increased DMI and potentially increased milk yield during early lactation.

CARBOHYDRATE NUTRITION

Because of the focus on increasing energy content of the diet fed during the close-up period, most research on energy content of the close-up diet has related to varying the non-fibre carbohydrate (NFC) content of the diet. A commonly held notion in the dairy industry is that diets higher in NFC content than traditional dry cow diets must be fed prior to calving in order to promote the development of ruminal papillae for adequate absorption of VFA produced during ruminal fermentation. This notion was based on one experiment in which dry cows were adapted from a diet containing a large amount of poor quality forage to a diet containing a much larger proportion of grain (Dirksen, Liebich and Mayer, 1985). However, Andersen, Sehested and Ingvartsen (1999) reported that transition cows subjected to dietary changes much more typical of those in modern nutritional management systems did not have meaningful changes in ruminal epithelia during the transition period. Regardless of the effect on rumen epithelia, feeding diets containing higher proportions of NFC should promote adaptation of rumen microbes to NFC levels typical of diets fed during lactation and provide increased amounts of propionate to support hepatic gluconeogenesis and microbial protein (providing the diet contains sufficient rumen degradable protein) to support protein requirements for maintenance, pregnancy, and mammogenesis.

Several investigators have examined the effects of NFC level in the close-up diet on metabolism and performance, and results from seven experiments are summarized in Table 1. Although the range of NFC concentrations investigated in these experiments was large, when compared with a low NFC diet, feeding a high NFC diet virtually always resulted in higher prepartum DMI and frequently resulted in one or more positive effects on energy metabolism or production (Table 11.1).

Table 11.1 Effect of NFC[1] concentration in the prepartum diet on metabolism and performance.

| Experiment | NFC, g/kg of DM | | Effect of high NFC |
	Low	High	
Grum, Drackley, Younker, LaCount and Veenhuizen, 1996	180	280	↑ prepartum DMI; ↑ prepartum insulin
Minor, Trower, Strang, Shaver and Grummer, 1998	240	440	↑ prepartum DMI; ↑ milk yield[2]
Mashek and Beede, 2000	350	380	↓ prepartum BHBA; ↑ prepartum insulin; ↑ milk yield[3]
Keady, Mayne, Fitzpatrick and McCoy, 2001	130	280	↑ prepartum DMI
Holcomb, Van Horn, Head, Hall and Wilcox, 2001	250	300	↑ prepartum DMI; ↓ peripartum NEFA
Doepel, Lapierre and Kennelly, 2002	240	300	↑ postpartum DMI; ↓ peripartum NEFA; ↓ liver triglycerides[4]
Rabelo, Rezende, Bertics and Grummer, 2003	380	450	↑ prepartum DMI; ↑ postpartum DMI[4] (d 1 to 20)

[1] Non-fibre carbohydrate; 100 – [(NDF-NDICP) + CP + EE + Ash]; NRC, 2001
[2] Cows fed higher NFC prepartum continued on higher NFC diet postpartum
[3] Higher NFC increased milk yield only in 3rd lactation and greater cows
[4] Statistical trend.

The NFC content of the diet is only one factor that affects performance.

Fermentability of NFC at a fixed carbohydrate concentration has also been examined in diets fed during both the prepartum and immediate postpartum periods. Dann, Varga and Putnam (1999) reported that increasing the fermentability of NFC prepartum by replacing cracked corn with steam-flaked corn (390 g/kg total NFC content of the diet) tended to increase prepartum DMI, postpartum milk yield, and plasma insulin concentrations during the immediate postpartum period; NEFA concentrations were decreased during the prepartum period by increasing intake of fermentable carbohydrate. Ordway, Ishler and Varga (2002) fed diets containing approximately 360 g/ kg of NFC during the prepartum period and replaced ground shelled corn (27 g/kg diet DM) with sucrose. Sucrose tended to increase plasma glucose concentration during the prepartum period, but did not affect peripartum performance or concentrations of NEFA during either the prepartum or postpartum periods.

Most of the experiments described above confounded NFC content and energy concentration of the prepartum diet – i.e., the increase in NFC content of the diet simultaneously increased the energy content of the diet, and, given that cows typically consumed more of the higher NFC diet, they also consumed more energy during the prepartum period. We were interested in exploring the concentration of NFC in the diet independent of energy content of the prepartum diet and, more specifically, the impact of deriving energy from starch-based NFC compared with other carbohydrate sources (Smith, Waldron, Overton, Drackley and Socha, 2002; 2003). Of particular interest was the effect of feeding a diet high in non-forage fibre sources (NFFS; i.e., beet pulp, soybean hulls) as a source of highly-digestible NDF on performance and metabolism during the periparturient period. The high NFC diet contained 1.59 Mcal/kg of NE_L (approximately 10.4 MJ/kg of ME), 400 g/kg of NFC, and 280 g/kg of starch; the high NFFS diet contained 1.54 Mcal/kg of NE_L (approximately 10.1 MJ/kg of ME), 340 g/kg of NFC, and 180 g/kg of starch. Dry matter intake did not differ between the two treatments during the prepartum and postpartum periods, and other aspects of performance and metabolism, including the dynamics of insulin action and glucose disposal in response to a glucose challenge, were virtually unaffected by the source of dietary carbohydrate in the prepartum diet.

Although we did not assess directly the impact of energy supply in this experiment (Smith *et al.*, 2002; 2003); results imply that the generally positive effects on performance and metabolism of feeding diets during the prepartum period that are moderately higher in NFC content are linked to energy supply from carbohydrate rather than NFC content of the diet per se. Therefore, the specific NFC content of the diet prepartum diet may have received unwarranted focus in research and also practical diet formulation in the dairy industry. This speculation is consistent with results reported recently by Pickett, Cassidy, Tozer and Varga (2003), who measured positive effects on metabolism and performance when NDF from forage was replaced by NDF from NFFS in diets fed during the prepartum period.

In summary, available data do not clearly support a single strategy for approaching carbohydrate nutrition of transition cows during the prepartum period. However, most studies report one or more positive outcomes when higher NFC diets are compared with a lower NFC control diet. This conclusion is consistent with the data of Hayirli *et al.* (2002) who reported that prepartum DMI was positively correlated with NFC content of the prepartum diet. The only guideline provided by the Dairy NRC (2001) for carbohydrate nutrition of dry cows was that NFC content of the close-up diet should not exceed 430 g/kg of diet DM. This recommendation is consistent with data indicating that feeding diets during the close-up period containing high concentrations of NFC (430 to 450 g/kg) appeared to accentuate the decrease in DMI occurring in the days preceding parturition (Minor *et al.*, 1998; Rabelo *et al.*, 2003). Based upon the available information, our current recommendation is to formulate diets for cows during the close-up period that contain moderately high concentrations (340 to 360 g/kg) of starch-based NFC sources, balancing the remainder of the carbohydrate portion of the diet with NDF from forage or non-forage fibre sources (see Table 11.2 for detailed nutrient guidelines for close-up cows).

Table 11.2 General goals for diet formulation for close-up cows.

	Standard	*Anionic*
• NE$_L$, Mcal/kg	1.50 to 1.54	
• ME, MJ/kg	9.8 to 10.1	
• Metabolizable protein, g/d	1100 to 1200	
• NFC, g/kg	340 to 360	
• Starch, g/kg	180 to 200	
• Dietary Ca, g/d	100	140
• Dietary Ca, g/kg	9.0	10.2
• Dietary P, g/kg	3.0 to 3.5	
• Mg, g/kg	4.0 to 4.2	
• Cl, g/kg	3.0	8.0 to 1.2
• K, g/kg	< 13	< 13
• Na, %	1.0 to 1.5	
• S, %	2.0	3.0 to 4.0
• Vitamin A (IU/d)	100000	100000
• Vitamin D (IU/d)	30000	30000
• Vitamin E (IU/d)	1800	1800

Dynamics of dry matter intake in transition cows

The dynamics of DMI in transition cows has implications for carbohydrate composition of the diet during the prepartum period. Controlled experiments have shown that cows fed ad libitum during the close-up period decrease their voluntary DMI beginning approximately ten days before expected calving, with a decrease from d 21 prepartum to d 1 prepartum in the order of 30%. Grummer (1995) summarized data from several experiments and reported that DMI on d 21 postpartum was correlated with DMI on d 1 prepartum. This led many investigators to focus on elucidating factors affecting DMI of dairy cows during the close-up period. Hayirli *et al.* (2002) summarized animal and dietary factors accounting for variation in DMI during the close-up period from a large number of studies. Not surprisingly, as prepartum body condition score increased, prepartum DMI decreased. Although DMI as a proportion of BW was comparable for cows averaging a 2.8 or 3.6 body condition score (1 to 5 scale), cows with an average body condition score of 4.4 had a pronounced decrease in prepartum DMI.

Despite the widely held belief that increased total DMI during the close-up period will increase DMI during the postpartum period and will increase overall transition cow success (Grummer, 1995; Hayirli *et al.*, 2002), a growing body of evidence supports the possibility that the shape of the prepartum DMI curve (i.e., the rate and extent of decrease of DMI prior to calving) might be a more meaningful predictor of overall transition health and performance. Recently, Mashek and Grummer (2003) examined the relationships among total DMI from d 21 to d 1 prepartum, change in DMI between d 21 and d 1 prepartum, and postpartum performance and metabolic indices. They reported that postpartum DMI and milk production were more strongly correlated with total prepartum DMI. As total prepartum DMI increased, postpartum DMI and milk yield also increased. However, meaningful metabolic indices (i.e., postpartum plasma NEFA concentrations and liver triglyceride accumulation) were more strongly correlated with the change in DMI from 21 d to 1 d prepartum; as the change in DMI decreased, postpartum plasma NEFA and liver triglycerides also decreased.

The effect of the magnitude of decrease in prepartum DMI has been studied in vivo by investigators seeking to determine the potential to restrict energy intake of dry cows in order to perhaps precondition metabolism to negative energy balance. Across several experiments, cows fed balanced prepartum diets restricted to below calculated energy requirements (usually about 80% of predicted requirements) did not decrease DMI during the days preceding parturition. These animals increased their postpartum DMI and milk yield at faster rates than cows consuming the same diets ad libitum (Douglas,

Drackley, Overton and Bateman, 1998; Holcomb *et al.*, 2001; Agenas, Burstedt and Holtenius, 2003). Furthermore, feed-restricted cows had greater insulin sensitivity (Holtenius, Agenas, Delavaud and Chilliard, 2003) and typically had blunted peripartal NEFA curves compared with those fed ad libitum (Douglas *et al.*, 1998; Holcomb *et al.*, 2001; Holtenius *et al.*, 2003).

Collectively these data are intriguing, but all experiments were conducted with cows that were individually housed and fed. Achieving uniform restricted intake in a typical group-fed situation on commercial farms, where social interactions and competition will predominate, is difficult to achieve. Simple replacement of forage with straw in an effort to provide a more bulky diet decreased insulin concentrations (Rabelo, Bertics, Mackovic and Grummer, 2001) and might not allow for comparable microbial adaptation to a higher energy diet during the prepartum period. Despite these findings, anecdotal reports from the dairy industry in the US indicate transition success in response to feeding a diet containing large amounts of chopped straw (3 to 4 kg of DM) prepartum, in a herd experiencing a relatively high incidence of postpartum displaced abomasum. The advantages of such a diet are that straw typically is low in potassium, which should mitigate the risk of hypocalcaemia. The high degree of bulkiness should ensure high rumen fill, thereby decreasing the risk of displaced abomasum and potentially limiting energy intake. The disadvantages of feeding this type of diet are the increased risk of sorting by cows and the relatively higher level of feeding management required to ensure consistent intake of a high-bulk diet. Controlled research has not been conducted to determine whether the shape of the prepartum DMI curve is flatter when cows are fed a high-bulk diet, thus mimicking the restricted-fed scenario described above. Furthermore, controlled research has not been conducted to determine whether feeding a high-bulk diet is more advantageous than a well-managed higher NFC diet as recommended above for nutritional management of transition cows.

Shortened dry periods and implications for transition cow management

Several years ago, Dr. Kermit Bachman and colleagues at the University of Florida began challenging the concept of the traditional 50 to 60-day dry period. Studies conducted by this group (summarized in Table 11.3) and several studies conducted at various universities in the US (Annen, Collier, McGuire, Vicini, Ballam and Lormore, 2003; Grummer and Rastani, 2003; Fernandez, Ryan, Galton and Overton, 2004) indicate that subsequent cow performance and metabolism during the next lactation is virtually unaffected by decreasing the length of the dry period to 30 to 40 d, particularly for multiparous cows. This has been an interesting area of research because field experience is generating knowledge at least at the pace of university-based research. My current personal recommendation is 40-d dry periods for all cows that continue to produce sufficient quantities of milk to merit continued milking.

Table 11.3 Summary of results from studies of shortened dry period conducted by the University of Florida.

Reference			305-day milk yield (kg)	
	Days dry	Number of cows	Adjusted[1]	Mature equivalent[1]
Bachman, 2002	34	15	9799	9125
	57	19	9978	8986
Schairer, 2001	32	10	11635	11008
	61	9	10222	10529
Gulay, Hayen, Bachman,	31	28	9580	9586
Belloso, Liboni and	61	27	9836	9700
Head, 2003				

[1] adjusted for previous lactation milk yield; [2] adjusted for parity

Shortening the length of the dry period has obvious implications for nutritional management systems for transition cows on commercial dairy farms. In situations where two-group nutritional systems are impractical, a shorter dry period allows implementation of a more suitable one-group nutritional scheme. As described above, two-group nutritional schemes for dry cows in a 60-d dry period scenario are preferred because of both economics and potential negative effects of overfeeding cows during the early dry period. However, in my opinion the moderate NFC (340 to 360 g/kg) close-up diet can be fed throughout a 40-d dry period, regardless of DCAD. I do not believe that the far-off group will disappear on most commercial farms because many cows will continue to have dry periods of 50 d or greater; however, it potentially becomes a smaller group. Alternatively, dairy producers can elect to feed a far-off diet from 40 to 20 d before expected parturition and then move cows to a close-up group until parturition.

Summary and implications

Available data support two-group nutritional strategies for dry cows, with diets containing moderately high concentrations (340 to 360 g/kg) of NFC from starch-based sources combined with additional use of non-forage fibre-based carbohydrate sources in order to provide sufficient substrates for oxidative and glucogenic metabolism and to promote synthesis of microbial protein in the rumen. Feeding diets containing NFC in excess of 380 g/kg DM might increase total prepartum DMI, but might accentuate the rate of decrease in DMI prior to calving, thereby leading to metabolic disorders and periparturient problems. Shortened dry periods facilitate adoption of a one-group dry cow nutritional management scheme in some situations, although university-based research and field experience will continue to provide insight into optimal management strategies for these cows.

References

Agenas, S., Burstedt, E. and Holtenius, K. (2003) Effects of feeding intensity during the dry period. 1. Feed intake, body weight, and milk production. *Journal of Dairy Science*, 86:870-882.

Andersen, J. B., Sehested, J. and Ingvartsen, K. L. (1999) Effect of dry cow feeding strategy on rumen pH, concentration of volatile fatty acids, and rumen epithelium development. *Acta Agriculturae Scandinavica, Section A, Animal Science.* 49:149-155.

Annen, E. L., Collier, R. J., McGuire, M. A., Vicini, J. L., Ballam, J. M. and Lormore, M. J. (2003) Preliminary study on the effect of continuous somatotropin treatment on dry period requirement in dairy cattle: production, health, and reproduction. *Proceedings Cornell Nutrition Conference for Feed Manufacturers*, Syracuse, NY pp. 41-50.

Bachman, K. C. (2002) Milk production of dairy cows treated with estrogen at the onset of a short dry period. *Journal of Dairy Science*, 85:797–803.

Bennink, M. R., Mellenberger, R. W., Frobish, R. A. and Bauman, D. E. (1972) Glucose oxidation and entry rate as affected by the initiation of lactation. *Journal of Dairy Science*, 55:712.(Abstract)

Burhans, W. S., Bell, A. W., Nadeau, R. and Knapp, J. R. (2003) Factors associated with transition cow ketosis incidence in selected New England herds. *Journal of Dairy Science*, 86(Supplement 1):247. (Abstract)

Contreras, L. L., Ryan, C. M. and Overton, T. R. (2004). Effects of dry cow grouping strategy and prepartum body condition score on performance and health of transition dairy cows. *Journal of Dairy Science*, 87:517-523.

Corbett, R. B. (2002) Influence of days fed a close-up dry cow ration and heat stress on subsequent milk production in western dairy herds. *Journal of Dairy Science*, 85(Supplement 1):191-192. (Abstract)

Dann, H. M., Litherland, N. B., Underwood, J. P., Bionaz, M. and Drackley, J. K. (2003) Prepartum nutrient intake has minimal effects on postpartum dry matter intake, serum nonesterified fatty acids, liver lipid and glycogen contents, and milk yield. *Journal of Dairy Science*, 86(Supplement 1):106. (Abstract)

Dann, H. M., Varga, G. A. and Putnam, D. E. (1999) Improving energy supply to late gestation and early postpartum dairy cows. *Journal of Dairy Science*, 82:1765-1778.

Dirksen, G., Liebich, H. and Mayer, K. (1985) Adaptive changes of the ruminal mucosa and functional and clinical significance. *Bovine Practitioner*, 20:116-120.

Doepel, L., Lapierre, H. and Kennelly, J. J. (2002) Peripartum performance and metabolism of dairy cows in response to prepartum energy and protein intake. *Journal of Dairy Science*, 85:2315-2334.

Domecq, J. J., Skidmore, A. L., Lloyd, J. W. and Kaneene, J. B. (1997) Relationship between body condition score and milk yield in a large dairy herd of high yielding Holstein cows. *Journal of Dairy Science*, 80:101-112.

Douglas, G. N., Drackley, J. K., Overton, T. R. and Bateman, H. G. (1998) Lipid metabolism and production by Holstein cows fed control or high fat diets at restricted or ad libitum intakes during the dry period. *Journal of Dairy Science*, 81(Supplement 1):295.(Abstract)

Drackley, J. K. (1999) Biology of dairy cows during the transition period: the final frontier? *Journal of Dairy Science*, 82: 2259-2273.

Emery, R. S., Liesman, J. S. and Herdt, T. H. (1992) Metabolism of long-chain fatty acids by ruminant liver. *Journal of Nutrition*, 122:832-837.

Fernandez, J., Ryan, C. M., Galton, D. M. and Overton, T. R. (2004) Effects of dry period length on performance and health of dairy cows during the subsequent lactation. *Journal of Dairy Science*, 87(Supplement 1):345.(Abstract)

Garnsworthy, P. C. and Jones, G. P. (1987) The influence of body condition at calving and dietary protein supply on voluntary food intake and performance in dairy cows. *Animal Production*, 44:347-353.

Garnsworthy, P. C. and Topps, J. H. (1982a) The effect of body condition of dairy cows at calving on their food intake and performance when given complete diets. *Animal Production*, 35:113-119.

Garnsworthy, P. C. and Topps, J. H. (1982b) The effects of body condition at calving, food intake, and performance in early lactation on blood composition of dairy cows given complete diets. *Animal Production*, 35:121-125.

Godden, S. M., Stewart, S. C., Fetrow, J. F., Rapnicki, P., Cady, R., Weiland, W.,

Spencer, H. and Eicker, S. W. (2003) The relationship between herd rbST-supplementation and other factors and risk for removal for cows in Minnesota Holstein dairy herds. *Proceedings Four-State Nutrition Conference*, LaCrosse, WI. MidWest Plan Service publication MWPS-4SD16. pp. 55-64.

Goff, J. P., and Horst, R. L. (1997) Physiological changes at parturition and their relationship to metabolic disorders. *Journal of Dairy Science*, 80:1260-1268.

Grummer, R. R. (1995) Impact in changes in organic nutrient metabolism on feeding the transition cow. *Journal of Animal Science*, 73:2820- 2833.

Grum, D. E., Drackley, J. K., Younker, R. S., LaCount, D. W. and Veenhuizen, J. J. (1996) Nutrition during the dry period and hepatic lipid metabolism of periparturient dairy cows. *Journal of Dairy Science*, 79:1850-1864.

Grummer, R. R. and Rastani, R. (2003) Why re-evaluate length of dry period? *Journal of Dairy Science*, 86:153-154 (Abstract)

Gulay, M. S., Hayen, M. J., Bachman, K. C., Belloso, T., Liboni, M. and Head, H. H. (2003) Milk production and feed intake of Holstein cows given short (30 d) or normal (60 d) dry periods. *Journal of Dairy Science*, 86:2030–2038.

Hayirli, A., Grummer, R. R., Nordheim, E. V. and Crump, P. M. (2002) Animal and dietary factors affecting feed intake during the prefresh transition period in Holsteins. *Journal of Dairy Science*, 85:3430-3443.

Holcomb, C. S., Van Horn, H. H., Head, H. H., Hall, M. B. and Wilcox, C. J. (2001) Effects of prepartum dry matter intake and forage percentage on postpartum performance of lactating dairy cows. *Journal of Dairy Science*, 84:2051-2058.

Holtenius, K., Agenas, S., Delavaud, C. and Chilliard, Y. (2003) Effects of feeding intensity during the dry period. 2. Metabolic and hormonal responses. *Journal of Dairy Science*, 86:883-891.

Keady, T.W.J., Mayne, C. S., Fitzpatrick, D. A. and McCoy, M. A. (2001) Effect of concentrate feed level in late gestation on subsequent milk yield, milk composition, and fertility of dairy cows. *Journal of Dairy Science*, 84:1468-1479.

Mashek, D. G. and Beede, D. K. (2000) Peripartum responses of dairy cows to partial substitution of corn silage with corn grain in diets fed during the late dry period. *Journal of Dairy Science*, 83:2310-2318.

Mashek, D. G. and Beede, D. K. (2001) Peripartum responses of dairy cows fed energy-dense diets for 3 or 6 weeks prepartum. *Journal of Dairy Science*, 84:115-125.

Mashek, D. G. and Grummer, R. R. (2003) The ups and downs of feed intake in prefresh cows. *Proceedings Four-State Nutrition Conference*, LaCrosse,

WI. MidWest Plan Service publication MWPS-4SD16. pp. 153-158.

Minor, D. J., Trower, S. L., Strang, B. D., Shaver, R. D. and Grummer, R. R. (1998) Effects of nonfiber carbohydrate and niacin on periparturient metabolic status and lactation of dairy cows. *Journal of Dairy Science*, 81:189-200.

National Research Council. (2001) *Nutrient requirements of dairy cattle. 7th revised edition*. National Academy Press, Washington, DC.

Ordway, R. S., Ishler, V. A. and Varga, G. A. (2002) Effects of sucrose supplementation on dry matter intake, milk yield, and blood metabolites of periparturient Holstein dairy cows. *Journal of Dairy Science*, 85:879-888.

Overton, T.R. (1998) Substrate utilization for hepatic gluconeogenesis in the transition dairy cow. *Proceedings Cornell Nutrition Conference for Feed Manufacturers*, Cornell Univ., Ithaca, NY, pp. 237-246.

Overton, T. R., Drackley, J. K., Douglas, G. N., Emmert, L. S. and Clark, J. H. (1998) Hepatic gluconeogenesis and whole-body protein metabolism of periparturient dairy cows as affected by source of energy and intake of the prepartum diet. *Journal of Dairy Science*, 81(Supplement 1):295.(Abstract)

Overton, T. R. and Waldron, M. R. (2004). Nutritional management of transition dairy cows: Strategies to optimize metabolic health. *Journal of Dairy Science*, 87(E. Supplement):E105-E119.

Pickett, M. M., Cassidy, T. W., Tozer, P. R. and Varga, G. A. (2003) Effect of prepartum dietary carbohydrate source and monensin on dry matter intake, milk production and blood metabolites of transition dairy cows. *Journal of Dairy Science*, 86(Supplement 1):10.(Abstract)

Rabelo, E., Bertics, S. J., Mackovic, J. and Grummer, R. R. (2001) Strategies for increasing energy density of dry cow diets. *Journal of Dairy Science*, 84:2240-2249.

Rabelo, E., Rezende, R. L., Bertics, S. J. and Grummer, R. R. (2003) Effects of transition diets varying in dietary energy density on lactation performance and ruminal parameters of dairy cows. *Journal of Dairy Science*, 86:916-925.

Reynolds, C. K., Aikman, P. C., Lupoli, B., Humphries, D. J. and Beever, D. E. (2003) Splanchnic metabolism of dairy cows during the transition from late gestation through early lactation. *Journal of Dairy Science*, 86:1201-1217.

Seal, C. J. and Reynolds, C. K. (1993) Nutritional implications of gastrointestinal and liver metabolism in ruminants. *Nutrition Research Reviews*, 6:185-208.

Schairer, M. L. (2001) *Estrogen treatments for the initiation of dryoff in dairy cows*. M.S. Thesis. University of Florida, Gainesville.

Smith, K. L., Waldron, M. R., Overton, T. R., Drackley, J. K. and Socha, M. T. (2002) Performance of dairy cows as affected by prepartum carbohydrate source and supplementation with chromium throughout the periparturient period. *Journal of Dairy Science*, 85(Supplement 1):23. (Abstract)

Smith, K. L., Waldron, M. R., Overton, T. R., Drackley, J. K. and Socha, M. T. (2003) Metabolism of dairy cows as affected by prepartum dietary carbohydrate source and supplementation with chromium throughout the periparturient period. *Journal of Dairy Science*, 86(Supplement 1):106. (Abstract)

Treacher, R. J., Reid, I. M. and Roberts, C. J. (1986) Effect of body condition at calving on the health and performance of dairy cows. *Animal Production*, 43:1-6.

12

RECENT ADVANCES IN THE USE OF DIETARY CATION-ANION DIFFERENCE (DCAD) FOR TRANSITION DAIRY COWS

W.K. SANCHEZ[1] AND D.K. BEEDE[2]
[1]Diamond V Mills, Cedar Rapids, Iowa, USA; [2]Michigan State University, East Lansing, Michigan, USA

Introduction

Dietary cation-anion difference (DCAD) can affect two major biological systems that are often challenged in transition dairy cows. These involve calcium (Ca) metabolism and acid-base homeostasis. Calcium is critical for normal cell function and contractile functions of the musculature of the gastrointestinal tract, uterus, and mammary gland; and, maintenance of normal acid-base status is one of the highest biological priorities. Formulating the appropriate DCAD to achieve specific physiological responses has become a useful tool to improve both Ca metabolism and acid-base status of transition dairy cows. This chapter includes a review of recent research on DCAD for transition cows. Recommendations for applying this information also are included. The chapter is divided into sections that focus separately on pre- and post-partum cows. Areas where additional information is needed are highlighted.

Background on the DCAD Concept

Over 80 years ago, Shohl and Sato (1922) first proposed that minerals affected acid-base status. Shohl (1939) proposed that maintenance of normal acid-base equilibrium required excretion of excess dietary cations and anions. He hypothesized that consumption of either, excess mineral cations relative to anions, or excess anions relative to cations, resulted in acid-base imbalances in the animal.

Once animal nutritionists began to test these hypotheses, mineral interrelationships were found to affect numerous metabolic processes. Leach

(1979) and Mongin (1980) reviewed related literature and concluded that mineral interrelationships had profound influences on animal biology. They theorized that for an animal to maintain its acid-base homeostasis, input and output of acidity had to be maintained. It was shown that net acid intake was related to the difference between dietary intake of cations and anions. The monovalent macromineral ions, Na, K and Cl were the most influential elements in the relationship (Mongin, 1980).

Nutrient metabolism results in the degradation of nutrient precursors to acids and bases. During normal metabolism the flux of resulting H^+ is great. In typical rations fed to dairy cows, inorganic cations (eg., Na, K, Ca, and Mg) exceed dietary inorganic anions (eg., Cl, S, and P) by several millequivalents (meq) per day. Carried with excess dietary inorganic cations are organic anions, which can be metabolized to HCO_3^-. Therefore, a diet with excess inorganic cations relative to inorganic anions is alkaline and a diet with excess inorganic anions relative to cations is acidogenic.

PHYSIOLOGICAL MECHANISMS

Research has been limited on the physiology of how DCAD affects blood Ca. Numerous studies have been conducted relating acid-base variables indirectly to blood Ca concentrations, bone resorption markers and indicators of Ca absorption. Researchers from the USDA/ARS National Animal Disease Center, Ames, Iowa, USA (J.P. Goff and R.L. Horst) have been studying the physiological mechanisms elicited by DCAD. Their research has addressed chiefly hormonal regulation mechanisms. When blood Ca declines in the transition cow, parathyroid hormone (PTH) is secreted and begins by acting on bone cells to release Ca. PTH also initiates the renal synthesis of the active vitamin D hormone (D_3). Vitamin D_3 facilitates increased absorption of calcium from the intestine. Collectively these responses are designed to maintain blood calcium. However, the drain on the blood calcium pool of dairy cows producing an abundance of colostrum proceeds at a faster rate than these hormonal responses can compensate and many cows suffer from associated problems (i.e., clinical milk fever, displaced abomasum, and retained foetal membranes; Goff and Horst, 2003).

Goff *et al.* (1991) proposed a theory that reducing the DCAD concentration in the diet somehow enhanced the sensitivity of or to the PTH hormonal response. Their team used blood concentrations of Ca, PTH and vitamin D from numerous cows fed diets with either low or high DCAD to relate vitamin D and blood Ca. They found that responsiveness to vitamin D was better in cows fed diets low in DCAD. Goff *et al.* (2004) used these data to test the hypothesis that cows fed a positive DCAD are less responsive to PTH than are cows fed a diet with low or negative DCAD.

In a recently reported study, Goff *et al.* (2004) used 16 late gestation Jersey cows fed diets with either a Low (13 g K and 11 g Cl, per kg DM) or High DCAD (27 g K and 4 g Cl, per kg DM) for 2 weeks. Parathyroid hormone was injected every 3 hours as samples were collected for analysis of vitamin D and Ca concentrations in blood. The PTH caused a rise in blood Ca over the 48-hour injection period. However, as hypothesized, the dietary treatments led to different responses. Cows fed the diet with Low DCAD had a more rapid rise in blood Ca during the first 6 hours of the injection period. Cows fed the diet with the High DCAD did not have a significant rise in blood Ca until 21 hours. Plasma vitamin D concentrations also were affected by dietary treatment. Vitamin D increased sharply by 6 hours and peaked at 21 hours in cows fed the Low DCAD, whereas cows fed the High DCAD diet did not respond as quickly or as consistently. This research directly tested the hypothesis that cows fed anionic diets respond better to the PTH signal. The findings provide further understanding and insight into the mechanisms behind the DCAD response. The DCAD affects the ability of the cow to respond to hypocalcaemia signals within the critical first few hours of the decline in blood Ca.

Cation-anion interrlationships

Leach (1979) and Mongin (1980) reviewed nutritional concepts related to cation-anion interrelationships. Historically, nutritionists intuitively knew it was difficult to evaluate the effect of one macromineral without considering the influences of others. Early theories evaluated total ash, mineral element ratios, and differences between two or among more of the macromineral elements.

ACID OR ALKALINE ASH

Nutritionists first investigated the alkalinity and acidity of the diet under the acid- or alkaline-ash concept (Shohl, 1939). It was recognized that food for humans had either an acid or alkaline ash. When food is metabolized in the body, organic anions, such as acetate, citrate, and malate are oxidized. Inorganic cations originally associated with these organic anions remain. Because organic anions can buffer H^+ ions generated through metabolism, a food with a large amount of organic anions (and thus inorganic cations) was considered alkaline. The pH of the ash represented the acid or alkaline nature of the food.

DIETARY CATION-ANION DIFFERENCE (DCAD)

Blood pH is determined ultimately by the number of cation and anion charges (valance charges) in blood. Cations and anions originating from the diet and absorbed into the blood have marked effects on blood pH. If more anions than cations enter the blood from the digestive tract, blood pH will decrease. Mongin (1980) was one of the first to propose a three-way interrelationship among dietary Na, K and Cl. He proposed that the sum of Na plus K minus Cl (in meq/100 g dietary DM) could be used to predict net acid intake. This equation commonly has been referred to as the dietary cation-anion balance (Tucker *et al.*, 1988) or dietary electrolyte balance (West *et al.*, 1991). In 1991 Sanchez and Beede introduced the term *cation-anion difference* to represent, more accurately, the mathematical calculation used and to avoid the erroneous connotation that the mineral cations truly are balanced with the mineral anions in the diet, or that they necessarily should be.

Expressed in its fullest form, the macromineral DCAD expression can be written as:

Full Expression: meq [(Na + K + Ca + Mg) - (Cl + S + P)]/100 g dietary DM.

A problem with including the multivalent macromineral elements (Ca, Mg, P, and S) in the DCAD expression for ruminants, relates to the variable and incomplete absorption and bioavailability of these ions compared with Na, K and Cl. Sodium, K, and Cl are believed to be nearly 100% absorbed and available, whereas absorption of the multivalent cations and anions is less than 100% and is variable depending on element and mineral contributing the source element (NRC, 2001). There also might be some confusion about using the molecular forms of the divalent anions [i.e., sulphates (SO_4) and phosphates (PO_4)] in the expression instead of the individual elements (i.e., S and P). The reason the elements are used is that only the elements and not the molecular variants are measured when feedstuffs are analyzed. Nonetheless, it is known that elemental S *per se* does not affect acid-base status. It also is known that a considerable portion of the S analyzed in feedstuffs is from SO_4, the form reactive in acid-base chemistry.

The expression used most often in non-ruminant nutrition and the literature is the monovalent cation-anion difference expressed as:

DCAD3:   meq (Na + K - Cl)/100 g dietary DM.

This expression was considered superior for non-ruminant nutritionists because it comes closest to representing feed ions that are completely dissociated and solubilized from their respective salts, and absorbed into the body (Mongin, 1980). Throughout the remainder of this chapter this three-element expression will be referred to as DCAD3.

Because of the additional common use of sulphate salts in rations of late-pregnant dairy cows (i.e., the last 2 to 4 weeks before parturition) to reduce DCAD, the expression that has gained most acceptance in ruminant nutrition is:

DCAD4: meq [(Na + K) - (Cl + S)]/100 g dietary DM.

Throughout the rest of this chapter this expression will be referred to as DCAD4.

Changes in macromineral recommendations in NRC (2001)

Before discussing recent research addressing the DCAD concept, we shall discuss recent changes in feeding recommendations for the individual macromineral elements. Authors of NRC (2001) Nutrient Requirements of Dairy Cattle (7th Revised Edition) made two significant changes compared with previous editions.

1. Macromineral requirements for most elements were estimated using the factorial method, taking into account an estimate of the net (absorbed) requirement for maintenance, growth, pregnancy and lactation.

2. To determine how much of the mineral element was needed in a particular diet, an absorption coefficient for each element was estimated to calculate the total amount needed in the ration to meet the absorbed mineral element requirement.

The factorial method was used for Ca and P in NRC (1989), but was extended to Na, K, and Mg and six of the trace elements in the 2001 edition. These changes help provide more accurate estimates of macromineral requirements of dairy cattle and may give insight into the animal responses observed to differing DCAD. Unfortunately, limited data are available to assign absorption coefficients to mineral elements in most forage and byproduct feedstuffs (which may have highly variable absorption coefficients), and the absorption coefficients assigned to inorganic sources are based on limited data.

Because of differences in absorption coefficients, there has been considerable discussion about assigning absorption coefficients within the DCAD expression (Goff *et al.*, 1997; NRC, 2001). There is general agreement that S is only about 60% of the relative value (relative acidogenicity) of Cl, but there are too few and conflicting data to assign absorption coefficient values to other elements. Fixed values within the overall DCAD expression for a di*et al*so do not allow weighting to be placed on individual elements from individual feed ingredients; presumably this is where much of the variation originates. For example, inorganic versus organic forms of mineral elements and acid-base responses of animals. Most nutritionists in the field are continuing to use the four-element expression (DCAD4) and then use tools such as urine pH (see below) to make adjustments to DCAD to achieve desired biological responses.

Recent research on DCAD with prepartum dairy cows

The transition period is defined here as between 3 weeks pre- and 3 weeks postpartum. The time from 3 weeks prepartum to parturition, is an extremely critical time for the modern, high performance dairy cow; in large part setting her entire lactational performance (Wang, 1990).

During the early 1970s researchers discovered that feeding a diet that was acidic (e.g., silages treated with inorganic acids) caused the concentration of blood Ca to increase in periparturient cows (Ender and Dishington, 1970). This led to more research (Block, 1984; Oetzel, 1988; Beede *et al.*, 1992) and to the eventual practice of feeding a diet with more anions relative to cations in the late prepartum period to help reduce milk fever and related metabolic problems. The increased blood Ca in these cows not only prevented them from going down with milk fever, but also reduced problems such as retained placenta and displaced abomasum (Oetzel, 1988; Massey *et al.*, 1993). These problems were implied to be associated with a Ca deficiency that prevented muscles from contracting. Therefore, nutritionists began feeding diets with fewer cations than anions (DCAD4) to prepartum cows to help improve Ca status around the time of colostrum formation and calving when Ca is often deficient. These diets are described as anionic diets or diets with a low or negative DCAD.

Milk fever has been estimated to affect 5 to 7% of high producing adult dairy cows in the United States (Jordan and Fourdraine, 1993). In addition, the prevalence of subclinical hypocalcaemia was 67% for multiparous Holstein dairy cows following calving (Beede *et al.*, 1992). Research indicated that cows with clinical milk fever produced about 14% less milk in the subsequent lactation, and their productive life is reduced compared with non-milk fever cows (Block, 1984; Curtis *et al.*, 1984). Furthermore, cows that experience milk fever have an increased risk of ketosis, mastitis (especially coliform mastitis), dystocia, left displaced abomasum, retained placenta, and milk fever in the subsequent lactation (Curtis *et al.*, 1984; Wang, 1990; Oetzel, 1988). Guard (1996) estimated that the cost associated with a single case of milk fever was approximately $334 US, when considering lost production and income, and veterinary and treatment costs.

THE PRACTICE OF FEEDING ANIONIC SALTS

The term "anionic salts" is a misnomer, because a true salt electrochemically is neither anionic nor cationic – it is neutral. Nonetheless, it is a term used quite widely in the feed and dairy industries in the United States. Feeding anionic salts or feeding a diet with a negative DCAD4 to late pregnant dry cows has become a common practice in dairies that can accommodate multiple groups of dry cows. Feeding late pregnant (the last 2 to 4 weeks before calving) dry cows less milliequivalents (meq) of Na and K relative to Cl and S (i.e., a low or negative DCAD) increases blood Ca at calving in response to changes in acid-base status. Studies showed that when dry cows were fed diets with negative DCAD, milk fever cases were reduced dramatically.

THE LATEST RESEARCH

Optimal DCAD. Controlled experiments have not determined optimal DCAD, which probably varies with breed of cow, feeding management, facilities, and other factors that might affect acid-base status. Recommended DCAD4 targets have ranged from about -15 to +15 meq/100 g dietary DM (Byers, 1992; Goff, 2000; Beede *et al.*, 1995). The negative DCAD concentrations are probably lower than needed, but can provide a margin to account for varying K concentrations in feeds and K consumed from pasture or free-choice hay. Diets with negative DCAD concentrations, however, have led to practical problems in the field, most often associated with reduced intakes (Oetzel and Barmore, 1992). The combined effects of DCAD on intake, blood Ca and urine pH are needed to find an optimal DCAD.

Researchers from the University of Idaho fed a wide range of DCAD in an attempt to quantify the relationships among DCAD, intake, blood Ca, and urine pH (Giesy *et al.*, 1997). They used four non-pregnant non-lactating Holstein cows given a Ca-chelating agent (Na_2-EDTA) intravenously to simulate hypocalcaemia. The DCAD4 was formulated to be +30, +10, -10, and −30; the actual treatments averaged +27.6, +11.9, -6.4 and -25.3 meq/ 100 g dietary DM. Sources of anions used to lower DCAD included ammonium chloride, ammonium sulphate, and magnesium sulphate.

Results from this study are shown in (Table 12.1). There was a quadratic effect of DCAD on dry matter intake, but the most negative DCAD treatment was the only treatment that significantly depressed intake. Decreasing DCAD increased blood ionized Ca linearly, but much of the benefit appeared between +27.6 and +11.9. The implications of these results are that incremental decreases in DCAD result in a linear increase in the freely available form of ionized Ca in the blood. There is a point when the diet becomes too low in DCAD and dry matter intake suffers. In many cases, simply lowering dietary K might increase blood Ca concentrations enough to control severe hypocalcaemia without resulting in clinical transition disorders. In other cases, anionic salts might be needed. Commercial anion products designed to maintain intakes might be valuable, particularly when forages with high concentrations of dietary K are fed. Information in Table 1 helps provide better predictability of the response that DCAD has on the combined effects of blood Ca, intake and urine pH. In practice, feed intake depression should be avoided and can become a concern somewhere between −5 and −25 meq of DCAD4.

Table 12.1 Effect of dietary cation-anion difference (DCAD) on dry matter intake, blood ionized Ca and urinary pH (Giesy *et al.*, 1997).

| | DCAD4[1] Treatment | | | | |
	+27.6	+11.9	-6.4	-25.3	*SEM*
Dry matter intake, kg/day	11.1	11.2	11.3	10.5	0.1[Q]
Blood Ionized Ca[2], mg/dl	2.82	3.64	3.62	3.99	0.20[L]
Urinary pH	8.38	7.65	6.42	6.04	0.46[L]

[1] DCAD4 expressed as meq $[(Na + K) - (Cl + S)]/100$ g DM.
[2] Samples collected after Na-EDTA administration.
[Q, L] Superscripts indicate a quadratic (Q) or linear (L) effect ($P < 0.01$) of DCAD on the response variable.

Using urine pH to monitor effectiveness of DCAD. Upon feeding supplemental anions to alter systemic acid-base status, urine pH changes quickly (within 2 to 4 days; Goff *et al.*, 1998). Monitoring urine pH, therefore, can be a useful tool to determine whether or not the particular ration is having the desired physiological effects. Recent research has better quantified the relationship between DCAD and urine pH.

Giesy *et al.* (1997; discussed in the previous paragraph), also found a linear relationship between DCAD and urine pH. Urine pH (before infusion of Na_2-EDTA) was 8.38, 7.65, 6.42, and 6.04 for the +27.6, +11.9, -6.4 and -25.3 meq, respectively (Figure 12.1). Because urine pH was incrementally affected by all treatments it should be a useful predictor of dietary DCAD.

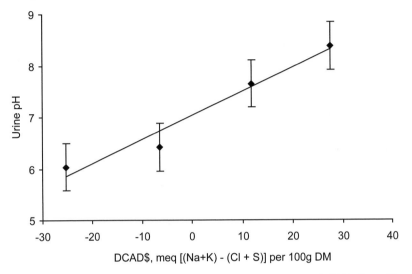

Figure 12.1 Urine pH response to varying dietary cation-anion difference (DCAD) (from Giesy *et al.*, 1997).

This linear relationship between DCAD and urine pH was similar to that found in an experiment with periparturient primi- and multiparous cows (Moore *et al.*, 1997). In that study late pregnant cows were fed for 21 days on diets with DCAD of +15, 0, and -5 meq (Na + K) - (Cl + S)/100g dietary DM. Supplemental anions were provided from calcium chloride, magnesium sulphate, and magnesium chloride. Total dietary Ca concentrations varied (4.4, 9.7, and 15 g Ca per kg DM) with the three decreasing DCADs, respectively. The source of supplemental Ca was from increasing calcium chloride and calcium carbonate in the 0 and -5 meq diets. Urine pH of late pregnant cows immediately before calving was 8.0 for +15 meq, 7.0 for 0 meq, and 6.2 for -5 meq.

Oetzel (2000) measured urine pH responses in groups of cows on commercial farms. With late pregnant cows fed diets over a wide range of DCAD, urine pH responded in a curvilinear fashion. From –10 to +10 meq urine pH had a steep positive slope, but above +10 meq urine pH became relatively constant (Figure 12.2). Charbonneau *et al.* (2004) conducted a meta analysis on the effect of DCAD on urine pH using 22 studies with 48 treatment groups and reported a similar non-linear response.

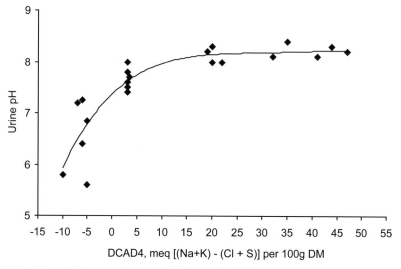

Figure 12.2 Effect of dietary cation-anion difference (DCAD) on urine pH of late-prepartum dairy cows. Each point is the mean urine pH of a group of cows fed a diet of known mineral element content. (Adapted from Oetzel, 2000).

From the literature it appears that over a DCAD range of -10 to +20 meq/100 g dietary DM urine pH will vary from 6.0 to 8.5. Variation in responses among reports might be due to other dietary factors (e.g., other cations and anions not in the DCAD expression, buffering capacity of the diet, exogenous acids such as lactic acid from silage, time of feeding, and feed intake differences). Because dietary DCAD4 is affected by the potential variation in four mineral elements and highly variable feed intakes during transition, the practical application is to use average urine pH values from several cows to predict both the actual dietary DCAD and the level of available Ca in blood.

Specific ion effects. One major implication of the latest research is that Cl ions have a greater relative impact than sulphate ions on acid-base status. Goff *et al.* (2004) recently demonstrated this in a series of studies when they fed Cl or SO_4 ions from different sources on an equimolar basis. All measures of acid-base status showed that Cl ions were more acidogenic than SO_4 ions. The most likely cause is differences between mineral elements in dissociation within and absorption from the digestive tract. Chloride is a strong monovalent ion that ionizes rapidly and dissociates from its salt form for easy absorption. Sulphate is much slower to dissociate and is absorbed less readily. This research supports the need for fine-tuning the DCAD expression based on more accurate information about absorption coefficients. Until those coefficients become available, readers can use this information to help trouble-shoot implementation of different formulations of DCAD and to look specifically at ion sources when animal responses are not as expected.

How much dietary Ca should be fed? In trials where negative DCAD aided prevention of milk fever, dietary concentrations of Ca typically were above those needed to meet requirements (typically between 12 and 15 g Ca/kg DM compared with recommendation for requirement of about 5 g/kg DM). The basis for these higher Ca concentrations is that negative DCAD is thought to increase urinary excretion of Ca, so more dietary Ca should be fed to replenish the blood pool of Ca. In cows infused with EDTA to cause hypocalcaemia, Schonewille et al. (1999) demonstrated that urinary Ca concentrations are much higher in cows fed diets with negative DCAD. Therefore, if a negative DCAD diet was fed with low dietary Ca (i.e., 5 to 10 g Ca/kg DM), chronic conditions of low or negative Ca balance could be detrimental. Therefore, many dairy nutritionists continue to formulate diets with Ca concentrations higher than needed to meet actual requirements.

However, in a series of recent studies (Rodriguez *et al.*, 1996; Rodriguez *et al.* 1997; Chan *et al.*, 2001; and Beede *et al.*, 2001) researchers found no benefit on health and lactational responses of periparturient cows by varying prepartum dietary Ca (between about 5 and 20 g Ca/kg DM) with negative DCAD. In some instances diets with elevated dietary Ca decreased DMI. For example, increasing dietary Ca by supplementation with calcium carbonate reduced DMI of late pregnant dairy cows (Beede *et al.*, 2001). Therefore, very high dietary Ca concentrations (i.e., concentrations that would provide greater than 150 g Ca/day) are not recommended. These studies indicate that approximately 100g dietary Ca is adequate when diets have a DCAD of about –5 meq/100 g dietary DM. However, these studies have relatively short-term data collection periods after parturition (10 to 70 days after calving). Until studies become available with longer postpartum data collection, most researchers continue to feed and recommend between 120 to 150 g Ca/day for cows fed diets with negative DCAD. When positive but relatively low DCAD (i.e., less than 20 meq/100 g) is used, nutritionists and researchers typically recommend feeding a reduced amount of dietary Ca (i.e., 50 to 100 g/day). When DCAD is more negative, a dietary Ca of 100 to 150 g/ day is typically fed. These recommendations also depend on the source of Ca used. For example, Ca from alfalfa forage (with a high amount of oxalate-bound Ca) is less available than Ca from carbonate forms, and Ca from calcium carbonate is less bioavailable than Ca from calcium chloride (NRC, 2001).

Negative DCAD and pregnant heifers. One question when feeding supplemental anions (negative DCAD) before calving is the potential negative effects on pregnant heifers. In nearly all DCAD studies, heifers beginning their first lactation were not included. One of the main questions is should dairy producers feed anion-supplemented diets to primiparous cows before calving? Moore *et al.* (2000) provided insight into this question by characterizing blood Ca, metabolic disorders, and production responses in cows and heifers (primiparous animals) fed varying DCAD before calving (Table 12.2). Both cows and heifers had increased blood Ca concentrations at calving when fed prepartum diets with lower DCAD. Within the experimental range of DCAD, the responses appear to be linear suggesting that each incremental drop in DCAD led to an increase in blood Ca. These responses are similar to those found in other studies using only multiparous cows. Dry matter intake before calving decreased by 1.5 kg/d for multiparous cows fed the anion-supplemented diet prepartum (-15 meq/100 g dietary DM). However, there was a trend (P = 0.15) for a large increase in fat-corrected milk production (4.3 kg/d) during the first 2 to 10 weeks of lactation. Therefore, feeding a negative DCAD to late pregnant multiparous cows was beneficial as it improved periparturient blood Ca status and subsequent milk production.

Table 12.2 Effect of varying dietary cation-anion difference (DCAD) on urine pH, Ca intake, energy balance and milk production in multiparous and primiparous dairy cows. (Adapted from Moore *et al.*, 2000.)

| | | *DCAD4* | | *Effect (P <)[1]* | |
	+*15*	*0*	*-15*	*C1*	*C2*
Multiparous					
Cows Urinary pH	7.95	7.32	6.01	**0.01**	**0.01**
Blood ionized Ca at calving, mg/dl	3.67	3.85	4.35	**0.03**	**0.04**
Prepartum DMI, kg/day	14.5	14.4	13.0	0.61	0.5
Energy Balance, MJ/day	35.2	34.5	25.1	0.63	0.51
3.5% FCM, kg/day	43.9	45.4	48.2	**0.15**	0.25
Primiparous					
Cows Urinary pH	8.03	7.37	6.42	**0.01**	**0.01**
Blood ionized Ca at calving, mg/dl	4.44	4.57	4.62	**0.02**	0.48
Prepartum DMI, kg/day	10.5	9.6	8.0	**0.02**	**0.05**
Energy Balance, MJ/day	15.7	11.0	0.4	**0.05**	**0.06**
3.5% FCM, kg/day	32.8	31.5	31.3	0.27	0.87

[1] Contrasts: C1: control vs. 0 and -15 DCAD; C2: 0 vs. -15 DCAD
[2] DCAD4 expressed as meq $[(Na + K) - (Cl + S)]/100$ g DM.

However, the situation with primiparous cows was much different to that observed with multiparous cows (Moore *et al.*, 2000). Overall, no statistical difference was found between cows and heifers. However, this might have been due to low replication. There was no improvement from feeding diets with lower DCAD to heifers, but there were negative effects on feed intake, energy balance and milk production (Table 12.2). No difference was noted in the cases of metabolic disorders due to treatment (probably due to small numbers of animals per treatment; 21 cows and 34 heifers). However, the total cases of transition related disorders (abomasal displacement, ketosis, metritis, milk fever, and retained placenta) were 7, 4, and 1 for multiparous cows fed diets with DCAD of +15, 0, and –15 meq/100 g DCAD, respectively. In contrast, the total cases of transition related disorders for first lactation cows were 8, 9, and 7 for diets with +15, 0, and –15 meq/100 g DCAD, respectively. These results support the notion (and recommendation by some nutritionists) that, whenever practically possible, late pregnant heifers should be separated and not fed supplementary anions. Some commercial anion products have been marketed that apparently do not have as negative effects on feed intake as directly-supplemented anionic salts in totally mixed rations, so these might be beneficial when heifers must be housed with multiparous cows prepartum. These data certainly point to the need for additional research on the effects of diets with negative DCAD on heifers.

Recent research on DCAD for early lactation cows

EFFECT OF DCAD ON LACTATION

Considerable research has been conducted on the effects of DCAD in mid and late lactation, but little information exists on the effects of DCAD in early lactation (Sanchez, 1999; Beede, 2003; Wu and Murphy, 2004). Numerous studies have shown relationships between DCAD and acid-base balance. Cows in early lactation often experience acid-base challenges. When fed diets with abundant highly fermentable carbohydrates, cows are prone to ruminal acidosis and challenges to systemic acid-base status. Conditions that are caused by or related to ruminal acidosis include reduced rumen function, abomasal displacement, reduced rumination, irritation to the ruminal wall, decreased feed consumption, low milk fat concentration, laminitis and reduced milk production; all potential maladies of the transition cow. For a high producing dairy cow, a positive DCAD presumably is efficacious to counter or neutralize ruminal and blood acids that are produced by accelerated and greater digestion and metabolism than would be typical in nonlactating or lower yielding cows.

The requirements for the individual mineral elements, Na, K, and Cl, have been defined using the factorial method, summing the grams of mineral element needed for maintenance, lactation, growth, and pregnancy divided by the absorption coefficient for that particular element (NRC, 2001). The dietary recommendation for S was set simply at 2 g S/ kg DM because insufficient information was available to use the factorial approach. As a point of reference for the remainder of the discussion in this chapter, based on current total dietary requirements (g/day) and recommendations for ration concentrations (%) of K, Na and Cl for lactating cows with milk yields ranging from 25 to 55 kg/day, the calculated DCAD3 is about +29 meq/100 g dietary DM; DCAD4 is about +16 meq/100g DM. These values are 3 to 4 meq/100 g dietary DM greater than those calculated from recommendations based on NRC (1989). Higher concentrations have been proposed recently. Wu and Murphy (2004) conducted a meta-analysis, primarily with data from cows in mid lactation, and found that milk yield, 4% fat-corrected milk yield, and dry matter intake peaked at +34, +49, and +40 meq DCAD3, respectively.

RESEARCH ON DCAD IN EARLY LACTATION

Delaquis and Block (1995) conducted one of the first studies with cows in early lactation. They measured responses by 12 Holstein cows during each of three stages of lactation. Within early (25 to 50 days in milk (DIM)), mid (107 to 137 DIM), and late (162 to 234 DIM) stages of lactation, two DCAD4 treatments were fed: +5.6 versus +25.8; +14.0 versus +37.3; and, +20.0 versus +37.5 meq/100 g dietary DM, respectively. Only DMI, milk yield and composition responses are addressed here. Daily DMI of cows increased with high versus low DCAD4 within early lactation (16.2 versus 15.2 kg/day) and, mid lactation (17.0 versus 15.6 kg/day). In late lactation, DMI for low and high DCAD4 were not different (16.8 versus 17.8 kg/day), supporting the notion that earlier lactation cows respond more to increasing DCAD. Milk yield responses were similar to DMI, with cows on low and high DCAD4 treatments yielding 18.3 versus 19.5 kg/day in early lactation, 18.2 versus 18.9 kg/day in mid lactation; no difference was detected in late lactation (14.9 versus 15.3 kg/day). Significant, but generally small, differences were noted in milk protein and lactose concentrations and yields with increasing DCAD4 in early and mid-lactation. The DMI and milk yield responses in early and mid-lactation to increasing DCAD4 are not surprising. Also, the lowest DCAD4 in both stages was quite low (5.5 [early] or 14.0 [mid] meq/ 100 g) and would be expected to decrease performance based on results of Tucker *et al.* (1988) and Sanchez *et al.* (1994a,b).

Although lactational responses to increasing DCAD4 were noted in early and mid-lactation in the above experiment, overall DMI and milk yield of cows were low compared with most modern Holstein cows. Thus, the stage of lactation data and differences in this experiment do not provide much useful insight as to whether or not early lactation cows (generally presumed to have higher milk yields, nutrient demands, and metabolic challenges) might benefit from higher DCAD4.

Roche *et al.* (2003 a,b) working in Australia and New Zealand studied DCAD for early lactation dairy cows in pasture-based systems. In their first experiment, cows were fed individually a ration of 5 kg of dry rolled barley plus ad libitum pasture forage (cut-and-carried for the experiment); a typical ration for early lactation cow in south-eastern Australia. The DCAD4 was varied by drenching individual cows twice-daily after milking with appropriate amounts of magnesium sulphate, magnesium chloride, and (or) sodium bicarbonate (Roche *et al.*, 2003b). The final DCAD concentrations of the experimental treatments were +21, +52, +102, and +127 meq/100 g dietary DM. Five cows received each treatment. As DCAD4 increased above +21 meq/100 g, DMI tended to decline ($P < 0.1$) and average daily body-weight change, and milk protein production declined ($P < 0.05$). However, concentrations of milk fat, protein, or lactose were unaffected by DCAD. There was a non-significant trend (25.4, 24.6, 24.7, 23.2 kg/cow per day) for reduction in milk yield as DCAD4 increased from +21, +52, +102, and +127 meq/100 g DM, respectively. Milk protein yield declined nearly 20% as DCAD4 increased from +21 to +127 meq/100 g dietary DM.

In their most recent experiment, Roche *et al.* (2003a) evaluated lactational performance of early lactation cows in a pasture-based system in New Zealand, where DCAD might range from 0 to +100 meq/100 g DM depending upon the particular pasture and fertilization scheme. Holstein-Friesian cows (n = 36) were grazed together and forage intake was estimated for individual cows. Average basal concentrations of K, Na, Cl, S (g/kg DM) and DCAD4 were 37.4, 3, 11, 3.6, and +55 meq/100 g forage DM during the 5-week study. One of four experimental treatments was delivered twice-daily by drench to individual cows randomly assigned to receive supplements containing varying amounts of sodium bicarbonate and calcium chloride to alter DCAD4. The actual final DCAD4 treatments (from pasture intake plus drench) were +23, +45, +70, and +88 meq/100 g dietary DM). Dry matter intake of pasture (overall average = 17 kg/cow per day), yield of milk (overall average = 26 kg/cow per day), yield and concentrations of milk protein and lactose, body-weight gain, and BCS change were not affected by increasing DCAD4. There were significant linear increases in milk fat concentration (39.6 to 42.2 g/kg) and fat yield (10% increase overall) with increasing DCAD. Systemic acid-base status was affected, as reflected by increases in blood pH, bicarbonate, base excess, and urine pH, as DCAD increased. The authors concluded that overall lactational performance of early lactation cows was not affected over this wide range of DCAD4 in this pasture-based system. There was no suggestion from this study that increasing DCAD4 above +23 meq/100g of dietary DM was beneficial to overall lactational performance, except for the slight rise in milk fat concentration and fat yield with increasing DCAD4. However, the lack of response to higher DCAD concentrations in these studies also could have been due to the fact that the treatments were delivered via a drench rather than the diet.

Wildman *et al.* (2004) studied the effects of DCAD and crude protein in early lactation (47 DIM) and reported beneficial responses to increasing DCAD. Eight early lactation Holstein cows were used in a replicated 4 x 4 Latin square design in which two DCAD3 treatments (+25 and +50 meq/100 g dietary DM; S was not reported) were factored with two concentrations of crude protein (150 and 170 g/kg). Increasing DCAD3 from +25 to +50 increased DMI (17.8 versus 19.1 kg/day), 3.5% fat-corrected milk (FCM) yield (20.2 versus 23.7 kg/day), fat yield (0.6 versus 0.8 kg/day), and energy-corrected milk yield (20.5 versus 23.8 kg/day). Higher DCAD also increased blood K, bicarbonate, and pH.

Another report with high producing cows (averaging greater than 36 kg milk/day) in early lactation was presented by Mooney and Allen (2002). Raising DCAD with either $NaHCO_3$ or $KHCO_3$ increased milk yield by 0.6 kg/day, milk fat by 2.2 g/kg, milk lactose by 0.7 g/kg, and 4% FCM yield by 1.3 kg/day, and increased efficiency of FCM yield compared with NaCl and KCl treatments (Table 12.3). Their research also showed that the Cl treatments reduced time spent eating compared with HCO_3 treatments.

Table 12.3 Effect of $NaHCO_3$ and $KHCO_3$ (HCO_3) versus NaCl and KCl (Cl) on intake, milk production and feed efficiency.

	Treatment		
	HCO_3	Cl	P
Milk, kg/day	37.3	36.7	*
DMI, kg/day	28.0	27.9	
Milk fat, g/kg	38.9	37.7	**
4% FCM, kg/day	36.7	35.4	*
FCM/DMI	1.31	1.27	**

* P < 0.05; ** P < 0.01

Adapted from Mooney and Allen, 2002

Source of cation. DCAD can be raised by increasing Na and (or) K as the mineral cation source, or by reducing the amounts of anions present in the formulation. The question of whether the source of the cation makes any difference has not been investigated thoroughly for cows at any stage of lactation. The studies of Tucker *et al.* (1988) and West *et al.*, (1991) suggested that increasing either K or Na concentrations (typically by adding the bicarbonate salt of Na or carbonate salt of K), either of which changes DCAD, resulted in similar lactational responses.

There are results to suggest that a specific cation effect exists in early lactation cows. A study conducted in Israel during winter (Silanikove *et al.*, 1997) with high producing early lactation cows (yielding greater than 39 kg milk/day) showed that cows were losing K (Table 12.4) but retaining Cl at 2 weeks into lactation. These researchers conducted total collection trials with three pairs of multiparous cows fed 3 g Na, 9.5 g K, 5.7 g Cl per kg DM (S was not reported). Results of that study showed that cows 2 weeks in milk were losing 925 meq/day of K and retaining 171 meq of Cl (responses were significantly different from zero). Delaquis and Block (1995) also reported a reduction in K balance (120 versus 126 g/day) for cows fed +5.6 DCAD4 versus +25.8 meq/100 g dietary DM. More information is certainly needed on the specific effects of dietary K and Cl in very early lactation.

Table 12.4 Apparent mineral element retention by Holstein cows at 2 weeks in lactation. Values are in meq/day.

	Intake			Output				Apparent
Mineral	*Feed*	*Water*	*Total*	*Milk*	*Urine*	*Faeces*	*Total*	*retention*
Na	1786	262	2048	696	1015	458	2169	-122
K	3219	0	3219	1746	1796	611	4153	-925*
Cl	2620	484	3104	1252	1056	625	2933	171*

* Significantly different from 0; $P < 0.05$.
Adapted from Silanikove et al.,1997

Effect of supplemental K on Mg absorption

There is a potential negative effect of excess dietary K on Mg absorption. Researchers from the Netherlands (Schonewille *et al.*, 1999; Jittakhot *et al.*, 2004) demonstrated the dependency that K has on dietary Mg. In their first study they showed that Mg transport across the ruminal epithelium was depressed when either intrinsically high dietary K concentrations or supplemental K was fed. In the second study, supplemental K (76 versus 21 g/day) as $KHCO_3$ decreased absorption and urinary excretion of Mg. Feeding more Mg (69 versus 41 g/day) as MgO counteracted the suppressant effect of K on Mg absorption in cows and increased absorption of Mg. When feeding higher amounts of dietary K (i.e., above NRC, 2001 requirements) raising dietary Mg might be efficacious.

Summary

The body of research on the effects of DCAD on transition cows provides valuable information for practicing nutritionists as well as directions for additional research. Formulating diets with different concentrations of DCAD has become a popular and effective tool for both pre- and postpartum dairy cows. New information is available on the effect of DCAD concentration on dry matter intake, blood Ca and urine pH. Apparent differences in absorption coefficients and metabolic effects between the anions Cl and S and the cations K and Na have been addressed and concentrations of dietary Ca have been evaluated. This research also has made us aware of the potential detrimental effects of excess anions for primiparous cows and excess K on Mg absorption. New information also exists on the physiological mechanisms that demonstrate improved sensitivity of PTH function in cows fed reduced DCAD in late pregnancy. Research on DCAD for the postpartum cow is much more limited, but there are indications that very low DCAD3 (less than 20 meq/100 g dietary DM) can affect production and that individual ions might have specific effects related to deficiencies of K and retention of Cl ions.

More research certainly is warranted in this area of nutrition. For prepartum cows, long-term studies with differing concentrations of Ca and DCAD are needed to determine optimal concentrations. Research on the direct cause of acid-base effects on PTH sensitivity will allow more understanding and potential new and better approaches to controlling hypocalcaemia. For lactating cows more research is needed with early lactation and high producing cows, particularly those in the first 50 days of lactation. Additional research on specific ion effects with early lactation cows might also prove valuable. For both prepartum and postpartum transition animals, new information is needed on the absorbability of the divalent macromineral elements. As the genetic potential for higher milk yield continues, careful attention to the transition diet will become even more critical. Despite the relative lack of information on this topic especially for early lactation, high yielding cows, there are some valuable new practical findings that dairy nutritionists can use to fine-tune one more components of the diet and improve the overall health, production, and profitability of dairy cows that continue to be challenged by the rigors of the transition period.

References

Beede, D.K. (2003). Optimum dietary cation-anion difference for lactating cows. *Proc. Four State Dairy Management Conference*, LaCrosse, WI, p.109-122.

Beede, D. K., Risco, C.A., Donovan, G.A., Wang, C., Archibald, L.G., and Sanchez, W.K. (1992). Nutritional management of the late pregnant dry cow with particular reference to dietary cation-anion difference and calcium supplementation. *Proc. 24th Ann. Convention Am. Assoc. Bovine Practitioners*, Orlando, FL, p. 51.

Beede, D.K. (1995). Macromineral element nutrition for the transition cow: Practical implications and strategies. *Proc. Tri-State Nutrition Conf.*, Ft. Wayne, IN, p. 175.

Beede D.K., Pilbeam T.E., Puffenbarger S.M., and Tempelman R.J. (2001). Peripartum responses of Holstein cows and heifers fed graded concentrations of calcium (calcium carbonate) and anion (chloride) three weeks before calving. *J Dairy Sci.*, **84**(Suppl 1), 83.

Block, E. (1984). Manipulating dietary anions and cations for prepartum dairy cows to reduce incidence of milk fever. *J. Dairy Sci.*, **67**, 2939-2948.

Byers, D. I. (1992). Formulating anionic dry cow rations (Practice Tips). *Proc. 24th Ann. Convention Am. Assoc. Bovine Pract.* p. 149.

Chan, P.S., West, J.W., Bernard, J.K. (2001). Metabolic measures around parturition for late gestation cows supplemented with moderate and high dietary calcium during hot weather. *J. Dairy Sci.*, **84**(Suppl. 1), 83.

Charbonneau, E., Pellerin, D. and Oetzel, G.R. (2000). Effect of lowered prepartum DCAD on urinary pH. A meta-analysis. *J. Dairy Sci.*, **87**(Suppl. 1), 441.

Curtis, C.R., Erb, H.N., Sniffen, C.J., and Smith, R.D. (1984). Epidemiology of parturient paresis: predisposing factors with emphasis on dry cow feeding and management. *J. Dairy Sci.*, **67**, 817-25.

Delaquis A.M. and Block, E. (1995). Dietary cation-anion difference, acid-base status, mineral metabolism, renal function, and milk production of lactating cows. *J. Dairy Sci.*, **78**, 2259-2284.

Ender, F., and Dishington, I.W. (1970). Etiology and prevention of parturient paresis puerperalis in dairy cows. Page 71 in *Parturient Hypocalcemia*. J. J. B. Anderson, ed. Academic Press, New York.

Giesy, J.G., Sanchez, W.K., McGuire, M.A., Higgins, J.J., Griffel, L.A. and Guy, M.A. (1997). Quantifying the relationship of dietary cation-anion difference to blood calcium in cows during hypocalcemia. *J. Dairy Sci.*, **80**(Suppl. 1), 142.

Goff, J.P. (2000). Pathophysiology of calcium and phosphorus disorders. *Vet.*

Clin. N. America: Food Animal Practice. **16**, 319-337.

Goff, J. P., Horst, R.L., Mueller, F.J., Miller, J.K., Kiess, G.A. and Dowlen, H.H. (1991). Addition of chloride to a repartal diet high in cations increases 1, 25-Dihydroxyvitamin D response to hypocalcemia preventing milk fever. *J. Dairy Sci.*, **74**, 3863-3871.

Goff, J.P., and Horst, R.L. (1998). Effect of time after feeding on urine pH determinations to assess response to dietary cation-anion adjustment. *J. Dairy Sci.* **81**(Suppl. 1), 44.

Goff, J.P. and Horst, R.L. (2003) Milk fever control in the United States. *Acta Vet Scand* Suppl. **97**, 145-7.

Goff, J. P., Ruiz, R., and Horst, R.L. (2004). Relative acidifying activity of anionic salts commonly used to prevent milk fever. *J. Dairy Sci.*, **87**, 1245-1255.

Guard, C. (1996). Fresh cow problems are costly: culling hurts the most. Page 100 in *Proc. 1994 Ann. Conf. Vet. Cornell Univ.*, Ithaca, NY.

Jittakhot, S., Schonewille, J.T., Wouterse, H., Yuangklang, C. and Beynen, A.C. (2004). Apparent magnesium absorption in dry cows fed at 3 levels of potassium and 2 levels of magnesium intake. *J. Dairy Sci.*, **87**, 379-385.

Jonsson, G., and Pehrson, B. (1970). Trials with prophylactic treatment of parturient paresis. *Vet. Record.* **87**, 575.

Jordan, E.R., and Fourdraine, R.H. (1993). Characterization of the management practices of the top milk producing herds in the country. *J. Dairy Sci.* **76**, 3247-3256.

Joyce, P.W., Sanchez, W.K., and Goff, J.P. (1997). Effect of anionic salts in prepartum diets based on alfalfa. *J. Dairy Sci.* **80**, 2866-2875.

Leach, R.M. 1979. Dietary electrolytes: Story with many facets. *Feedstuffs.* April 30, p. 27.

Massey, C.D., Wang, C., Donovan, G.A., Beede, D.K. (1993). Hypocalcemia at parturition as a risk factor for left displaced abomasums in dairy cows. *J. Am Vet Med Assoc.*, **15**, 203(6)852-3.

Mongin, P. (1980). Electrolytes in nutrition: review of basic principles and practical application in poultry and swine. In *Third Ann. Int. Mineral Conf.* Orlando, FL. p.1.

Mooney, C.S. and Allen, M.S. (2002). Effect of dietary strong ions on milk yield, milk composition, and chewing activity in lactating dairy cows. *J. Dairy Sci.*, **85**, Suppl. 1, 109. Abstract.

Moore, S.J., VandeHaar, M.J., Sharma, B.K., Pilbeam, T.E., Beede, D.K., Bucholtz, H.F., Liesman, J.S., Horst, R.L., and Goff, J.P. (1997). Varying dietary cation anion difference (DCAD) for dairy cattle before calving. *J. Dairy Sci.* **80**(Suppl. 1), 170.

Moore, S.J., VandeHaar, M.J., Sharma, B.K., Pilbeam, T.E., Beede, D.K., Bucholtz, H.F., Liesman, J.S., Horst, R.L., and Goff, J.P. (2000). Effects of altering dietary cation-anion difference on calcium and energy metabolism in peripartum cows. *J. Dairy Sci.*, **83**, 2095-2104.

NRC (National Research Council). (1989). *Nutrient Requirements of Dairy Cattle.* 6th Revised Ed., National Academy Press, Washington, DC.

NRC (National Research Council). (2001). *Nutrient Requirements of Dairy Cattle.* 7th Revised Ed., National Academy Press, Washington, DC.

Oetzel, G.R., and Barmore, J.A. (1992). Palatability of anionic salts fed in a concentrate mix. *J. Dairy Sci.*, **75**(Suppl. 1), 297.

Oetzel, G.R., Olson, J.D., Curtis, C.R., Curtis, M.J., and Fettman, M.J. (1988). Ammonium chloride and ammonium sulfate for prevention of parturient paresis in dairy cows. *J. Dairy Sci.*, **71**, 3302-3309.

Oetzel, G.R. (2000). Management of milk fever and mineral disorders. *Vet. Clin. N. America: Food Animal Practice.* **16**, 369-377.

Roche, J.R., Petch, S., and Kay, J.K. (2003a). Effect of dietary cation-anion difference on the milk production of early lactation cows. *J. Dairy Sci.*, **86,** (Suppl. 1).

Roche, J.R., Dalley, D., Moate, P., Grainger, C., Rath, C.M., and O'Mara, F. (2003b). Dietary cation-anion difference and the health and production of pasture-fed dairy cows. 1. Dairy cows in early lactation. *J. Dairy Sci.*, **86**, 970-978.

Rodriguez, L.A., Pilbeam, T.E., and Beede, D.K. (1997). Effects of dietary anion source and calcium concentration on dry matter intake and acid-base status of nonlactating nonpregnant Holstein cows. *J. Dairy Sci.*, **80**(Suppl. 1), 241.

Rodriguez, L.A., Pilbeam, T.E., Ashley, R.W., Neudeck, S.L., Templeman, R.J., Davidson, J.A., and Beede, D.K. (1996). Concurrently lowering dietary cation-anion difference while raising calcium content prepartum reduces urine pH and peripartum subclinical hypocalcemia. *J. Dairy Sci.*, **79**, (Suppl. 1), 197.

Sanchez, W.K. and Beede, D.K. (1991). Interrelationships of dietary Na,K and Cl and cation-anion difference in lactation rations. *Proc. Florida Rum. Nutr. Conf.* University of Florida, Gainesville. p.31.

Sanchez, W.K., Beede, D.K., and Cornell, J.A. (1994a). Interactions of sodium, potassium, and chloride on lactation, acid-base status, and mineral concentrations. *J. Dairy Sci.*, **77**, 1661-1675.

Sanchez W.K., Beede D.K., Delorenzo M.A. (1994b). Macromineral element interrelationships and lactational performance: empirical models from a large data set. *J Dairy Sci.*, **77**, 3096-3110.

Sanchez, W.K. (1999). Another new look at DCAD for the postpartum dairy cow. *Proc. 1999 Mid-South Ruminant Nutrition Conf.* p.1.

Silanikove, N. E. Maltz, A. Halevi, and Shinder, D. (1997). Metabolism of water, sodium, potassium and chlorine by high yielding dairy cows at the onset of lactation. *J. Dairy Sci.*, **80,** 949-956.

Shohl, A.T. (1939). *Mineral Metabolism*. Reinhold Publishing Corp, New York.

Shohl, A.T. and Sato, A. (1922). Acid-base metabolism. I. Determination of base balance. *J. Biol. Chem.* **58**, 235.

Schonewille, J.T., Wouterse, H., Yuangklang, C., and Beynen, A.C. (1999). Effects of intrinsic potassium in artificially dried grass and supplemental potassium bicarbonate on apparent magnesium absorption in dry cows. *J. Dairy Sci.*, **82**, 1824-1830.

Tucker, W.B., Harrison, G.A., and Hemken, R.W. (1988). Influence of dietary cation-anion balance on milk, blood, urine, and rumen fluid in lactating dairy cattle. *J. Dairy Sci.*, **71**, 346-354.

Wang, C. (1990). *Influence of dietary factors on calcium metabolism and incidence of parturient paresis in dairy cows*. Ph.D. Diss., Univ. Florida, Gainesville.

West, J.W., Mullinix, B.G., and Sandifer, T.G. (1991). Changing dietary electrolyte balance for dairy cows in cool and hot environments. *J. Dairy Sci.*, **74**, 1662-1674.

West, J.W., Haydon, K.D., Mullinix, B.G., and Sandifer, T.G. (1992). Dietary cation-anion balance and cation source effects on production and acid-base status of heat-stress cows. *J. Dairy Sci.*, **75**, 2776-2786.

Wildman, C.D., West, J.W., and Bernard. J.K. (2004). Effect of dietary cation-anion difference and protein on milk yield and blood metabolites of lactating dairy cows. *J. Dairy Sci.*, **87**, (Suppl. 1), 270.

Wu, H. and Murphy, M.R. (2004). Dietary cation-anion difference effects on performance of and acid-base status of lactating dairy cows: A meta analysis. *J. Dairy Sci.*, **87**, 2222-2229.

13

METABOLIZABLE METHIONINE OPTIMIZATION OF DAIRY COW RATIONS

J.C. ROBERT
ADISSEO France SAS, 42 Avenue Aristide Briand, 92160 Antony

From crude protein to metabolizable amino acid requirements

PROTEIN

Before the 1980s, crude protein or apparent digestible protein were used for calculating ruminant   rations. With increasing genetic merit of animals and increasing productivity, particularly for dairy cows in the 1980s, a new approach was proposed for estimating requirements and supplies of dietary protein taking into account the digestive physiology of ruminants. Metabolizable protein (MP) was used as common basic concept, defined as the true protein being digested post ruminally and the component amino acids absorbed in the small intestine. Microbial protein produced in the rumen, ruminally undegraded feed crude protein (RUP) and endogenous crude protein provide the supply of MP to the intestine. Therefore, systems such as the French PDI system (Vérité and Peyraud, 1989) were proposed to account for protein supplies and requirements in terms of Digestible (Metabolizable) Protein.

AMINO ACIDS

Absorbed amino acids and not protein per se are the required nutrients. Amino acids (AA) are provided by ruminally synthesised microbial protein, RUP and endogenous secretions. Proteins from these sources must be digested in the small intestine to release AA for absorption. They are the building blocks for synthesis of tissue and milk protein. Of approximately twenty AA found in animal tissue and milk proteins, 10 are considered to be "essential".

223

Essential AA (EAA), unlike "non essential" (NEAA), either cannot be synthesised by animal tissues or if they can, not in amounts sufficient to meet metabolic needs. When addressing AA nutrition of ruminants, quantitative and qualitative components should be treated as a whole and not as alternatives. Adequate supplies of all AA, EAA and NEAA have to be supplied at the metabolic level. This is achieved by using MP systems such as PDI. The qualitative supply concerns EAA as each EAA is required in different amounts and an ideal profile of absorbed EAA exists for maintenance, growth and lactation. This concept has been documented in swine (Henry, 1993; NRC, 1998) and poultry (Leclercq, 1996; NRC, 1994). These ideal EAA profiles remain to be established for dairy cattle, but it is known that two factors account for most of the variation in AA profiles of duodenal protein. These are the proportional contribution that RUP makes to total protein passage and the AA composition of that RUP (Rulquin and Vérité, 1993; Rulquin, Guinard and Vérité, 1998; NRC, 2001). When EAA are absorbed in the correct profile, their efficiency of use for protein synthesis is maximized and, on the contrary, this efficiency is less than maximum if the EAA profile is less than ideal. In this case, the EAA in shortest supply relative to requirements, the first limiting AA, will determine the extent of protein synthesis. Effectively, considering intestinal AA profiles when formulating diets may allow a higher level of animal productivity, particularly in terms of milk protein synthesis, and provides the opportunity to reduce the amount of RUP theoretically needed.

LIMITING AMINO ACIDS FOR DAIRY CATTLE

Limiting AA are those which are in shortest supply relative to requirements. Methionine has been identified as first limiting, but lysine and more recently, possibly, histidine have also been suggested. Schwab, Satter and Clay (1976), using post ruminal AA infusions, demonstrated that methionine and lysine were likely to be the two most limiting AA in dairy rations for milk protein secretion. Many subsequent studies, using post ruminal infusions of individual AA or ruminally protected methionine and lysine, have shown that in the vast majority of rations, methionine and lysine are the first two limiting AA (Rulquin and Vérité, 1993). As might expected, the extent and sequence of limitation appear to be affected primarily by the amount of by-pass protein (RUP) in the diet and its AA composition. Methionine has been shown to be first limiting for milk protein production when dairy cattle were fed high forage or soyabean hull-based diets and intake of RUP was low. Methionine has also been identified as first limiting for dairy cows fed a variety of diets in which most of the supplemental RUP was provided by soyabean protein, animal derived proteins or a combination of the two. When soyabean meal made the major contribution to by-pass protein, methionine was clearly the first limiting AA (Armentano, Bertics and Ducharme, 1993; Robert, Sloan and Denis, 1996; Robert, Sloan and Lahaye, 1995; Rulquin and Delaby, 1994) and there is no further improvement in milk protein content with additional metabolizable lysine supply (Sloan, 1993). Lysine has been identified clearly as first limiting for milk protein synthesis when maize protein was the major constituent of by-pass protein supply (Rulquin, Le Hénaff and Vérité , 1990; Robert, Sloan and Nozière, 1996). Methionine and lysine have been identified as co-limiting when cows were fed maize silage based diets with little protein supplementation (NRC, 2001). Co-limitation may also exist with rations composed of maize grain as the only energy source and soyabean meal as the main protein source (Koch, Whitehouse, Garthwaithe, Wasserstrom and Schwab, 1996). More recently, histidine has been shown to be more limiting than lysine or methionine when cows are fed rations based on grass silage plus barley and oats, with or without feather meal as the sole source of RUP (Kim, Choung and Chamberlain, 1999; Vanhatalo, Huhtanen, Toivonen and Varvikko, 1999; Korhohen, Vanhatalo, Varvikko and Huhtanen, 2000), but its role as limiting milk protein synthesis is questionable as it has been mainly seen to increase milk yield and its precise requirement has not been established (Rulquin, Vérité, Guinard-Flament and Pisulewski, 2001).

AMINO ACID REQUIREMENTS OF LACTATING DAIRY COWS

Methionine and lysine requirements

A fundamental progress was accomplished by INRA in France from 1993 to 1998 using initially the framework of PDI system to evolve to a metabolizable (digestible) AA system (Rulquin, Guinard and Verité, 1998; Rulquin and Vérité, 1993; Rulquin, Pisulewski, Verité and Guinard, 1993). This system is a tool for formulating the dairy cow ration in terms of metabolizable methionine and lysine.

Digestible (metabolizable) AA, especially digestible methionine (MetDI) and digestible lysine (LysDI) values for raw materials, have been estimated using a systematic methodological approach (Rulquin, Guinard and Vérité, 1998). MetDI and LysDI requirements have been calculated by establishing exponential representations of milk protein yield and content responses to increasing doses of MetDI and LysDI obtained from literature (163 diets) (Rulquin, Pisulewski *et al.*, 1993). The INRA system is based on the ideal protein concept, and supplies and requirements have been expressed in terms of profiles, as a proportion of digestible (metabolizable) protein allowed by the energy supplied to the animal (PDIE). The exponential relationships established for methionine have been obtained with variable MetDI and constant LysDI and they are only valid for LysDI values greater than 6.5%. For lower LysDI concentrations there is no response to increasing supply of MetDI (Rulquin, Pisulewski *et al.*, 1993). The MetDI and LysDI requirements have been established respectively at 2.5 and 7.3 % of PDIE by these authors.

NRC (2001) proposed requirements for methionine and lysine in MP for lactating dairy cows. The methodology used by NRC was the "indirect" dose response approach described by Rulquin, Pisulewski *et al.* (1993). Milk protein secretion responses to increased doses of MetDI were used. The database included data from 27 trials and 87 observations, methionine being supplied by abomasum or duodenum continuous infusions or by feeding ruminally protected forms. The final regression analysis was limited to the trials where LysDI was 6.50 minimum. Several models were tested and the rectilinear one was accepted as the final one. The requirement estimated as the value at the breakpoint was 2.4 MetDI for maximal yield and content of milk protein. The same methodology applied to lysine led to a requirement of 7.2 LysDI with a limit value of 1.95 for MetDI. These requirement values, 2.4 for MetDI and 7.2 for LysDI, proposed by NRC (2001) are similar to those obtained by Rulquin, Pisulewski *et al.* (1993).

The Cornell Net Carbohydrate and Protein System (CNCPS) for evaluating cattle diets, and its associated AA sub-model, is the most dynamic of the factorial models described to date (O'Connor, Sniffen, Fox and Chalupa, 1993). The EAA requirements of Holstein cows have been estimated by Schwab (1996) using the CNCPS. Requirements were expressed on the basis of both daily amounts (g/day) and as profiles (each EAA as a % of total EAA). The authors estimated requirements as 5.2 % of total EAA for methionine and 16.3 for lysine, which roughly correspond to 2.4-2.5 for MetDI and 7.7-7.9 for LysDI.

Schwab (1996) reviewed the results of specific dose response studies (direct dose-response approach) and estimated requirements (% of total EAA) of 5.3 for methionine and 15 for lysine. As indicated by Sloan (1997), a precise estimate of MetDI requirements has been much more difficult to establish than LysDI requirements using this method. In some trials the marginal response to additional MetDI remained positive and linear between 1.5 and 2.4 MetDI, with a LysDI supply of 7.3 (Pisulewski, Rulquin, Peyraud and Vérité, 1996; Socha, Schwab, Putnam, Whitehouse, Kierstead and Garthwaite, 1994a, b). In another trial (Socha *et al.*, 1994c), no response to increased MetDI was found. Where LysDI was only 6.5, no significant increase in milk protein secretion was found with additional MetDI, whereas simultaneously increasing LysDI and MetDI from 6.50/1.80 to a minimum of 6.80/2.15, increased milk protein secretion by 37g/head/day (Sloan, Robert and Lavedrine, 1994).

Recently, Lapierre, Pacheco, Berthiaume, Ouellet, Schwab, Holtrop and Lobley (2004) found discrepancies between AA duodenal flux estimated by NRC (2001) and measured portal flux. New equations for estimating AA flows at the duodenum were proposed. The authors suggested that variable rather than fixed factors for transfer efficiencies should be incorporated into future predictive models. Hanigan (2004) also suggested that our current feeding systems could be improved by inclusion of a more accurate consideration of the metabolic complexities of the splanchnic tissues.

However the values of 2.5% and 7.3% of PDIE proposed by Rulquin, Pisulewski et al. (1993) currently appear to be good estimates of MetDI and LysDI requirements respectively. For application to formulating for digestible amino acids for dairy cows, a reasonable compromise between animal performance optimization and cost has led to practical recommendations of 2.2 and 7.0 % of PDIE, respectively for MetDI and LysDI requirements.

How to balance dairy cow rations in terms of metabolizable methionine

FEW RAW MATERIALS ARE GOOD SOURCES OF MetDI

Using the Metabolizable AA system proposed by INRA (Rulquin, Pisulewski *et al.*, 1993; Rulquin *et al.*, 1998) MetDI and LysDI supplies can be calculated from the composition of the ration and compared to the needs of dairy cows. For that purpose INRA published databases for all commonly used raw materials (Rulquin, Guinard, Vérité and Delaby, 1993; Sauvant, Perez and Tran, 2003).

Few raw materials are good sources of MetDI and in practical formulation it is impossible to meet both MetDI and LysDI constraints without using synthetic amino acid products: for example maize gluten meal is rich in MetDI (2.5 % PDIE), but very poor in LysDI (3.7 % PDIE); fish meal is rich in MetDI and LysDI (2.8 and 7.8 respectively), but animal protein is currently banned in Europe. These extreme values are linked to the low rumen degradability of protein in the raw material and to its specific AA profile; in this case, the by-pass contribution will also be important, linked to the protein content of the raw material. As an example, maize grain is an exception among cereals: its N-rumen degradability is very low (0.4) and its metabolizable AA contribution will be high with high rates of maize inclusion in the diet. In this case, maize contributes to a LysDI deficit (5.8 % PDIE) with a MetDI (2.0 % PDIE) lower than requirements. These values will be exacerbated in the case of maize gluten meal with a higher protein content and a lower N-rumen degradability (0.28). This is not the case for maize gluten feed, where N-rumen degradability is high (0.7). Forages have high rumen degradability so their metabolizable AA profiles are very close to that of rumen microorganisms, with relatively low values for MetDI (1.9 to 2.0 % PDIE) and high LysDI values (6.8 to 7.0). The raw materials that can most influence metabolizable AA profile are concentrated sources of by-pass proteins. Soyabean meal is a balanced source of LysDI (7.0 % PDIE), but other protein meals of vegetable origin tend to be low in both MetDI and LysDI (Table 13.1).

Table 13.1 Potential concentration of LysDI and MetDI in commonly used raw materials for dairy cow rations.

	g/kg DM		*g/100 g PDIE*			
	LYSDI	*METDI*	*LYSDI*	*METDI*		
Grass silage	**4.8**	**1.3**	**7.04**	**1.96**		all forage sources have
Maize silage	**4.7**	**1.3**	**6.92**	**1.98**	→	similar LysDi and
Alfalfa	**4.9**	**1.2**	**6.94**	**1.74**		MetDi values
Sugar beet pulp	8.2	2.0	7.78	1.89		
Wheat	7.2	2.1	6.58	1.93		
Maize grain	**7.0**	**2.4**	**5.83**	**2.01**	→	all maize grain and
Maize gluten feed	**7.4**	**2.4**	**5.95**	**1.95**		byproducts are poor in
Maize gluten meal	**19.1**	**12.7**	**3.71**	**2.48**		LysDi
Soyabean meal	16.9	3.6	7.02	1.51		
Rapeseed meal	10.5	3.1	6.83	2.00		
Bloodmeal	**40.0**	**5.2**	**8.04**	**1.04**	→	a rich source of Lysine
Feathermeal	**17.8**	**4.3**	**3.08**	**0.74**	→	poor source of lys & met
Fishmeal	**32.2**	**11.4**	**7.82**	**2.76**	→	good but expensive source of lys & met

Many European rations are deficient in MetDI, but not in LysDI, particularly if maize silage and soyabean meal are used. To reach the recommended MetDI and LysDI levels in practical formulation, a combination of raw materials can be used for LysDI supply and good quality protected methionine for MetDI supply.

METHIONINE SUPPLEMENTATION OF THE RATION

Technologies proposed

Balancing MetDI deficient rations consists of supplementing with protected methionine. Different technologies have been proposed:

- surface coating with a fatty acid / pH sensitive polymer mixture (Smartamine ™, SmM- Adisseo)

- surface coating or matrices involving fat or saturated fatty acids, ethyl cellulose and minerals (Mepron® M85-Degussa)

- liquid source of the hydroxy analogue of methionine (DL, 2-hydroxy-4 (methylthio) butanoic acid, HMB) (Rhodimet ™ AT88-Adisseo) (Alimet®-Novus)

- and more recently the isopropyl ester of HMB (HMBi, Metasmart ™-Adisseo).

Why we need accurate methionine bioavailability values of protected products

These products will be used to balance the ration in terms of metabolizable methionine, the objective being to supply extra methionine to the mammary gland for increasing casein synthesis. To determine what dose of protected product to supply in rationing, and to include it at an economic level, a precise methionine bioavailability value is needed. Methionine bioavailability is defined as the proportion of methionine in the product that will be absorbed through the gut wall into the bloodstream of an animal.

How to measure methionine bioavailability of protected products

Different methods to estimate methionine bioavailability from rumen protected methionine have been proposed:

- in vitro tests: - pH lab tests apply to pH sensitive coating
  - rumen simulation techniques, only measure rumen resistance

- in situ tests: - rumen nylon bag
  - mobile nylon bag, allowing measurement of both rumen resistance and apparent digestibility

- digestive tract flow measurements: duodenal, ileal, faecal.

These methods only provide, in the best case, a value for potential to be absorbed; some, e.g. digestive tract flow studies, are cumbersome and costly.

Blood tests have been developed to determine a more complete evaluation of bioavailability:

- the blood tests proposed can be qualitative, measuring changes in blood methionine concentration after protected methionine supply for 5 days (Blum, Bruckmaier and Jans, 1999; Sudekum, Wolfram, Ader and Robert, 2004).

- Rulquin and Kowalczyk (2001; 2003) proposed a quantitative blood method using methionine absorbed as a function of plasma methionine concentration. This relationship was established through measurement of changes in plasma methionine concentration after infusing increasing methionine doses into the duodenum for 4 days (saturation method).

- Blood kinetics tests have been proposed as quantitative methods by determining the area under the curve (AUC) described by blood methionine concentrations after a spot dose supply of a fixed quantity of methionine equivalent to the protected product. The AUC is compared to the AUC of a reference product with known methionine bioavailability (SmM). In this case, the relationship between AUC and bioavailability is assumed to be linear (Robert, Sloan, Etave and Bouza, 2001).

A more elaborate quantitative blood kinetics method was proposed (Robert, Etave, d'Alfonso and Bouza, 2001), using the relationship between spot digestible methionine supply and plasma methionine response in terms of AUC. The exponential relationship found was used to establish a reverse calibration curve which allows prediction of the amount of metabolizable methionine delivered by a rumen protected methionine according to its measured AUC value. Bioavailability was calculated as the ratio of metabolizable methionine to methionine supplied by the product (Figures 13.1 and 13.2).

Methionine bioavailabilities of different technologies

We reviewed the literature on bioavailability of methionine in the three main technologies actually available.

Consistent results were obtained by the different methods of evaluation for pH sensitive   coating (SmM) (Robert, 1992; Robert, Williams and Bouza, 1997; Robert, d'Alfonso, Etave, Depres and Bouza, 2002; Rulquin and Kowalczyk, 2003), in agreement with the value of 80 % generally used in practice.

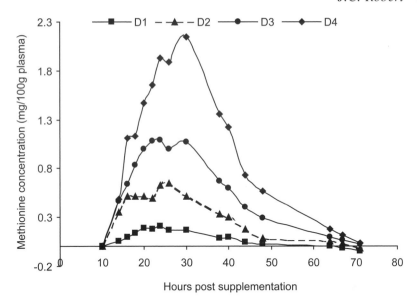

Figure 13.1 Plasma methionine concentration kinetics for increasing metabolisable methionine pulse supplies (D1 = 7; D2 = 18; D3 = 29; D4 = 39 g metabolisable methionine (from Robert *et al.* 2001)

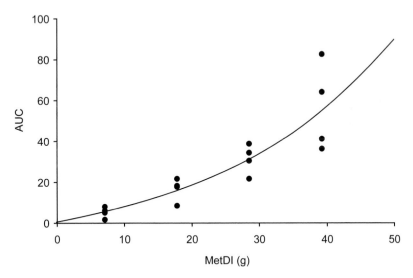

Figure 13.2 Modelling area under curve (AUC) response to increasing doses of metabolisable methionine (AUC = 15.94 (1-exp (0.038 metab.met)); SED = 9.89) (Robert *et al.*, 2001)

For Mepron® M85, contradictory results were observed: methionine bioavailability varied from 25% (Koenig and Rode, 2001) to around 50 % (Overton, LaCount, Cicela, Clark, 1996; Berthiaume, Lapierre, Stevenson, Coté and McBride, 2000) with in situ mobile-bag evaluations. A digestive tract flow experiment gave values around 50 % (Berthiaume, Dubreuil, Stevenson, McBride and Lapierre, 2001) and blood tests gave values between 20 and 35 % (Robert, Williams and Bouza, 1997; Koenig and Rode, 2001). Recently, Olley, Ordway, Whitehouse, Schwab and Sloan (2004) measured methionine bioavailabilities for Mepron® M85 of 41 based on change in plasma methionine concentration and 22 % based on methionine plus cystine concentrations.

For HMB, conflicting results were also observed by different authors using different methodologies. Measurements using in vitro rumen simulation techniques produced rumen resistance values between 22 and 43 %, depending on retention time (Vasquez-Anon, Cassidy, McCullough and Varga, 2001). Rumen by-pass evaluated in vivo was between 40 and 50 %, with 12 to 45 % measured at the duodenum (Koenig, Rode, Knight, McCullough, Andrews and Kung, 1996; Koenig, Rode, Knight, McCullough, 1999; Koenig, Rode, Knight, Vazquez-Anon, 2002). The authors suggested that differences between ruminal and duodenal measurements were due to omasum and/or abomasum absorption. Blood kinetics measurements show very low bioavailability values: around 3% (Robert, Williams and Bouza, 1997). Recently, Noftsger, St-Pierre and Firkins (2004) measured 5% of HMB by-passing the rumen in lactating dairy cows receiving a practical HMB supplementation dose included in a total mixed ration.

Possible origin of discrepancies

With all these results, some discrepancies could be due to bias linked to the specific methodology used. For example, there is a risk of overestimating bioavailabity with in sacco tests due to not taking into account mechanical aspects of digestion or/and uncertainty about rumen residence time chosen for calculations. For digestive tract flow estimates, the mode of calculation of rumen by-pass rate can lead to different evaluations. For example, for HMB, a fractional rate constant calculation (Koenig *et al.*, 1996; 1999; 2002) generally gives results greater than those obtained with a concentration method. This latter mode of calculation leads to rumen by-pass values of 5 to 10 %, in agreement with results of blood kinetics tests (Graulet, personal communication).

A new source of metabolizable methionine: HMBi, the isopropyl ester of HMB (Metasmart™,   Adisseo)

HMB blood increase following HMBi feeding

A very fast appearance of HMB in the peripheral blood was observed after

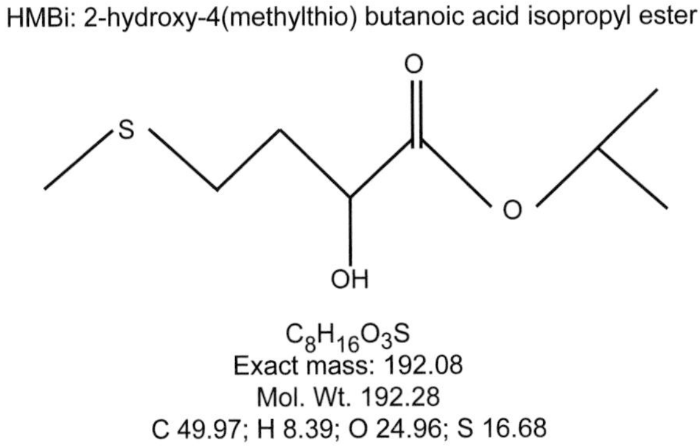

HMBi: 2-hydroxy-4(methylthio) butanoic acid isopropyl ester

$C_8H_{16}O_3S$
Exact mass: 192.08
Mol. Wt. 192.28
C 49.97; H 8.39; O 24.96; S 16.68

Figure 13.3 Molecule identification – HMBi - Metasmart.

supplying a spot dose of HMBi into the rumen of non-pregnant dry cows (Robert, Richard, d'Alfonso, Ballet and Depres, 2001). Before HMB administration, plasma concentrations were zero; 10 minutes after administration, plasma HMB increased to 1.4 mg per 100g and continued to increase progressively to reach a peak value of 3 mg/100g at 1h15 mn. Plasma HMB concentrations subsequently decreased progressively to reach a basal value 8h after HMBi administration. This observation of a very fast absorption was confirmed in several trials, with HMB maximum plasma concentrations at short times (1 to 2 h) after supply into the rumen (Robert, *et al* unpublished; Nozière, Richard, Graulet, Durand, Rémond and Robert, 2004).

Blood methionine increase following HMBi feeding

Plasma methionine concentrations peaked shortly after HMB concentrations. Methionine appeared in blood following HMB and in greater concentrations: peak at 4 hours versus 1 to 2 hours, and peak concentrations of 3.2 mg versus 1.5 mg (Robert, Richard, d'Alfonso *et al.*, 2001). The blood kinetics of HMB and methionine were different: a sharp increase and decrease for HMB and larger spread over time for methionine. It is likely that these observations correspond to HMB conversion into methionine: this conversion at the metabolic level has been widely studied in monogastrics, where HMB passes through an intermediary compound, 2-keto-4 (methylthio) butanoic acid, thereafter converting to L Methionine by transamination (Dupuis, Saunderson, Puigserver and Brachet, 1989). This conversion would mainly take place in the liver and kidneys in poultry (Saunderson, 1985), but other tissues and particularly intestinal mucous could be involved (Dupuis *et al.*, 1989). In ruminants, Belasco (1972) observed, in vitro, the ability of enzymatic systems in liver and kidney microsomes to transform HMB into methionine.

Wester, Vasquez-Anon, Parker, Dibner, Calder and Lobley (2000a and b) have shown in sheep that HMB supplied post-ruminally is largely metabolized into methionine by extra-hepatic tissues.

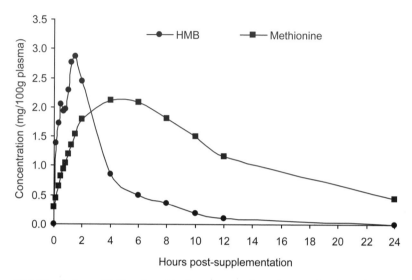

Figure 13.4 Blood plasma HMB and methionine concentrations after a pulse dose of HMBi (69g) into the rumen of cows .

HMBi Absorption mechanism

Comparisons between HMBi and SmM have been made, in terms of time for methionine to appear and to peak in peripheral blood after spot supplies into the rumen of non-pregnant dry cows. For SmM, the time for methionine appearance in the blood was 8 hours with a peak around 20 hours; for HMBi, the time for HMB and methionine to appear in peripheral blood was much shorter: 10 mn. or less with HMB and methionine peaking respectively at 1h30 mn. and 4h. (Robert, Richard and Bouza, 2001; Robert, d'Alfonso *et al.*, 2002). SmM is known to be inert in the rumen and to deliver methionine to the small intestine where it is absorbed, which explains the long time before appearance in the blood. From the previous observations, HMBi is not absorbed at the intestinal level and the most likely site of absorption is the rumen wall. This was verified using the emptied-washed rumen technique of Kristensen, Gäbel, Pierzynowsky and Danfaer (2000) on a non-pregnant dry Jersey cow receiving a spot dose of HMBi (Graulet, Richard and Robert, 2004). Rapid and important increases of HMB and methionine concentrations in peripheral blood were clearly observed. This was not the case with DL methionine or HMB. Applying the blood kinetics method (AUC) led to a bioavailability determination of 91% for the methionine equivalent of HMBi, demonstrating clearly the high potential of HMBi to be absorbed through the rumen wall in the absence of microorganisms. Rumen absorption of HMBi was also confirmed using catheterization of both ruminal and jugular veins of two rumen-functional Holstein cows; HMBi supplied as a spot dose into the rumen was only detected as traces at the ruminal vein level and not at all at the jugular vein. In contrast, large and rapid increases in HMB and methionine concentrations were observed at both sites following HMBi supplementation. Higher plasma concentrations of HMB were observed in the ruminal vein (8.5 mg/100 g), compared with the jugular vein (5.0 mg/ 100 g), confirming ruminal absorption of HMBi. The concomitant presence of HMB and isopropanol in the ruminal vein suggests that HMBi is hydrolyzed in the rumen wall. Nozière *et al.* (2004) have also observed absorption of HMBi through the rumen wall of sheep and hydrolysis to HMB and isopropanol at this site.

"Methionine bioavailability" of HMBi

"Methionine bioavailability" of HMBi was determined by the quantitative blood kinetics method (Robert, Richard and Bouza, 2001; Robert, d'Alfonso *et al.*, 2002). Values obtained were 50 % for the liquid form of HMBi and 56 % for a powder presentation (30 % of liquid HMBi on a carrier). From a survey of all measurements performed (Robert, unpublished), the mean bioavailability of the liquid presentation of HMBi was 51% (17 determinations with a standard deviation of 8). For the powder presentation the bioavailability was 55 % (8 determinations with a standard deviation of 4). Schwab, Whitehouse, McLaughlin, Kadaryia, St-Pierre, Sloan and Robert (2001), using responses in milk protein concentration to incremental doses of HMBi or SmM, obtained a bioavailability around 50%. Guyot, Robert and Sauvant (2004) performed a meta-analysis of seven production experiments analysing response in milk protein concentration to increasing HMBi, SmM and HMB doses, and concluded a 50 % bioavailability for methionine equivalent of HMBi.

HMBi in the rumen

In vitro trials have shown that the portion HMBi remaining in the rumen is hydrolysed to HMB and isopropanol (Robert, Ballet, Richard and Bouza, 2002a). This result is in agreement with those of Digenis, Amos, Mitchell, Swintowsky, Yang, Schelling and Parish (1974) and Patterson and Kung (1988) for other methionine and HMB chemical derivatives. In vitro trials have also shown that the HMB derived from HMBi is used as a substrate by rumen microorganisms and stimulates rumen fermentation (Robert, Ballet, Richard and Bouza, 2002b; Robert, Paquet, Richard and Bouza, 2003). This utilization of HMB by rumen microorganisms is widely illustrated in the literature (Schwab, 1998). The HMBi that is not absorbed through the rumen wall remains in the rumen and is used by rumen micro-organisms. This is supported by the results of Noftsger, St-Pierre and Firkins (2004), who found no HMBi in the omasum of lactating dairy cows receiving 22 g/day of HMBi incorporated into a total mixed ration, and only small amounts of HMB (2-3 % of the HMB contained in the HMBi fed).

Conclusion

The 50 % bioavailability of methionine equivalent of HMBi results from its high potential to cross the rumen wall versus the activity of rumen microorganisms to hydrolyse and use it.

Influence of metabolizable methionine optimization of dairy rations on lactation performance

Production responses to metabolizable methionine, and factors causing variation in response, were reviewed by Schwab (1989), Rulquin (1992), Sloan (1997) and NRC (2001). Most of the trials examined included either post ruminal amino acid infusion or supplementation with pH sensitive coated methionine (Smartamine™ M, SmM). Numerous trials refer to both MetDI and LysDI, but in this review we have tried to separate specific effects of MetDI. As previously mentioned, the MetDI effect will be limited by inadequate supply of LysDI (LysDI<6.5).

POST RUMINAL AMINO ACID INFUSION OR SUPPLEMENTATION WITH pH SENSITIVE COATED METHIONINE

Milk protein secretion

Balancing MetDI in dairy rations has a positive effect on milk protein secretion. The main parameters improved are milk protein concentration and milk protein yield. The consistent increases in milk protein secretion, particularly in milk protein concentration, observed as a result of improvements in MetDI supply are due to an increase in milk casein since the proportion of casein within milk crude protein or milk true protein is increased (Pisulewski *et al.*, 1996). As a consequence, cheese making properties are increased and particularly cheese yield (Hurtaud *et al.*, 1995). Long term studies show a consistent response to MetDI supply in terms of milk protein yield over the full lactation. Brunschwig and Augeard (1994) observed a 60 g/cow/day improvement in milk protein yield throughout lactation. Milk protein concentration increased by 1 g/kg in early lactation, and the response increased to 2 g/kg at the 20th week of lactation, whilst milk yield decreased due to stage of lactation. Similar results were obtained by Pabst (unpublished), Polan *et al.*(1991) and Rogers *et al.*(1989).

Milk yield

Milk yield responses to MetDI and LysDI optimization are more common in cows during early lactation than in mid or late lactation. These milk volume improvements led to more marked increases in terms of milk protein yield. Socha *et al.* (1994d) observed an increase in milk volume of up to 3.5 kg/ day in early lactation and milk protein yield was increased by an average of 80 g/cow/day.

Robert, Sloan and Bourdeau (1994) observed a large increase in milk protein yield and concentration (+ 1.6 g/kg) during the first 84 days of lactation after balancing the ration for MetDI. This was associated with an increase in milk volume of 0.5 to 1 kg/cow/day. Thiaucourt (1996), in a field trial involving 2000 cows, confirmed that balancing MetDI improved milk yield during the first 100 days of lactation (+ 2.2 kg/day) and also increased milk protein concentration (+ 1.3 g/kg).

Dry matter intake (DMI)

According to the literature, balancing MetDI of dairy rations has very little influence on dry matter intake, except in specific circumstances when MetDI and LysDI supply are balanced during the pre-calving period (Socha *et al.*, 1994d; Robert, unpublished; Brunschwhig, Augeard, Sloan and Tanan, 1995). In these trials, the quantity of MetDI supplied before parturition was the same as that calculated to meet requirements after parturition and a negative effect on DMI was observed during early lactation. As emphasized by Sloan (1997), the problem was initiated pre-calving, with a large decrease in DMI when MetDI was increased greatly above requirements (3.1 % instead of 2.4 % of PDIE) and when the LysDI to MetDI ratio was also unbalanced (2.7 versus the recommended 3.0 (NRC, 2001)). As a consequence of these observations, it appears particularly important to balance MetDI and LysDI profiles before calving.

Influence of energy and protein on milk performance response

In terms of energy, no interaction was observed within the range tested - energy intakes between 90 and 110 per cent of recommendations (Rulquin and Delaby, 1994b; Brunschwig *et al.*, 1995). An exception was the counterbalancing effect of MetDI and LysDI against the milk fat depression observed when fat was fed to supply extra energy (Bertrand, Pardue and Jenkins, 1996; Chow, DePeters and Baldwin, 1990).

Greatest production responses were observed with normal levels of crude protein in the ration (140 to 180 g/kg). The relative balance of AA absorbed by the dairy cow determines the response, rather than the total quantity of AA absorbed. Where different protein levels were tested (100 versus 105 % of recommendations for PDI) and the AA profile of protein entering the small intestine for each level was similar, milk protein responses to extra supply of MetDI and LysDI were of the same magnitude (Sloan, Robert and Mathé, 1989; Socha *et al*, 1994; Colin-Schoellen, Laurent, Vignon, Robert and Sloan, 1995).

Efficiency of feed nitrogen and energy utilization

Sloan (1997) reviewed this question in depth and underlined the fact that the proportions of extra MetDI appearing in milk protein averaged only 10%. This probably reflects the law of diminishing returns. MetDI supplementations were generally made to reach the final requirements of cows. Methionine is also used for maintenance functions and is involved in numerous metabolic processes.

For energy, Sloan (1997) made the assumption that energy yielding nutrients both of dietary origin and from body reserves are utilized much more efficiently for milk production when rations are balanced for MetDI and LysDI. The improvement in energy status of animals receiving a ration supplemented with bioavailable methionine could be due to the key role of methionine at the liver level. This is of a particular importance in early lactation when energy balance is negative. MetDI supply increased the quantity of very low density lipoproteins (VLDL), the lipoprotein complex essential for transporting mobilized triglycerides from the liver towards peripheral tissues, particularly the mammary gland, to provide energy nutrients for lactation (Bauchart, Durand, Gruffat and Chilliard, 1998). Positive effects on reproductive performance have also been observed with balancing MetDI of dairy rations (Robert, Mathé, Bouza, Valentin and Demirdjian, 1996; Thiaucourt, 1996). This could be linked to better utilization of energy leading to a more favourable energy status for successful reproduction.

In conclusion, MetDI supply appears not only as a positive factor improving milk performance (milk protein secretion and milk yield in early lactation), but also has a positive influence in terms of reducing metabolic disorders and improving the energy status of dairy cows, particularly in early lactation.

ETHYL CELLULOSE AND FAT COATED METHIONINE (MEPRON® M85)

Few studies have been published concerning the influence of Mepron M85 supplementation on milk production.

Overton, Emmert and Clark (1998), in a long-term study involving 44 multiparous dairy cows receiving an alfalfa plus corn silage ration, observed a trend to increase crude protein and casein concentrations of milk with a supplementation of 20 g/day of Mepron M85. Mepron M85 increased plasma concentrations of methionine, decreased histidine and tended to decrease arginine, lysine and ornithine. In a short-term trial involving 16 dairy cows in mid lactation, Leonardi, Stevenson and Armentano (2003), tested two levels of crude protein (161 and 188 g/kg) with or without supplemental methionine as Mepron M85 (17 to 19 g/head/day). For the ration without Mepron M85 supplementation, the LysDI was 6.4 and MetDI was 1.75 (NRC 2001). No interaction between CP level and methionine supplementation was observed. Milk protein concentration increased significantly from 31.7 to 32.6 g/kg with the addition of Mepron M85. Methionine supplementation did not affect N excretion in urine and faeces. Soder and Holden (1999) tested three levels of Mepron M85 (0-15 and 30 g/day) on 6 cows in a short-term trial, the main objective being to measure the influence of supplementation on immunity and as a second objective to measure the influence on milk production. No effect on milk yield or milk composition was observed, although serum methionine concentrations increased with Mepron M85 supply.

It is difficult to draw conclusions from these results given the small number of trials reported.

HMB SUPPLEMENTATION OF DAIRY RATIONS

All the above results were obtained with MetDI supplied by coated protected methionine. An abundant literature, reviewed by Schwab (1998), also investigated the potential influence of a chemical derivative of methionine, the hydroxy analogue of methionine (HMB). The main response recorded was an increase in milk fat concentration and only three positive results were seen in milk protein (Hutjens and Shultz, 1971; Van Hellemond and Sprietsma, 1977; Bargghava, Otterby, Murphy and Donker, 1977). Recently four trials were performed to compare the influence of HMB, HMBi and SmM on lactation performance in dairy cows, (Rulquin and Delaby, 2003; Schwab, Whitehouse, McLaughlin, Kadaryia, St-Pierre, Sloan, Gill and Robert, 2001; Sylvester, St-Pierre, Sloan, Beckman and Noftsger, 2003; Noftsger *et al*, 2004). No improvement in milk protein concentration was observed with HMB, but improvements were observed in response to SmM and HMBi. These results appear to be in agreement with a low methionine bioavailability value for HMB. When positive results were observed, particularly the effect on milk fat content and yield, these seem to be linked to HMB utilization by rumen microorganisms (Schwab et al, 1998). Additional in vitro and in vivo experiments are needed to provide guidance as to the type of diet and feeding programs that will benefit most from HMB supplementation.

HMB + SmM SUPPLEMENTATION OF DAIRY RATIONS

A combination of HMB and pH sensitive coated methionine (SmM) was tested by Noftsger and St-Pierre (2003) in a production trial starting at 4 weeks after calving and lasting for 12 weeks. The authors compared 4 treatments combining crude protein level and low or high estimated digestibility of RUP (LoDRUP and HiDRUP) and supplementation with HMB plus SmM for one of the treatments. Treatments were: 1) 183 g CP/kg + LoDRUP; 2) 183 g CP/kg + HiDRUP; 3) 169 g CP/kg + HiDRUP and 4) 170 g CP/kg + HIDRUP + HMB + SmM. Milk yield (40.8, 46.2, 42.9, 46.6 kg/day), milk protein concentration (29.5, 29.8, 29.9, 30.9 g/kg) and milk fat concentration (34.2, 36.4, 36.6, 37.3 g/kg) were measured for the four treatments respectively. Milk urea N and blood urea N were lower with the lower CP diets. Efficiency of N use for milk protein production was higher with the higher digestibility RUP, especially for the treatment supplemented with methionine. Supplementing the HiDRUP source with rumen available and rumen escape sources of methionine (HMB + SmM) resulted in maximum milk and protein production and maximum N efficiency, indicating that post ruminal digestibility of RUP and amino acid balance were more important than total RUP supply. Estimates of environmental efficiency indicate that lowering CP and balancing AA can sharply decrease the amount of N released into the environment. It should be noted that the combination tested in this trial is similar to HMBi because it supplied both MetDi and rumen available HMB.

Sloan (1997) also emphasised the potential of amino acid supplementation to reduce nitrogen pollution. However, to reduce dietary CP concentrations by as much as is achieved in pig nutrition, more attention needs to be paid not only to lysine and methionine but also to other potentially limiting amino acids. Such an optimization of amino acids in the ration needs information on requirements for limiting amino acids other than lysine and methionine. Lapierre (2004) recommends that in future the supply of individual AA will be balanced according to the needs of the dairy cow based on understanding the various metabolic controls that exist within the animal. Such an approach would allow a reduction in N intake without a major impact on milk production, but with a substantial reduction in urinary N excretion and negative environmental impacts.

HMBI (ISOPROPYL ESTER OF HMB)

Because HMBi provides bioavailable methionine to dairy cows, milk protein concentration should be enhanced when HMBi is fed.

Four short-term and two-long term trials were carried out. Cows received rations based on maize silage and soyabean meal or mixed rations based on grass, maize silage and soyabean meal. All rations were estimated to be deficient in MetDI (1.7 to 1.8% of PDIE) but provided adequate LysDI (6.7 to 7.1% of PDIE).

The short-term trials were carried out after peak lactation in early or mid lactation, with latin square or cross over designs, and for periods of 2 weeks (Noftsger *et al.*, 2004; Rulquin and Delaby, 2003; Schwab *et al.*, 2001). Milk protein concentration increased by 0.9 to 1.4 g/kg. No change in DMI, milk yield or fat concentration was observed.

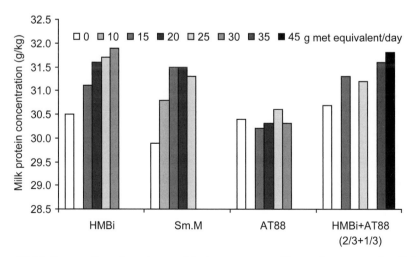

Figure 13.5 Influence of supplementary methionine source on milk protein concentration (Adapted from Schwab *et al.*, 2001).

The long-term trials were carried out over 120 days starting at the third week of lactation (Jurjanz, unpublished) and 100 days from the fourth week of lactation (Sylvester et al, 2003) on dairy cows with milk yields of 30 and 41.5 kg/day respectively. Jurjanz (unpublished) tested a supplementation of 27 g HMBi/head/day supplying 9.5 g/head/day MetDI versus a control ration, on 36 dairy cows. No significant change was observed in DMI, milk yield, fat concentration or fat yield. Milk protein concentration was significantly increased (+ 1.1 g/kg) and lactose content decreased (- 1.1 g/kg, p <0.05). Milk protein nitrogen and casein nitrogen increased significantly (+ 0.2 g/l, p < 0.1 and +0.2 g/l, p < 0.05 respectively); NPN concentration and proportion in milk decreased significantly (-19 mg/l, p < 0.05 and -0.6 %, p < 0.01 respectively) with HMBi treatment versus Control. Sylvester *et al.*, (2003) compared four treatments: control, HMBi (0.15 % in the ration), HMB (0.1 %) and a combination of HMBi and HMB (0.15 % + 0.05 % respectively) using 60 dairy cows. HMB supplementation had no effect on milk yield or composition, with only lactose production showing an improvement. Significant (p < 0.05) increases in milk yield (+ 2.9 kg/day), true protein concentration (+ 1.5 g/kg), true protein yield (+ 115 g/day), fat yield (+ 165 g/day) and lactose yield (+ 182 g/day) were observed with HMBi. No significant interaction was observed between HMB and HMBi on any of the production traits. HMBi supply reduced the amount of N excreted in urine by increasing the amount of N secreted in milk.

The results of these production trials were used in a meta-analysis of the influence of different sources of methionine on milk protein concentration (Guyot, Robert and Sauvant, 2004). This meta-analysis was carried out in order to predict the increase in milk protein content (MPC, g/kg) in relation to the quantity of estimated MetDI given to cows (MMQ; g/head/day). Regressions were calculated taking into account the methionine contents of each product (70% for HMBi, 78% for SmM and 88 % for HMB) and their respective bioavailability (50% for HMBi, 80% for SmM and 40% for HMB). The response curves to graded amounts of MetDI from HMBi and SmM were similar (not significantly different). The relationship was: MPC= 0.357 *ln MMQ (R^2= 0.77; RSD = 0.11 g/kg). This result confirmed the 50% bioavailability value for HMBi. In this meta-analysis, HMB did not increase milk protein concentration. The equation was: MPC= 0.0022 *ln MMQ (R^2= 0; RSD: 0.15 g/kg).

A field trial of HMBi (Metasmart [TM]) was carried out in France on 20 farms with 35 to 110 Holstein cows each (average 60) (Guyot T., unpublished), during the 2002-2003 winter. Mean milk production was 8500 kg per year, milk protein concentration ranged from 31 to 34 g/kg, and milk fat concentration ranged from 37 to 43 g/kg. Diets were mainly maize silage and soyabean meal, but half of the farms also included grass silage in the diets (10 to 50 % of the forage DMI). The experimental design was a change-over with successive periods of control and HMBi supplementation (25 to 30 g/head/day) of one month each. One to two HMBi supplementation periods was conducted before cows were turned out to spring pasture. Milk yield and composition (protein and fat) were individually measured twice per month by the official milk recording organisation. Statistical analysis used a model that included farm and production characteristics for each cow to separate stage of lactation and HMBi effects. HMBi supplementation significantly improved milk protein concentration on 16 farms. For all farms, milk protein improved by 0.8 g/kg on average. Significant improvements in milk protein varied from 0.6 to 1.4 g/kg, depending on farm. Milk fat concentration increased overall by an average of 1.3 g/kg with HMBi supplementation. The response varied, with significant increases ranging from 1 to 2.9 g/kg for 13 farms. No difference was observed in milk yield between the two treatments. Further analysis was carried out to identify the main causes of variation. When MetDI was lower than 1.83, the milk protein increase was 1g/kg and when LysDI was higher than 6.87, the milk protein improvement was 1.1 g/kg. Responses were highest on farms where the control milk protein concentration was greater than 32.5 g/kg. These farms showed a milk protein improvement of 1.2 g/kg on average. Probably AA were not the first limiting factors on farms with a relatively low basal milk protein concentration. These results confirm that different factors, especially base levels of MetDI and LysDI, must be taken into account before feeding supplemental AAs. The increase in milk fat concentration observed in this trial could be due to the HMB coming from HMBi and remaining in the rumen. If this improvement in fat concentration is confirmed, it should also be taken into account when assessing the overall need to supplement with HMBi.

HMBI IN PELLETED FEED

Unlike coated products (SmM and Mepron M85 for example), HMBi can be incorporated into pelleted feeds. Stability of HMBi in pelleted feeds has been established under industrial conditions, when added either as a liquid or as a powder (Table 13.2). The molecule mixes well and provides homogeneous feed as well as allowing pellets of adequate hardness (Dollat, unpublished). It can be easily analysed in feeds. For example, in four production trials (Rulquin, unpublished; Rulquin and Delaby, 2003; Jurjanz, unpublished; Schwab *et al.*, 2001), HMBi supplementation significantly improved milk protein concentration whether HMBi was fed in a meal or in pelleted feed.

Table 13.2 Effect of pelleting of HMBi in liquid (L) or solid (S) form on recovery in two different types of diet: concentrate (CA) or complete feed (CF) (Dollat. J.M.. unpublished).

Mixture	HMBi added to meal (mg/kg)	HMBi measured in meal (mg/kg)	HMBi expected in pellets (mg/kg)	HMBI measured in pellets (mg/kg)	Recovery (%)
CA/HMBiL(2,5)	2295	2290	2217	2350	**106**
CA/HMBiL(1)	932	950	925	910	**98**
CA/HMBiS(1)	804	920	897	860	**96**
CF/HMBiL(1)	958	800	755	770	**102**
CF/HMBiS(1)	827	940	887	880	**99**

Practicalities of metabolizable methionine optimization of dairy cow rations - economics

Obviously the first steps in ration formulation are to balance the ration in terms of net energy, metabolizable protein and to optimize rumen function. Subsequently, using existing metabolizable AA systems, the basal MetDI and LysDI supplies can be calculated from composition of the ration and compared with dairy cow requirements. Very few raw materials are good sources of MetDI, and in practical formulation it is impossible to meet both the MetDI and the LysDI constraints without using synthetic amino acid products. To reach the recommended MetDI and LysDI levels raw materials for LysDI supply can be combined with good quality protected methionine for MetDI supply. This is especially important and highlights the crucial point of using a reliable estimation of bioavailability.

Good quality rumen-protected products are much more cost-effective than feed raw materials for supplying MetDI (metabolizable methionine). For example, with Smartamine™ M, the ingredient cost is well below 0.02 $ per gram of MetDI, compared with 0.07 for blood meal, 0.05 for soyabean meal and 0.04 for fish meal.

In terms of value, increases in milk protein are the main benefit, depending on the milk protein payment to the producer. Other improvements come into play as well, such as milk yield in early lactation, improvement in N efficiency, reduction of N excretion, lower milk urea nitrogen, reduction of metabolic problems and energy status improvement favouring reproduction.

Today, numerous practical applications are in place, primarily in cheese-producing countries, or regions, where milk protein is well-valued, particularly with Smartamine ™ M. In typical MetDI- deficient European dairy rations, SmM is often fed at 12 g/animal/day (up to 18 g for high-producing cows or very deficient rations). This supplies a little over 7 g MetDI (up to 11 g for the highest doses) allowing MetDI requirement (2.2 % of PDIE) to be met. The LysDI requirement (7 % of PDIE) is satisfied through formulating with adequate raw materials.

SmM, however, cannot be included into pellets, unlike HMBi. The characteristics of the product allow HMBi to be incorporated into pelleted feed. Taking into account the HMBi product characteristics and its methionine equivalent bioavailaibility, the recommended dosage is 25-30 g/head/day to balance the MetDI of the ration. HMBi is also a source of rumen-available HMB and thus acts on ruminal fermentation. Indirect improvements in milk yield and milk fat can result from this ruminal HMB supply. This has to be taken into account according to production objectives and context, milk fat quotas for example, when choosing between a protected methionine that is inert in the rumen (SmM) and a partly ruminal fermentation enhancer (HMBi).

Conclusion

The concept of ideal AA profile is now well documented for ruminants, particularly for dairy cows. Methionine and lysine have been identified as first limiting for milk protein synthesis. Metabolizable AA systems have been proposed (Rulquin *et al.*, 1993, 1998; NRC, 2001). These systems are tools for formulating dairy cow rations in terms of metabolizable methionine (MetDI) and metabolizable lysine (LysDI). To reach the recommended MetDI and LysDI levels in practical formulation a judicious combination of raw materials for LysDI supply combined with a good quality protected methionine for MetDI supply is the best solution. The typical ruminally protected methionine is the coated form like Smartamine™ M. A new molecule which can be incorporated into pelleted feed is proposed, HMBi, which is a chemical derivative of the hydroxy analogue of methionine, acting both as metabolizable methionine and a ruminal fermentation enhancer. The cost of ruminally protected methionine should be considered against its potential to improve milk protein yield and other aspects of milk performance. Increases in milk protein concentration from 1 to 2 g/kg and milk protein yield from 60 to 100 g/head/day are attainable over a complete lactation and milk yield can increase by up to 2.5 kg in early lactation. Other improvements should also be taken into account, such as improvement in N efficiency, reduction of metabolic problems, and energy status improvement favouring reproduction.

The challenges for future research could concern:

• refining MetDI and LysDI recommendations for diverse situations (stage of lactation, milk yield potential, lactation number)

- determining requirements for other EAA to improve N efficiency and further reduce dietary CP concentrations and N pollution

- enhancing current AA systems to be more flexible by taking into account some dynamic aspects at the digestive and metabolic levels. However, current AA systems are robust, simple and effective enough to be used with satisfaction in a wide variety of dietary situations.

References

Armentano, L.E., Bertics, S.J. and Ducharme, G.A. (1993) Lactation responses to rumen protected methionine or methionine with lysine, in diets based on alfalfa haylage. *Journal of Dairy Science*,**78**, (suppl. 1) 202.

Bauchart, D., Durand, D., Gruffat, D. and Chilliard, Y. (1998) Mechanisms of liver steatosis in early lactation cows – effect of hepatoprotector agents. pp 1-9. *Proc. Cornell Nutr. Conf. Feed Manuf., Cornell Univ.* Ithaca NY

Belasco, I.J. (1972) Stability of methionine hydroxy analog in rumen fluid and its conversion in vitro to methionine by calf liver and kidney . *Journal of Dairy Science,* **55**, 775-784.

Berthiaume, R., Dubreuil, P., Stevenson, M., McBride, B.W. and Lapierre, H. (2001) Intestinal disappearance and mesenteric and portal appearance of amino acids in dairy cows fed ruminally protected methionine . *Journal of Dairy Science*, **84**, 194-203.

Berthiaume, R., Lapierre H., Stevenson, M., Coté, N. and McBride, B.W. (2000) Comparison of the in situ and in vivo intestinal disappearance of ruminally protected methionine. *Journal of Dairy Science*, **83**, 2049-2056.

Bertrand, J.A., Pardue, F.F. and Jenkins, T.C. (1996) Effect of protected amino acids on milk production and composition of Jersey cows fed whole cottonseed. *Journal of Dairy Science*, **79,** (Suppl. 1) 5022.

Bhargava P.V., Otterby D.E., Murphy, J.M. and Donker, J.D. (1977) methionine hydroxy analog in diets for lactating cows. *Journal of Dairy Science*, **60**, 1594-1604.

Blum, J.W., Bruckmaier, R.M. and Jans, F. (1999) Rumen protected methionine fed to dairy cows : bioavailability and effects on plasma amino acids pattern and plasma metabolites and insulin concentrations. *Journal of Dairy Science*, **82**, 1991-1998

Brunschwig, P. and Augeard, P. (1994) Acide aminé protégé: effet sur la production et la composition du lait des vaches sur régime ensilage de maïs. *Journées de Recherches sur l'alimentation et la nutrition des Herbivores* 16-17 mars - INRA Theix.

Brunschwig, P., Augeard, P., Sloan, B.K. and Tanan, K. (1995) Feeding of protected methionine from 10 days pre-calving and at the beginning of lactation to dairy cows fed a maize silage based ration. *Rencontres Recherches Ruminants*, **2**, 249.

Brunschwig, P., Augeard, P., Sloan, B.K., and Tanan, K. (1995) Supplementation of maize silage or mixed forage (maize and grass silage) based rations with rumen protected methionine for dairy cows. *Annales Zootechnie*, **44**, 380.

Casper, D.P., Schingoethe, D.J., Yang, C.M.J. and Mueller, C.R. (1987) Protected Methionine supplementation with extruded blend of soybeans and soybean meal for dairy cows. *Journal of Dairy Science*, **70**, 321-330.

Chow, L.M., DePeters, E.J. and Baldwin, R.L. (1990) Effect of rumen-protected methionine and lysine on casein in milk when diets high in fat or concentrate are fed. *Journal of Dairy Science* **73**, 1051-1061.

Colin-Schoellen, O., Laurent, F., Vignon, B., Robert, J.C. and Sloan, B.K. (1995) Interactions of ruminally protected methionine and lysine with protein source or energy level in the diet of cows. *Journal of Dairy Science*, **78**, 2807-2818.

Digenis, G.A., Amos H.E., Mitchell, G.E., Swintowsky, J.V., Yang, K., Schelling, G.T. and Parish, R.C. (1974) Methionine substitute in ruminant nutrition. II. Stability of non nitrogenous compounds related to methionine during in vitro incubation with rumen micro-organisms. *Journal of Pharmaceutical Sciences*, **63**, 751-754.

Dupuis,L., Saunderson, C.L., Puigserver, A. and Brachet, P. (1989) Oxidation of methionine and 2-hydroxy-4-methylthiobutanoic acid stereoisomers in chicken tissues. *British Journal of Nutrition*, **62**, 63-75.

Graulet, B., Richard, C. and Robert, J.C. (2004) The isopropyl ester of methionine hydroxy-analogue is absorbed through the rumen wall in the cow. *Journal of Animal and Feed Science*, **13**, (Suppl. 1) .

Guyot, T., Robert, J.C. and Sauvant, D. (2004) Meta-analysis of the influence of different sources of methionine on the milk protein content. *Journal of Dairy Science*, 87, (Suppl. 1) 162.

Hanigan, M.D. (2004) Quantitative aspects of splanchnic metabolism in the lactating ruminant. *Proceedings of the British Society Animal Science,* 258.

Henry, Y. (1993) Affinement du concept de la protéine idéale pour le porc en croissance. *INRA Productions Animales*, **6**, 199-212.

Hutjens, M.F. and Schultz, LH. (1971) Addition of soybeans or methionine hydroxy analog to high-concentrate rations for dairy cows. *Journal of Dairy Science*, **54**, 1637-1644.

Hurtaud, C., Rulquin, H. and Vérité, R. (1995) Effect of rumen protected methionine and lysine on milk composition and on cheese yielding capacity. *Annales Zootechnie*, **44**, 382.

Kim, C.H., Choung, J.J. and Chamberlain, D.G. (1999) Determination of the first limiting amino acid for milk production in dairy cows consuming a diet of grass silage and cereal based supplement containing feather meal. *Journal of the Science of Food and Agriculture*, **79**, 1703-1708.

Koch, K.L., Whitehouse, N.L, Garthwaite, B.D., Wasserstom, V.M. and Schwab, C.G. (1996) Production responses of lactating Holstein cows to rumen stable forms of lysine and methionine. *Journal of Dairy Science*, **79**, (Suppl. l) 24.

Koenig, K.M. and Rode, L.M. (2001) Ruminal degradability, intestinal disappearance and plasma methionine response of rumen- protected methionine in dairy cows. *Journal of Dairy Science*, **84**, 1480-1487.

Koenig, K.M., Rode, L.M., Knight, C.D., McCullough, P.R., Andrews, K.A. and Kung, L. Jr (1996) Ruminal escape and subsequent absorption of Alimet in the digestive tract of lactating cows. *Journal of Animal Science*, **74**, (Suppl. 1), 266.

Koenig, K.M., Rode, L.M., Knight, C.D. and McCullough, P.R. (1999) Ruminal escape, gastrointestinal absorption, and response of serum methionine to supplementation of liquid methionine hydroxy analog in dairy cows. *Journal of Dairy Science*, **82**, 355-361.

Koenig, K.M., Rode, L.M., Knight, C.D. and Vazquez-Anon, M. (2002) Rumen degradation and availability of various amounts of liquid methionine hydroxy analog in lactating dairy cows. *Journal of Dairy Science*, **85**, 930-938.

Korhohen, M., Vanhatalo, A., Varvikko, T. and Huhtanen, P. (2000) Responses to graded post ruminal doses of histidine in dairy cows fed grass silage diets. *Journal of Dairy Science*, **83**, 2596-2608.

Kristensen, N.B., Gäbel, G., Pierzynowski, S.G. and Danfaer, A. (2000) Portal recovery of short-chain fatty acids infused into the temporarily-isolated and washed reticulo-rumen of sheep. *Bristish Journal of Nutrition*, **84**, 477-482.

Lapierre, H.(2004) The route of absorbed nitrogen to milk protein. *Proceedings of the British Society Animal Science,* 255.

Lapierre, H., Pacheco, D., Berthiaume, R., Ouellet, D.R., Schwab, C.G., Holtrop, G. and Lobley, G.E. (2004) What is the true supply of amino acids? *Journal of Dairy Science*, **87**, (Suppl. 1), 240.

Leclercq, B. (1996) Les rejets azotés issus de l'aviculture: importance et progrès envisageables. *INRA Productions Animales*, **9**, 91-101.

Leonardi, C., Stevenson, M. and Armentano, L.E. (2003) Effect of two levels of crude protein and methionine supplementation on performance of dairy cows. *Journal of Dairy Science*, **86**, 4033-4042.

National Research Council (1994) *Nutrient requirements of Poultry.* 9th rev. ed.

Natl. Acad. Press, Washington, DC.

National Research Council (1998) *Nutrient requirements of Swine.* 10th rev. ed. Natl. Acad. Press, Washington, DC.

National Research Council (2001) *Nutrient requirements of dairy cattle.* 7th rev. ed. Natl. Acad. Press, Washington, DC.

Noftsger, S. and St-Pierre, N.R. (2003) Supplementation of Methionine and selection of highly digestible rumen undegradable protein to improve nitrogen efficiency for milk production. *Journal of Dairy Science*, **86**, 958-969.

Noftsger, S., St-Pierre, N.R. and Firkins, J.L. (2004) Determination of undegradability and ruminal effects of HMB, HMBi, and DL-Met in lactating cows. *Journal of Dairy Science*, **87,** (Suppl.1), 218.

Nozière, P., Richard, C., Graulet, B., Durand, D., Rémond, D. and Robert, J.C. (2004) Investigation of the site of absorption and metabolism of HMBi and HMB on sheep. *Journal of Dairy Science*, **87,** (Suppl. 1), 220.

O'Connor, J.O., Sniffen, C.J., Fox, D.G. and Chalupa, W. (1993) A net carbohydrate and protein system for evaluating cattle diets: IV Predicting amino acid adequacy. *Journal of Dairy Science*, **71**, 1298-1311.

Olley, J.B., Ordway, R.S., Whitehouse, N.L., Schwab, C.G. and Sloan, B.K. (2004) Use of changes in plasma sulphur amino acid concentrations to compare the ability of methionine (Met) products to provide absorbable Met to lactating dairy cows fed a Met-adequate diet. *Journal of Dairy Science*, **87,** (Suppl. 1) 162.

Overton, T.R., Emmert, L.S. and Clark, J.H. (1998) Effects of source of carbohydrate and protein and rumen-protected methionine on performance of cows. *Journal of Dairy Science*, **81**, 221-228.

Overton, T.R., LaCount, D.W., Cicela, T.M. and Clark, J.H. (1996) Evaluation of a ruminally protected methionine product for lactating dairy cows. *Journal of Dairy Science*, **79**, 631-638.

Patterson, J.A. and Kung, L.Jr. (1988) Metabolism of DL-methionine and methionine analogs by rumen microorganisms. *Journal of Dairy Science*, **71**, 3292-3301.

Pisulewski, P.M., Rulquin, H., Peyraud, J.L. and Vérité, R. (1996) Lactational and systemic responses of dairy cows to post-ruminal infusions of increasing amounts of methionine. *Journal of Dairy Science*, **79**, 1781-1791.

Polan, C.E., Cummins, K.A., Sniffen, C.J., Muscato, T.V., Vicini, J.L., Crooker, B.A., Clark, J.H., Johnson, D.G., Otterby, D.E., Guillaume, B., Muller, L.O., Varga, G.A., Murray, R.A. and Peirce-Sandner, S. (1991) Responses of dairy cows to supplement rumen-protected forms of methionine and lysine. *Journal of Dairy Science*, **74**, 2997-3013.

Robert, J.C. (1992) Tests in vitro. Tests in sacco. Acides aminés protégés dans le rumen et disponibles dans l'intestin. Une technologie RPAN originale. *Journées AFTAA-CAAA, Tours.*

Robert, J.C., Ballet, N., Richard, C. and Bouza, B. (2002a) Ruminal metabolism of 2-hydroxy-4 (methylthio) butanoic acid isopropyl ester (HMBi). *Journal of Dairy Science*, **85**, (Suppl. 1), 240.

Robert, J.C., Ballet, N., Richard, C. and Bouza, B. (2002b) Influence of 2-hydroxy-4(methylthio) butanoic acid isopropyl ester (HMBi) on the digestibility of organic matter and energy value of corn silage measured in vitro. *Journal of Dairy Science*, **85**, (Suppl. 1), 240.

Robert, J.C., d'Alfonso, T., Etave, G., Depres, E. and Bouza, B. (2002) Quantifying the metabolisable methionine contribution of a liquid or powder presentation of 2-hydroxy-4(methylthio) butanoic acid isopropyl ester (HMBi). *Journal of Dairy Science*, **85**, (Suppl. 1), 71.

Robert, J.C., Etave, G., d'Alfonso, T. and Bouza, B. (2001) A blood kinetics methodology to measure bioavailability of rumen protected methionine sources for ruminants. *Journal of Dairy Science*, **84**, (Suppl. 1), 281.

Robert, J.C., Mathé, J., Bouza, B., Valentin, S. and Demirdjian, S. (1996) The effect of protected methionine supplementation on dairy cow fertility. *Journal of Dairy Science*, **79**, (Suppl. 1), 147.

Robert, J.C., Paquet, S., Richard, C. and Bouza, B. (2003) Influence of 2-hydroxy-4 (methylthio) butanoic acid isopropyl ester (HMBi) concentration on in vitro estimated organic matter digestibility of diets varying in proportion of corn silage relative to concentrate. *Journal of Dairy Science*, **86**, (Suppl. 1), 276.

Robert, J.C., Richard, C. and Bouza, B. (2001) Influence of monomer or dimer forms of isopropyl ester of HMB, on the supply of metabolisable methionine to the blood of ruminants. *Journal of Dairy Science*, **84**, (Suppl. 1), 281.

Robert, J.C., Richard, C., d' Alfonso, T., Ballet, N. and Depres, E. (2001) Investigation of the site of absorption and metabolism of a novel source of metabolisable methionine: 2-hydroxy-4 (methylthio) butanoic acid isopropyl ester. *Journal of Dairy Science*, **84**, (Suppl. 1), 35.

Robert, J.C., Sloan, B.K. and Bourdeau, S. (1994) The effect of supplementation of corn silage plus soybean meal diets with rumen protected methionine on the lactational performance of dairy cows in early lactation. *Journal of Dairy Science*, **77**, (Suppl. 1), 349.

Robert, J.C., Sloan, B.K. and Denis, C. (1996) The effect of graded amounts of rumen protected methionine on lactational responses in dairy cows. *Journal of Dairy Science*, **79**, (Suppl. 1), 256.

Robert, J.C., Sloan, B.K., Etave, G. and Bouza, B. (2001) Influence of length and ramification of the alcohol radical of esters of methionine and of 2-

hydroxy-4 (methylthio) butanoic acid on methionine bioavailability. *Journal of Dairy Science*, **84**, (Suppl. 1), 34.

Robert, J.C., Sloan, B.K. and Lahaye, F. (1995) Influence of increasing doses of intestinal digestible methionine (MetDI) on the performance of dairy cows in mid and late lactation. *IVth International Symposium on the Nutrition of Herbivores,* Clermont-Ferrand, France.

Robert, J.C., Sloan, B.K. and Nozière, P. (1996) The effect of graded amounts of rumen protected lysine on lactational performance in dairy cows. *Journal of Dairy Science*, **79**, (Suppl. 1), 257.

Robert, J.C., Williams, P.E.V. (1997) Influence of forage type on the intestinal availability of methionine from a rumen-protected form. *Journal of Dairy Science*, **80**, (Suppl. 1), 248.

Robert, J.C., Williams, P.E.V. and Bouza, B. (1997) Influence of source of methionine and protection technology on the post ruminal delivery and supply to the blood of dairy cows of an oral supply of methionine. *Journal of Dairy Science*, **80**, (Suppl. 1), 248.

Rogers, J.A., Peirce-Sandner, S.B., Papas, A.M., Polan, C.E., Sniffen, C.J., Muscato, T.V., Staples, C.R. and Clark, J.H. (1989) Production responses of dairy cows fed various amounts of rumen-protected methionine and lysine. *Journal of Dairy Science*, **72**, 1800-1817.

Rulquin, H. (1992) Intérêts et limites d'un apport de méthionine et de lysine dans l'alimentation des vaches laitières. *INRA Productions Animales*, **5**, 29-36.

Rulquin, H. and Delaby, L. (1994a) Lactational responses of dairy cows to graded amounts of rumen-protected methionine. *Journal of Dairy Science*, **77**, (Suppl. 1), 345.

Rulquin, H. and Delaby, L. (1994b) Effects of energy status on lactational responses of dairy cows to rumen-protected methionine. *Journal of Dairy Science*, **77**, (Suppl. 1), 346.

Rulquin, H. and Delaby, L. (1997) Effects of the energy balance of dairy cows on lactational responses to rumen-protected methionine. *Journal of Dairy Science*, **80**, 2513-2522.

Rulquin, H. et Delaby, L. (2003) Détermination de la biodisponibilité chez la Vache laitière de deux analogues de la méthionine pouvant supporter le passage en filière. In *Rencontres Recherches Ruminants*, **10**, 397.

Rulquin, H., Guinard, J. and Vérité, R. (1998) Variation of amino acid content in the small intestine digesta of cattle : development of a prediction model. *Livestock Production Science*, **53**, 1-13.

Rulquin, H., Guinard, R., Vérité, R. and Delaby, L. (1993) Teneurs en Lysine (LysDI) et Méthionine (MetDI) digestibles des aliments pour ruminants. *Journées AFTAA-CAAA, Le Mans*.

Rulquin, H. and Kowalczyk, J. (2000) A blood procedure to determine bioavailability of rumen protected Met for ruminants. *Journal of Dairy Science*, **83**, (Suppl. 1), 268.

Rulquin, H. and Kowalczyk, J. (2003) Development of a method for measuring lysine and methionine bioavailability in rumen-protected products for cattle. *Journal of Animal and Feed Science*, **12**, 3, 465-474 .

Rulquin, H., Le Henaff, L. and Vérité, R. (1990) Effects on milk protein yield of graded levels of lysine infused into the duodenum of dairy cows fed diets with two levels of protein. *Reproduction Nutrition Development*, **30,** (Suppl. 2), 238.

Rulquin, H., Pisulewski, P.M., Vérité, R. and Guinard, J. (1993) Milk production and composition as a function of post-ruminal lysine and methionine supply: a nutrient response approach. *Livestock Production Science*, **37**, 69-90.

Rulquin, H. and Vérité, R. (1993) Amino acid nutrition of dairy cows : Productive effects and animal requirements. In *Recent Advances in Animal Nutrition - 1993* , pp 55-77. Edited by P.C. Garnsworthy and D.J.A. Cole. Nottingham University Press, Nottingham.

Rulquin, H., Vérité, R., Guinard-Flament, J. and Pisulewski P.M. (2001) Acides Aminés digestibles dans l'intestin . Origines des variations chez les ruminants et répercussions sur les protéines du lait. *INRA Productions Animales*, **14**, 201-210.

Saunderson, C.L. (1985) Comparative metabolism of L-methionine, DL-methionine and DL-2- hydroxy-4-methylthiobutanoic acid by broilers chicks. *British Journal of Nutrition*, **54**, 621-633.

Sauvant, D., Perez, J.M., Tran, G. (2003) Tables de composition et de valeur nutritive des matières premières destinées aux animaux d'élevage. INRA-AFZ- INA Paris Grignon

Soder, K.J. and Holden, L.A. (1999) Lymphocyte proliferation response of lactating dairy cows fed varying concentrations of rumen–protected methionine. *Journal of Dairy Science*, **82**, 1935-1942.

Schwab, C.G. (1989) Amino acids in dairy cow nutrition. pp 75-101. *Proceedings of Technical Symposium, in conjunction with California Animal Nutrition Conference*, Fresno, CA.

Schwab, C.G. (1996) Amino acid requirements of the dairy cow. pp 1-23. *Proceedings of Technical Symposium, in conjunction with California Animal Nutrition Conference*, Fresno, CA.

Schwab, C.G. (1998) Methionine analogs for dairy cows : a subject revisited. pp 1-24. *Proceedings of Technical Symposium, in conjunction with California Animal Nutrition Conference*, Fresno, CA.

Schwab, C.G., Satter, L.O. and Clay, A.B. (1976) Response of lactating dairy cows to abomasal infusion of amino acids. *Journal of Dairy Science,* **59**,

1254-1269.

Schwab, C.G., Whitehouse, N.L., McLaughlin, N.L., Kadariya, R.K., St-Pierre, N.R., Sloan, B.K., Gill, R.M. and Robert, J.C. (2001) Use of milk protein concentrations to estimate the methionine bioavailability of two forms of 2-hydroxy-4 (methylthio) butanoic acid (HMB) for lactating cows. *Journal of Dairy Science*, **84**, (Suppl. 1) 35.

Sloan, B.K. (1993) Ensilage de maïs/soja: quels acides aminés faut-il apporter pour augmenter le taux protéique chez la vache laitière? *Journées AFTAA-CAAA ,Tours.*

Sloan, B K. (1997) Developments in amino acid nutrition of dairy cows. In *Recent Advances in Animal Nutrition*-1997, pp. 167-198. Edited by P.C. Garnsworthy and J.Wiseman. Nottingham University Press, Nottingham.

Sloan, B.K., Robert, J.C. and Lavedrine, F. (1994) The effect of protected methionine and lysine supplementation on the performance of dairy cows in mid lactation. *Journal of Dairy Science*, **77**, (Suppl. l), 343.

Sloan, B.K., Robert, J.C. and Mathé, J. (1989) Influence of dietary crude protein content plus or minus inclusion of rumen protected amino acids (RAA) on the early lactation performance of heifers. *Journal of Dairy Science*, **72**, (Suppl. 1), 506.

Socha, M.T., Schwab, C.G., Putnam, D.E., Whitehouse, N.L., Kierstead, N.A. and Garthwaite, B.D. (1994a) Determining methionine requirements of dairy cows during peak lactation by post-ruminally infusing incremental amounts of methionine. *Journal of Dairy Science*, **77**, (Suppl. l), 350.

Socha, M.T., Schwab, C.G., Putnam, D.E., Whitehouse, N.L., Kierstead, N.A.and Garthwaite, B.D. (1994b) Determining methionine requirements of dairy cows during early lactation by post-ruminally infusing incremental amounts of methionine. *Journal of Dairy Science*,**77**, (Suppl.1), 246.

Socha, M.T., Schwab, C.G., Putnam, D.E., Whitehouse, N.L., Kierstead, N.A. and Garthwaite, B.D. (1994c) Determining methionine requirements of dairy cows during mid lactation by post-ruminally infusing incremental amounts of methionine. *Journal of Dairy Science*, **77**, (Suppl.l), 351.

Socha, M.T., Schwab, C.G., Putnam, D.E., Whitehouse, N.L., Kierstead, N.A. and Garthwaite, B.D. (1994d) Production responses of early lactation cows fed rumen-stable methionine or rumen-stable lysine plus methionine at two levels of dietary crude protein. *Journal of Dairy Science*, **77**, (Suppl. 1), 352.

Südekum, K.H., Wolfram, S., Ader, P. and Robert, J.C. (2004) Bioavailability of three ruminally protected methionine sources in cattle. *Animal Feed Science and Technology*, **113**, 17-25.

Sylvester, J.T., St-Pierre, N.R., Sloan, B.K., Beckman, J.L. and Noftsger, S.M. (2003) Effect of HMB and HMBi on milk production, composition, and N efficiency of Holstein Cows in early and mid-lactation. *Journal of Dairy Science*, **86**, (Suppl. 1), 60.

Thiaucourt, L. (1996) L'opportunité de la méthionine protégée en production laitière. *Bulletin des GTV* 2B, 45-52.

Van Hellemond, K.K. and Sprietsma, J.E. (1977) Effect of supplementation of methionine and methionine hydroxy analog on milk yield, milk fat and milk protein. *Z.Tierphysiol. Tierernährg. u. Futtermittelkde*, **39**, 109-115.

Vanhatalo, A., Huhtanen, P., Toivonen, F.J. and Varviko, T. (1999) Response of dairy cows fed grass silage diets to abomasal infusions of histidine alone or in combination with methionine and lysine. *Journal of Dairy Science*, **82**, 2674-2685.

Vasquez-Anon, M., Cassidy, T., McCullough, P. and Varga, G.A. (2001) Effects of Alimet on nutrient digestibility, bacterial protein synthesis, and ruminal disappearance during continuous culture. *Journal of Dairy Science*, **84**, 159-166.

Vérité, R. and Peyraud, J.L. (1989) Protein: the PDI system. In *Ruminant Nutrition: Recommended Allowances and Feed Tables – 1989*, pp 33-47.Edited by R.Jarrige, INRA, Libbey Eurotext, Paris, France.

Wester, T.J., Vasquez-Anon, M., Parker, D., Dibner, J., Calder, A.G. and Lobley G.E. (2000a) Metabolism of 2-hydroxy-4-methylthio butanoic acid (HMB) in growing lambs. *Journal of Dairy Science*, **83**, (Suppl. 1), 269.

Wester, T.J., Vasquez-Anon, M., Parker, D., Dibner, J., Calder, A.G. and Lobley, G.E. (2000b) Synthesis of methionine (Met) from 2-hydroxy-4-methylthio butanoic acid (HMB) in growing lambs.*Journal of Dairy Science*, **83**, (Suppl. 1), 269.

14

ALTERNATIVE STRATEGIES FOR MANIPULATING MILK FAT IN DAIRY COWS

RICHARD J. DEWHURST AND MICHAEL R. F. LEE
Institute of Grassland and Environmental Research, Plas Gogerddan, Aberystwyth SY23 3EB, United Kingdom

Introduction

Manipulation of the level and type of fat in milk remains an important target for dairy breeding and nutrition programmes. Over the last century, there were substantial increases in milk fat content as a result of genetic selection and increased understanding of diets that led to milk fat depression. Genetic advances were particularly marked within the British Friesian strain of Holsteins. The drive for higher milk fat content was associated with increased yields of cheese and butter and the central role of milk fat as an energy source in the human diet.

Concern about the hypercholesteraemic effects of medium-chain saturated fatty acids led to a growing demand for reduced-fat dairy products. However, this trend should now be challenged since recent advances have shown positive effects of some milk fat components for human health and there are epidemiological studies suggesting positive effects of consumption of dairy products in relation to cardio-vascular risk (Warensjö *et al.*, 2004). Indeed, as this chapter highlights, there are growing opportunities to enhance the beneficial components of milk and such products will require a continued focus on maintaining or increasing milk fat content.

On a shorter-time scale, and more pragmatically, the involvement of milk fat content in defining milk quotas in the UK means that farmers may need to reduce milk fat content dramatically and rapidly once their liabilities become clear towards the end of the quota year.

After some introductory remarks about milk fat composition, this chapter considers feeding strategies and mechanisms associated with total milk fat content. The paper then moves on to consider manipulation of proportions of the major fatty acids within milk fat. It is increasingly clear that fatty acids are themselves involved in the regulation of milk fat content, so the two areas are necessarily connected.

For each major fatty acid, we consider what is possible in terms of the highest concentrations in milk that have been achieved using post-rumen infusions, well-protected supplements, or extreme diets. We also consider what has been achieved through the use of more realistic feeding strategies: varying the levels and types of concentrates and forages that are used commercially. The final section on forage effects provides some new perspectives that might be useful in understanding mechanisms and defining possible new strategies. It is of interest because in the natural grazing situation, biohydrogenation - which is central to both the level and type of fat in milk - was evolved to deal with the ingestion of chloroplasts, which contain high levels of polyunsaturated fatty acids (PUFA) and produce oxygen. Biohydrogenation is the conversion of the PUFA, notably α-linolenic acid (C18:3) and linoleic acid (C18:2) in forages, through a series of interesting intermediates into more saturated forms such as vaccenic acid (*trans*-11 C18:1) and stearic acid (C18:0).

Milk fat composition

Milk fat generally contains 0.6 to 0.7 saturated fatty acids (SFA), 0.25 to 0.35 monounsaturated fatty acids (MUFA) and up to 0.05 polyunsaturated fatty acids (PUFA; Jensen, 2002). Linoleic, conjugated linoleic and α-linolenic acids are the main PUFA- typically 0.02, 0.005 and 0.005 of milk fatty acids. Oleic acid is normally two-thirds of the MUFA (0.2 of total fatty acids), with the remainder of the MUFA mainly *cis* and *trans* isomers of C18:1. The PUFA and MUFA are generally regarded as beneficial for human health, and there is recent evidence of beneficial effects of *trans*-11 C18:1 (*trans* vaccenic acid; TVA; Corl *et al.*, 2003; Lock *et al.*, 2004). The predominant SFA are myristic acid (C14:0; 0.08 to 0.14 of milk fatty acids), palmitic acid (C16:0; 0.22 to 0.35 of milk fatty acids) and C18:0 (0.09 to 0.14 of milk fatty acids). There are concerns about the effects of SFA on plasma cholesterol, though C18:0 is regarded as neutral (Yu *et al.*, 1995) in this regard and C16:0 may not be hypercholesteraemic if the diet contains recommended levels of C18:2 (Clandinin *et al.*, 2000). Certainly, myristic acid (C14:0) is regarded as more potent than palmitic acid (C16:0) in raising plasma lipids (Clandinin *et al.*, 2000). There is current research interest in levels of conjugated linoleic acid (CLA) in milk because of possible effects on a range of human conditions, including cancer and obesity (Belury, 2002). Milk generally contains only low levels of longer-chain PUFA, eicosapentaenoic acid (EPA; C20:5) and docosahexaenoic acid (DHA; C22:6), and there is still much debate about the value of the main omega-3 fatty acid in milk (C18:3) as a precursor of EPA and DHA - the elongation and desaturation pathways may be enhanced when these are deficient in the human diet (Jacobs *et al.*, 2004). This review will focus on beneficial fatty acids in milk, C18:3, CLA and TVA; generally the use of forages or oilseed-based products to increase levels of these fatty acids increases C18:0 and decreases levels of C16:0 and C14:0.

Nutritional regulation of milk fat content

Davis and Brown (1970) reviewed the early literature on milk fat depression and highlighted two main causes: increased readily digestible carbohydrate (coupled to low levels of effective fibre) and supplementation with plant and fish oils which contain high levels of PUFA. Griinari *et al* (1998) linked these effects and showed that dietary PUFA are required for low fibre diets to lead to milk fat depression. It was also clear that an interaction between diet and rumen function was involved in the mechanism. Despite this background, the mechanism(s) of milk fat depression are still not fully elucidated, though there have been substantial recent advances (Bauman and Griinari, 2003).

The fundamental problem in describing the cause of milk fat depression is one which is central to many issues in nutritional science - diets (or absorbed nutrients) are complex mixtures in which increases in one area are necessarily balanced by reductions in another (Parks, 1982). This is true at the level of diet components so that, for example, increasing levels of rapidly fermented carbohydrates tend to be associated with reduced levels of fibre. It is also true at the level of nutrients produced in the rumen. The fact that changes in proportions of volatile fatty acids (VFA) tend to be correlated, and that there were often correlations with levels of *trans* C18:1 fatty acids in milk, led to the proposal of a number of different mechanisms for milk fat depression through simple associations. Bauman and Griinari (2003) pointed out that differences in rumen VFA proportions with low fibre diets reflect increased propionate production rather that a decrease in production of acetate and butyrate, so the hypothesis that these are limiting for *de novo* fatty acid synthesis is weak. Similarly, these authors argue that the postulated effect of propionic acid (via insulin) is unlikely to be quantitatively significant.

It is now clear that some biohydrogenation intermediates of dietary PUFA are able to elicit milk fat depression: *trans*-10 C18:1 and *trans*-10, *cis*-12 CLA, and probably others. The evidence for *trans*-10 C18:1 is indirect, based on correlations with milk fat content (Griinari *et al.*, 1998), whilst the availability of a pure form of *trans*-10, *cis*-12 CLA meant that direct causality could be established (Baumgard *et al.*, 2001). *Trans*-10, *cis*-12 CLA production has been associated with high levels of concentrate (starch) feeding (Beaulieu *et al.* 2002; Wang *et al.* 2002), which reflects the nature of the major culturable rumen bacteria known to produce this isomer (*Megasphaera elsdenii*; Kim *et al.*, 2000). Bauman and Griinari (2003) showed that changes in levels/supply of *trans*-10 C18:1 and *trans*-10, *cis*-12 CLA are not always associated with milk fat content and cautioned that other biohydrogenation intermediates are probably involved. In particular, milk fat depression associated with feeding fish oil leads to increases in TVA and *cis*-9, *trans*-11 CLA which is not thought to cause milk depression, but had little effect on *trans*-10, *cis*-12 CLA which does. In this case, it is possibly *trans*-10 C18:1 that is responsible for milk fat depression: this isomer increased from zero to 1.6 g/100g of milk fatty acids when Offer *et al.* (1999) fed 250 g/d fish oil. Given the earlier problems with correlated variables and the complex interactions between fatty acids, particularly post-rumen, it is essential that further work in this area applies multivariate tools to describe inter-relationships between fatty acids (Fievez *et al.*, 2003; Cabrita *et al.*, 2003) and their effects (e.g. on milk fat content).

What is possible with concentrates?

MILK FAT

Sutton (1986) illustrated the general effect of forage to concentrate ratio on milk fat content. Milk fat content is relatively unaffected when concentrates make up less than 0.6 of diet DM, but there is a progressively increased risk of milk fat depression at higher levels of concentrate feeding. It is relatively easy to depress milk fat content by up to 0.5, from a typical 35 to 40 g/kg down to 20 g/kg, though this is neither predictable nor safe. A number of strategies can be used to reduce milk fat depression with high concentrate inclusions, including increasing the number of concentrate meals (Sutton *et al.*, 1985), and using concentrates based on maize which is more slowly fermented than barley (Sutton *et al.*, 1980) or digestible fibre sources such as sugar beet pulp (Sutton *et al.*, 1987). It had been assumed that these effects were related to variation in production of VFA in the rumen, though it now appears (Bauman and Griinari, 2003) that although correlated with VFA production, the effect is related to production of fatty acid biohydrogenation intermediates in the rumen.

The other concentrate-based strategy to decrease milk fat content is feeding PUFA- particularly from fish oil. The wide range of levels of milk fat depression observed in response to fish oil are shown in Table 14.1. The biohydrogenation theory of milk fat depression (Bauman and Griinari, 2003) provides a common explanation for all of the concentrate-based effect on milk fat content, since both starch and fish oil affect the production of biohydrogenation intermediates. Further work is needed to identify the precise isomers involved and modes of action.

The early experiment of Bines *et al.* (1978) showed the potential to increase milk fat content by feeding fat supplements which were 'protected' both from the action of the rumen and from action on the rumen. Well protected fats provide additional precursors for milk fat synthesis, whilst they do not lead to production of biohydrogenation intermediates associated with milk fat depression. However, this type of supplement also reduces milk protein and might not be suitable for producers where this affects milk price.

Recent research (Lock *et al.*, 2004) has shown potential to regulate milk fat content by inclusion of a rumen protected (lipid encapsulated) form of *trans*-10, *cis*-12 CLA- one of the biohydrogenation intermediates known to be involved in milk fat depression- in concentrates. The product (80 g/d) supplied 7.8 g/d of *trans*-10, *cis*-12 CLA, resulted in 0.15 g *trans*-10, *cis*-12 CLA per 100g of milk fat, and caused a 0.13 proportional reduction in milk fat content.

Mobilisation of body fat can make a significant contribution to milk fat content in early lactation (Bauman and Griinari, 2003). However, effects of manipulating body reserves in the dry period, whether by feeding better forages or additional concentrates, have not been consistent. This situation is probably complicated by the fact that cows with greater body reserves tend to have reduced (forage) intakes (Ingvartsen *et al.*, 1999).

MILK PUFA

Levels of PUFA in milk are usually low as a direct consequence of rumen biohydrogenation. A number of studies have shown that it is possible to obtain much higher levels if biohydrogenation can be avoided. Feeding high levels of encapsulated safflower oil led to milk with 35 g C18:2/100g fatty acids (Chilliard *et al.*, 2000), whilst feeding high levels of a product in which linseed oil was protected with formaldehyde-treated proteins led to milk with 20 g C18:3/100g milk fat (McDonald and Scott, 1977). Petit *et al.* (2002) obtained milk with 14 g C18:3/100g milk fat by duodenal infusion of linseed oil. A number of protected oilseed products have been fed to dairy cows with varying effects on C18:3. For example, Goodridge *et al.* (2001) obtained milk with 6.4 g C18:3/100g milk fat using a protected linseed product, whilst

Table 14.1 Effects of feeding unprotected fish oil on milk fat from dairy cows.

Experiment	Fish oil level (g/d)	Control				+ fish oil			
		Milk (kg/d)	Fat (g/kg)	c9t11 CLA (g/100g)	t10c12 CLA (g/100g)	Milk (kg/d)	Fat (g/kg)	c9t11 CLA (g/100g)	t10c12 CLA (g/100g)
Mattos et al. 2004	200	25.0	52.0	0.46	0	20.4	43.5	0.68	0.021
Abughazaleh et al. 2003	432	28.6	34.9	0.40	0	29.7	32.5	0.88	0.04
Petit et al. 2002	206	22.1	41.2			22.2	31.1		
Whitlock et al. 2002	432	32.1	35.1	0.60	0	29.1	27.9	2.03	0.03
Baer et al. 2001	2% of DM		33.7	0.66			22.9	2.43	
Donovan et al. 2000	612	31.7	29.7	0.60		27.4	23.0	1.90	
Jones et al. 2000	394	30.3	27.4	1.10*		31.6	22.5	2.24*	
Keady et al. 2000	450	22.5	42.3			25.7	32.5		
Offer et al. 1999	250	16.3	45.4			16.6	34.5		
Chilliard and Doreau 1997	300	26.5	38.6			28.0	25.3		
Cant et al. 1997	348	22.3	39.0			21.0	27.4		

*total CLA

a poorly-protected product fed by Petit *et al.* (2002) resulted in only 2.0 g C18:3/100g milk fat. Processing oilseeds is generally far less effective at increasing C18:2 and C18:3 in milk than feeding rumen protected lipid supplements (Kennelly, 1996).

The highest levels of the longer-chain PUFA, EPA and DHA, in milk have been achieved using post-rumen infusions of fish oil (4.1 and 2.3 g/100g: Hagemeister *et al.*, 1988) or fish oil protected from rumen biohydrogenation by formaldehyde-treated protein (1.3 and 2.2 g/100g respectively; Gulati *et al.*, 2003). There has been some confusion about the extent of biohydrogenation of EPA and DHA in the rumen, with some studies suggesting that it is related to concentration in the diet (Gulati *et al.*, 1999). The efficiency of transfer of EPA and DHA from diet into milk has been invariably low (0.02 to 0.04; Chilliard *et al.*, 2001).

MILK CLA

The highest recorded levels of CLA (5.1 g/100g of milk fatty acids) and TVA (17.0 g/100g of milk fatty acids) in milk from normally-fed cows were observed by Gulati *et al.* (2003), who fed 1.1 kg/day of a mixture of soyabean oil and fish oil (70/30) in which the rumen protection technology was weak. More typical levels of fish oil supplementation (250 g/day) led to CLA and TVA levels of 1.55 and 7.50 g/100g of milk fatty acids respectively (Offer *et al.*, 1999)

Abomasal infusion of 150 g/d of a CLA product with mixed isomers led to production of milk with 6.4 g CLA per 100 g milk fatty acids (Chouinard *et al.*, 1999). Abomasal infusion of a protected *trans*-10, *cis*-12 CLA product (14 g/d) resulted in milk fat containing 0.7 g of *trans*-10,*cis*-12 CLA per 100 g milk fatty acids (Baumgard *et al.*, 2001).

What is possible with forages?

MILK FAT

The association between milk fat content and forage to concentrate ratio has been described above. The association between milk fat content and dietary fibre, particularly in longer forms, is so well established that milk fat depression was used as the basis for defining target levels for physically effective NDF (peNDF) in dairy rations (Mertens, 1997). This association does not depend on any assumption about the mechanism of milk fat depression with low peNDF diets.

Feeding maize silage at high levels can lead to milk fat depression in comparison with grass silage feeding (Dewhurst *et al.*, 2001), though forage mixtures are the more normal practice in the UK. Phipps *et al.* (1995) evaluated the effects of a number of different forages included as one-third of the forage component along with grass silage. Replacing grass silage with maize silage, whole-crop wheat and fodder beet had little effect on milk fat content. Inclusion of brewers' grains reduced milk fat content and, again, this might be related to the effects of its high oil content on rumen biohydrogenation.

There is growing interest in the use of red clover silage in low-input and organic dairy systems because of its N-fixing abilities. There is some evidence of a reduction in milk fat content when red clover silage replaces grass or lucerne silage (Thomas *et al.*, 1985; Broderick *et al.*, 2000), though this is not always evident (Dewhurst *et al.*, 2003a; Al-Mabruk *et al.*, 2004). It seems likely that this effect can also be explained by the biohydrogenation hypothesis of Bauman and Griinari (2003). Biohydrogenation is reduced when feeding red clover silage (Dewhurst *et al.*, 2003b; Lee *et al.*, 2003) and there is evidence of increased production of *trans*-10,*cis*-12 CLA when red clover silage is fed (Dewhurst *et al.*, 2003b)- particularly when it is the sole feed (Lee *et al.*, submitted).

MILK PUFA (INCLUDING CLA)

Plants are the primary source of omega-3 fatty acids in both terrestrial and marine ecosystems. This section will focus on potential to exploit herbage PUFA as an alternative to dwindling and increasingly polluted marine sources (Jacobs *et al.*, 2004). The concentration of C18:3 in milk depends on levels of C18:3 in feed and the extent of biohydrogenation in the rumen. Whilst forages usually represent the main source of C18:3 in the diet, this tends to be counteracted by higher levels of biohydrogenation with increasing forage proportion (increasing pH) in the diet (Latham *et al.* 1972; Kalscheur *et al.* 1997; Kucuk *et al.* 2001).

A number of studies have identified the general relationship between levels of C18:3 in herbage and concentrations of C18:3 and CLA in milk. Thomson and Van Der Poel (2000) showed this effect in relation to changes in concentrations of fatty acids in grasses over the grazing season, whilst Chouinard *et al.* (1998) showed effects of stage of growth at which grass was cut for silage-making. However, Loyola *et al.* (2002) showed differences in the CLA content of milk from cows grazing different ryegrass cultivars, despite their similar fatty acid profiles.

Some of the most marked effects of forages on milk PUFA result from feeding fresh forage as opposed to conserved hay or silage; Tables 14.2 and 14.3 (updated from Dewhurst *et al.*, 2003c) provide a summary of reported effects of fresh herbage on C18:3 and CLA. Many of these effects reflect the loss of PUFA during field wilting (see below). The highest levels of C18:3 and CLA in milk from pasture-fed cows were 2.31 and 2.21 g/100g of milk fatty acids respectively. The level of C18:3 achieved with pasture feeding was only around one-tenth of that achieved under extreme conditions and one-third of levels achieved with a more normal level of protected fat supplement. The level of CLA in milk achieved by pasture feeding was comparable to levels achieved with moderate levels of fish oil (200-400 g/ day; Table 14.1), though pasture and fish oil effects may be additive since they represent different mechanisms (supply of precursors and altering microbial metabolism; Bauman and Griinari, 2003).

Table 14.2 Effect of the forage component of diets on the α-linolenic acid (C18:n-3) content (g/100g total fatty acids) of milk fat from dairy cows.

| | *Diets based on:* | |
	Fresh forage	*Conserved forage*
Timmen & Patton, 1986	0.84 (pasture)	0.36 (grass/wheat silage)
Aii *et al.* 1988	1.97 (grass)	1.46 (grass hay)
Aii *et al.* 1988	1.34 (grass)	1.13 (grass hay)
Hebeisen *et al.* 1993	2.31 (grass)	0.45 (conserved grass)
Kelly *et al.* 1998	0.95 (grass-white clover)	0.25 (maize and legume silages)
Dhiman *et al.* 1999	2.02 (grass-white clover)	0.81 (lucerne hay; grass-white clover)
White *et al.*, 2001	0.73 (grass-white clover)	0.37 (maize and lucerne silages)
Schroeder *et al.*, 2003	0.57 (winter oat pasture)	0.07 (maize silage)
Whiting *et al.* (2004)	1.13 (lucerne)	0.83 (lucerne silage)

A number of applied studies have investigated effects of restricting pasture and shown either no change or small decreases in concentrations of C18:3 and CLA in milk when pasture allocation is reduced (Stanton *et al.*, 1997; Stockdale *et al.*, 2003). Loor *et al.* (2003) showed 0.4 to 0.7 proportional increases in concentrations of C18:3, CLA and TVA in milk when cows were grazed for 7-8 hour periods in addition to being fed conserved forages in a TMR. Alpine pasture leads to production of milk (cheese) with enhanced levels of CLA and omega-3 PUFA (Innocente *et al.*, 2002; Haswirth *et al.*,

Table 3. Effect of the forage component of diets on the conjugated linoleic acid[*] content (g/ 100g total fatty acids) of milk fat from dairy cows.

	Diets based on: Fresh forage	Conserved forage
Timmen & Patton, 1986	1.34 (pasture)	0.27 (grass and wheat silages)
Precht & Molkentin, 1997	0.76 (grass)	0.38 (maize and grass silages)
Precht & Molkentin, 1997	1.05 (grass)	0.55 (grass silage; green maize)
Kelly *et al.* 1998	1.09 (grass-white clover)	0.54 (maize and legume silages)
Dhiman *et al.* 1999	2.21 (grass-white clover)	0.89 (lucerne hay; grass-white clover)
White *et al.*, 2001	0.66 (grass-white clover)	0.36 (maize and lucerne silages)
Schroeder *et al.*, 2003	1.12 (winter oat pasture)	0.41 (maize silage)

[*] generally *cis*-9, *trans*-11 C18:2

2004). Collomb *et al.* (2002) investigated these effects further and identified relationships between milk fatty acids and the species within herb-rich pastures. A number of these associations merit further attention in terms of potential mechanisms for increasing milk PUFA.

Our recent studies with red clover silage (Dewhurst *et al.*, 2003a,b; Lee *et al.*, 2003) have identified a substantial reduction in the extent of rumen biohydrogenation of C18:3 when feeding red clover silage instead of grass silage. This effect relates, in part at least, to the effect of polyphenol oxidase, which is activated when red clover tissue is damaged and which, for example, reduces the extent of lipolysis (Lee *et al.*, 2004). Biohydrogenation was also slightly reduced when feeding legume silages (white clover and lucerne) in the studies of Dewhurst *et al* (2003b) and Doyon *et al.* (2004), perhaps because of higher rumen passage rates.

There is limited evidence of potential for production of C20:5 by chain elongation from (forage) C18:3 in the mammary gland (e.g. Hebeisen *et al.*, 1993), though this does not appear to respond to increasing supply of the precursor (Petit *et al.*, 2002) and was not seen in all studies (Whiting *et al.*, 2004).

Making more use of forage fatty acids

A number of authors (Dewhurst *et al.*, 2001; Elgersma *et al.*, 2003a; Boufaied *et al.*, 2003) have recently commented on genetic variation in herbage fatty acid levels and we have identified QTL in a ryegrass mapping family (LB Turner and RJ Dewhurst, unpublished) which will facilitate rapid selection for high-lipid grasses. As with most herbage quality traits, there are substantial environmental effects and interactions that make exploitation more difficult. The most notable effects are the number and timing of cuts, and season (Saito *et al.*, 1969; Bauchart *et al.*, 1984; Dewhurst *et al.*, 2001; Boufaied *et al.*, 2003; Elgersma *et al.*, 2003a,b). Management practices that inhibit initiation of flowering increase fatty acid levels. (Bauchart *et al.*, 1984; Dewhurst *et al.*, 2002).

Oxidative loss of PUFA during field wilting represents a major loss to the food chain, with substantial losses of C18:3 during hay-making (Aii *et al.*, 1988) and modest losses during wilting prior to ensiling (Dewhurst and King 1998; Boufaied *et al.*, 2003). These losses are associated with the lipoxygenase system - a plant defence mechanism which is initiated in damaged tissues. Plant lipases release free C18:3 and C18:2 from damaged membranes (Thomas, 1986) and these are rapidly converted to hydroperoxy PUFA by the action of lipoxygenases (Feussner and Wasternack, 2002). The hydroperoxy PUFA are further catabolised to yield a range of volatile anti-microbial and anti-fungal compounds, such as leaf aldehydes and alcohols (e.g. Fall *et al.*, 1999). These compounds develop rapidly and provide the smell of freshly-cut herbage.

'Stay-green' grasses have provided one approach to reduce losses during wilting. We have investigated a 'stay-green' grass which lacks one of the enzymes involved in chlorophyll breakdown (Thomas and Smart, 1993) and so retains thylakoid membrane structure later in senescence. Stay-green grasses showed substantially reduced losses of fatty acids when artificially senesced (Harwood *et al.* 1982). Dewhurst *et al.* (2002) found a small reduction in losses of fatty acids during wilting, though the effect may have been limited by the rapid drying conditions.

Lipolysis is an essential pre-requisite for rumen biohydrogenation of fatty acids in the rumen. It is likely that reduced lipolysis offers at least part of the explanation for reduced rumen biohydrogenation of C18:3 with red clover silage in comparison with grass silage (Lee *et al.* 2003). Moate *et al.* (2004) suggested that lipolysis rate varied between 0.03 and 5.00 per hour across a wide range of feeds, with the lowest rate for cocksfoot hay. It was suggested that reduced lipolysis (and consequently reduced biohydrogenation potential) resulted from the difficulty of releasing intracellular lipids into the rumen fluid with this dry, fibrous feed.

A reduction of lipolysis, in the silo and/or rumen, as a consequence of the polyphenol oxidase system may be the basis for effects of red clover (silage) on lipid metabolism. Red clover shows considerable potential as a natural route for production of both omega-3 enriched milk fat and reduced-fat milk. Further work is needed to understand the basis of these effects and to enhance them.

Recent work at IGER (MRF Lee and RJ Dewhurst, unpublished) has investigated the effects of volatiles (aldehydes and alcohols) released from freshly damaged herbage on fatty acid biohydrogenation in an *in vitro* rumen system. The effect of these compounds on bacterial proliferation is well recognised in other ecosystems (Deng *et al.*, 1993). This study provided preliminary evidence of the toxicity of these compounds to rumen micro-organisms (a reduction in the synthesis of odd-chain fatty acids). Some of these compounds led to increased rumen biohydrogenation: possibly reflecting an association between the ability to handle these compounds and ability to biohydrogenate PUFA.

Conclusions

Dairying in the UK is tending to diverge, with some producers feeding high levels of concentrates and trying to produce commodity milk at low cost through diluting maintenance requirements. Other producers see potential in developing speciality niche production systems, linked to either product composition (e.g. omega-3 enriched milk) or system attributes (e.g. organic milk). Maintaining milk fat concentrations above statutory minima is mainly an issue for producers targeting maximum milk yields and low-cost production. The main causes of milk fat depression in these situations have been known for many years. Producers need to avoid feeding high levels of processed, starchy concentrates in large meals, maintain adequate levels of physically effective fibre, and avoid high levels of PUFA, particularly fish oil. Rumen-protected fats have a role in maintaining milk fat content in high-yielding cows, though these can lead to depression of milk protein content. There is often a trade-off between milk fat content and milk protein content - with strategies based on reducing levels of rapidly fermentable carbohydrates (starch) and increasing protected fat tending to increase milk fat content, but reduce milk protein content.

Earlier hypotheses about the role of VFA in milk fat depression now appear to be incomplete, which highlights the need to be careful about attributing effects to individual nutrients within complex and partially correlated mixtures. This caution should be applied to development of the new biohydrogenation hypothesis of milk fat depression since duodenal fatty acid profiles are notoriously complex and correlated. The biohydrogenation theory provides a unifying hypothesis for these effects which appear to result from the action of several intermediates of biohydrogenation - notably *trans*-10 C18:1 and *trans*-10, *cis*-12 CLA. There is some evidence that this mechanism is also relevant when feeding red clover silage and further work is justified to exploit this mechanism and facilitate production of natural low-fat milk from low-input and organic systems.

Paradoxically, opportunities to manipulate the fatty acid composition of milk fat are greatest with the use of specialised supplements. Use of protected post-rumen infusion or rumen protected fat supplements in a range of studies has resulted in milk with up to 35 g of C18:2, 20 g of C18:3, 6.4 g of CLA, 1.4 g of EPA and 2.2 g of DHA per 100g milk fatty acids. Overall, forages currently represent one of the best practical approaches to increasing *cis*-9, *trans*-11 CLA and TVA in milk. There is potential to increase delivery of omega-3 fatty acids from forage into milk through plant breeding and management designed to increase levels in herbage and to reduce losses during field wilting and rumen biohydrogenation. Thus low-input or organic milk might be associated with naturally increased levels of CLA or omega-3 fatty acids. However, concentrations achieved are much lower than those that have been achieved using protected lipids or post-rumen infusion.

References

AbuGhazaleh, A.A., Schingoethe, D.J., Hippen, A.R., Kalscheur, K.F. and Whitlock, L.A. (2002) Fatty acid profiles of milk and rumen digesta from cows fed fish oil, extruded soybeans or their blend. *Journal of Dairy Science*, **85**, 2266-2276.

Aii, T., Takahashi, S., Kurihara, M. and Kume, S. (1988) The effects of Italian ryegrass hay, haylage and fresh Italian ryegrass on the fatty acid composition of cows' milk. *Japanese Journal of Zootechnical Science*, **59**, 718-724.

Al-Mabuk, R.M., Beck, N.F.G. and Dewhurst, R.J. (2004) Effects of silage species and supplemental vitamin E on the oxidative stability of milk. *Journal of Dairy Science*, **87**, 406-412.

Baer, R.J., Ryali, A., Schingoethe, D.J., Kasperson, K.M., Donovan, D.C., Hippen, A.R. and Franklin, S.T. (2001) Composition and properties of milk and butter from cows fed fish oil. *Journal of Dairy Science*, **84**, 345-353.

Bauchart D., Verite, R. and Remond, B. (1984) Long-chain fatty acid digestion in lactating cows fed fresh grass from spring to autumn. *Canadian Journal of Animal Science,* **64(Suppl.)**, 330-331.

Bauman, D.E. and Griinari, J.M. (2003) Nutritional regulation of milk fat synthesis. *Annual Reviews of Nutrition*, **23**, 203-227.

Baumgard, L.H., Sangster, J.K. and Bauman, D.E. (2001) Milk fat synthesis in dairy cows is progressively reduced by increasing supplemental amounts of *trans*-10, *cis*-12 conjugated linoleic acid (CLA). *Journal of Nutrition*, **131**,1764-1769.

Beaulieu, A.D., Drackley, J.K. and Merchen, N.R. (2002) Concentrations of conjugated linoleic acid (*cis*-9, *trans*-11-octadecadienoic acid) are not increased in tissue lipids of cattle fed a high-concentrate diet supplemented with soybean oil. *Journal of Animal Science*, **80**, 847-861.

Belury, M.A. (2002) Dietary conjugated linoleic acid in health: physiological effects and mechanism of action. *Annual Review of Nutrition*, **22**, 505-531.

Bines, J.A., Brumby, P.E., Storry, J.E., Fulford, R.J. and Braithwaite, G.D. (1978) The effect of protected lipids on nutrient intakes, blood and rumen metabolites and milk secretion in dairy cows during early lactation. *Journal of Agricultural Science*, **91**, 135-150.

Boufaïed, H., Chouinard, P.Y., Tremblay, G.F., Petit, H.V., Michaud, R. and Bélanger, G. (2003) Fatty acids in forages. I. factors affecting concentrations. *Canadian Journal of Animal Science*, **83**, 501-511.

Broderick, G.A., Walgenbach, R.P. and Sterrenburg, E. (2000) Performance of lactating dairy cows fed alfalfa or red clover silage as the sole forage. *Journal of Dairy Science*, **83**, 1543-1551.

Cabrita, A.R.J., Fonseca, A.J.M., Dewhurst, R.J. and Gomes, E. (2003) Nitrogen supplementation of corn silages. 2. Assessing rumen function using fatty acid profiles of bovine milk. *Journal of Dairy Science*, **86**, 4020-4032.

Cant, J.P., Fredeen, A.H., MacIntyre, T., Gunn, J. and Crowe, N. (1997) Effect of fish oil and monensin on milk composition in dairy cows. *Canadian Journal of Animal Science*, **77**, 125-131.

Chilliard, Y. and Doreau, M. (1997) Influence of supplementary fish oil and rumen-protected methionine on milk yield and composition in dairy cows. *Journal of Dairy Research*, **64**, 173-179.

Chilliard, Y., Ferlay, A., Mansbridge, R.M. and Doreau M. (2000) Ruminant milk fat plasticity: nutritional control of saturated, polyunsaturated, *trans* and conjugated fatty acids. *Annales de Zootechnie*, **49**, 181-205

Chouinard, P.Y., Corneau, L, Kelly, M.L., Griinari, J.M. and Bauman, D.E. (1998) Effect of dietary manipulation on milk conjugated linoleic acid concentrations. *Journal of Animal Science*, **76 (Suppl.1)**, 233 (abstract).

Chouinard, P.Y., Corneau, L., Barbano, D.M., Metzger, L.E. and Bauman, D.E. (1999) Conjugated linoleic acids alter milk fat composition and inhibit milk fat secretion in dairy cows. *Journal of Nutrition*, **129**, 1579-1584.

Clandinin, M.T., Cook, S.L., Konrad, S.D. and French, M.A. (2000) The effect of palmitic acid on lipoprotein cholesterol levels. *International Journal of Food Sciences and Nutrition,* **51**, S61-S71.

Collomb, M., Bütikofer, U., Sieber, R., Jeangros, B. and Bosset, J-O. (2002) Correlation between fatty acids in cows' milk fat produced in the lowlands, mountains and highlands of Switzerland and botanical composition of the fodder. *International Dairy Journal*, **12**, 661-666.

Corl, B.A., Barbano, D.M., Bauman, D.E. and Ip, C. (2003) *Cis*-9, *trans*-11 CLA derived endogenously from trans-11 18:1 reduces cancer risk in rats. *Journal of Nutrition*, **133**, 2893-2900.

Davis, C.L. and Brown, R.E. (1970) Low-fat milk syndrome. In: *Physiology of Digestion and Metabolism in the Ruminant*. Ed. AT Phillipson. Pages 545-565. Oriel Press: Newcastle-upon-Tyne.

Deng, W., Hamilton-Kemp, T.R., Nielsen, M.T., Andersen, R.A., Collins, G.B. and Hildebrand, D.F. (2004) Effects of six-carbon aldehydes and alcohols on bacterial proliferation. *Journal of Agricultural and Food Chemistry*, **41**, 506-510.

Dewhurst, R.J. and King, P.J. (1998) Effects of extended wilting, shading and chemical additives on the fatty acids in laboratory grass silages. *Grass and Forage Science*, **53**, 219-224.

Dewhurst, R.J., Scollan, N.D., Youell, S.J., Tweed, J.K.S. and Humphreys, M.O. (2001) Influence of species, cutting date and cutting interval on the fatty acid composition of grasses. *Grass and Forage Science*, **56**, 68-74.

Dewhurst, R.J., Wadhwa D., Borgida L.P. and Fisher W.J. (2001) Rumen acid production from dairy feeds. 1. Effects on feed intake and milk production of dairy cows offered grass or corn silage. *Journal of Dairy Science*, **84**, 2721-2729.

Dewhurst, R.J., Moorby, J.M., Scollan, N.D., Tweed, J.K.S. and Humphreys, M.O. (2002) Effects of a stay-green trait on the concentrations and stability of fatty acids in perennial ryegrass. *Grass and Forage Science*, **57**, 360-366.

Dewhurst, R.J., Fisher, W.J., Tweed, J.K.S. and Wilkins, R.J. (2003a) Comparison of grass and legume silages for milk production. 1. Production responses with different levels of concentrate. *Journal of Dairy Science*, **86**, 2598-2611.

Dewhurst, R.J., Evans, R.T., Scollan, N.D., Moorby, J.M., Merry, R.J. and Wilkins, R.J. (2003b) Comparison of grass and legume silages for milk production. 2. *In vivo* and *in sacco* evaluations of rumen function. *Journal of Dairy Science*, **86**, 2612-2621.

Dewhurst, R.J., Scollan, N.D., Lee, M.R.F., Ougham, H.J. and Humphreys, M.O. (2003c). Forage breeding and management to increase the beneficial fatty acid content of ruminant products. *Proceedings of the Nutrition Society*, **62**, 329-336.

Dhiman, T.R., Anand, G.R., Satter, L.D. and Pariza, M.W. (1999) Conjugated linoleic acid content of milk from cows fed different diets. *Journal of Dairy Science*, **82**, 2146-2156.

Donovan, D.C., Schingoethe, D.J., Baer, R.J., Ryali, J., Hippen, A.R. and Franklin, S.T. (2000) Influence of dietary fish oil on conjugated linoleic acid and other fatty acids in milk fat from lactating dairy cows. *Journal of Dairy Science*, **83**, 2620-2628.

Doyon, A., Tremblay, G.F. and Chouinard, P.Y. (2004) Lactation performance and milk fatty acid profile of dairy goats fed four different forage species. *Journal of Dairy Science*, **87 (Suppl. 1),** 467 (abstract).

Elgersma, A., Ellen, G., van der Horst, H., Muuse, B.G., Boer, H. and Tamminga, S. (2003) Comparison of the fatty acid composition of fresh and ensiled perennial ryegrass (*Lolium perenne* L.), affected by cultivar and regrowth interval. *Animal Feed Science and Technology*, **108**, 191-205.

Elgersma, A., Ellen, G., van der Horst, H., Muuse, B.G., Boer, H. and Tamminga, S. (2003). Influence of cultivar and cutting date on the fatty acid composition of perennial ryegrass (*Lolium perenne* L.). *Grass and Forage Science*, **58**, 323-331.

Fall, R., Karl, T., Hansel, A, Jordan, A. and Lindinger, W. (1999) Volatile organic compounds emitted after leaf wounding: On-line analysis by proton-transfer-reaction mass spectrometry. *Journal of Geophysical Research-Atmospheres*, **104**, 15963-15974.

Feussner, I. and Wasternack, C. (2002). The lipoxygenase pathway. *Annual Review of Plant Biology*, **53**, 275-297.

Fievez, V., Vlaeminck, B., Dhanoa, M.S. and Dewhurst, R.J. (2003). Use of principal components analysis to investigate the origin of heptaecenoic and conjugated linoleic acids in milk. *Journal of Dairy Science*, **86**, 4047-4053.

Goodridge, J., Ingalls, J.R. and Crow, G.H. (2001) Transfer of omega-3 linolenic acid and linoleic acid to milk from flaxseed or Linola protected with formaldehyde. *Canadian Journal of Animal Science*, **81**, 525-532.

Griinari, J.M., Dwyer, D.A., McGuire, M.A., Bauman, D.E., Palmquist, D.L. and Nurmela, K.V.V. (1998) Trans-octadecenoic acids and milk fat depression in lactating dairy cows. *Journal of Dairy Science*, **81**, 1251-1261.

Gulati, S.K., Ashes, J.R. and Scott, T.W. (1999) Hydrogenation of eicosapentaenoic and docosahexaenoic acids and their incorporation into milk fat. *Animal Feed Science and Technology*, **79**, 57-64.

Gulati, S.K., McGrath, S., Wyn, P.C. and Scott, T.W. (2003) Preliminary results on the relative incorporation of docosahexaenoic and eicoapentaenoic acids into cows milk from two types of rumen protected fish oil. *International Dairy Journal*, **13**, 339-343.

Hagemeister H., Precht, D. and Barth, C.A. (1988) Zum transfer von omega-3-

fettsäuren in das milchfett bei kühen. *Michwissenschaft*, **43**, 153-158.

Harwood, J.L., Jones, A,W,H,M, and Thomas, H. (1982) Leaf senescence in a non-yellowing mutant of *Festuca pratensis*. III. Total acyl lipids of leaf tissues during senescence. *Planta*, **156**, 152-157.

Hauswirth, C.B., Scheeder, M.R.L. and Beer, J.H. (2004). High ω-3 fatty acid content in alpine cheese. The basis for an alpine paradox. *Circulation*, **109**, 103-107.

Hebeisen, D.F., Hoeflin, F., Reusch, H.P., Junker, E. and Lauterburg, B.H. (1993) Increased concentrations of omega-3 fatty acids in milk and platelet rich plasma of grass-fed cows. *International Journal of Vitaminology and Nutrition Research*, **63**, 229-233.

Ingvartsen, K.L., Friggens, N.C. and Faeverdin, P. (1999) Food intake regulation in late pregnancy and early lactation. *British Society of Animal Science Occasional Publication* Number 24, 37-54.

Innocente, N., Praturlon, D. and Corradini, C. (2002) Fatty acid profile of cheese produced with milk from cows grazing on mountain pastures. *Italian Journal of Food Science*, **14**, 217-224.

Jacobs, M.N., Covaci, A., Gherghe, A. and Schepen, P. (2004) Time trend investigation of PCBs, PBDEs, and oragnochlorine pesticicdes in selected n-3 polyunsaturated fatty acid rich dietary fish oil and vegetable oil supplements; nutritional relevance for human essential n-3 fatty acid requirements. *Journal of Agricultural and Food Chemistry*, **52**, 1780-1788.

Jensen, R.G. (2002) The composition of bovine milk lipids: January 1995 to December 2000. *Journal of Dairy Science*, **85**, 295-350.

Jones, D.F., Weiss, W.P. and Palmquist, D.L. (2000) Short communication: Influence of dietary tallow and fish oil on milk fat composition. *Journal of Dairy Science*, **83**, 2024-2026.

Kalscheur, K.F., Teeter, B.B., Piperova, L.S. and Erdman, R.A. (1997) Effect of forage concentration and buffer addition on duodenal flow of trans-$C_{18:1}$ fatty acids and milk fat production in dairy cows. *Journal of Dairy Science*, **80**, 2104-2114.

Keady, T.W.J., Mayne, C.S. and Fitzpatrick, D.A. (2000) Effects of supplementation of dairy cattle with fish oil on silage intake, milk yield and milk composition. *Journal of Dairy Research*, **67**, 137-153.

Kelly, M.L., Kolver, E.S., Bauman, D.E., Van Amburgh, M.E. and Muller, L.D. (1998) Effect of intake of pasture on concentrations of conjugated linoleic acid in milk of lactating cows. *Journal of Dairy Science*, **81**, 1630-1636.

Kennelly J.J. (1996) The fatty acid composition of milk fat as influenced by feeding oilseeds. *Animal Feed Science and Technology*, **60**, 137-152.

Kim, Y.J., Liu, R.H., Bond, D.R. and Russell, J.B. (2000) Effect of linoleic acid concentration on conjugated linoleic acid production by *Butyrivibrio fibrisolvens* A38. *Applied and Environmental Microbiology*, **66**, 5266-5230.

Kucuk, O., Hess, B.W., Ludden, P.A. and Rule, D.C. (2001) Effect of forage:concentrate ratio on ruminal digestion and duodenal flow of fatty acids in ewes. *Journal of Animal Science*, **79**, 2233-2240.

Latham, M.J., Storry, J.E. and Sharpe, M.E. (1972) Effect of low-roughage diets on the microflora and lipid metabolism in the rumen. *Applied Microbiology*, **24**, 871-877.

Lee, M.R.F., Martinez, E.M. and Scollan, N.D. (2003) Plant enzyme mediated lipolysis of Lolium perenne and Trifolium pratense in an in-vitro simulated rumen environment. *Aspects of Applied Biology*, **70**, 115-120.

Lee, M.R.F., Harris, L.J., Dewhurst, R.J., Merry, R.J. and Scollan, N.D. (2003) The effect of clover silages on long chain fatty acid rumen transformations and digestion in beef steers. *Animal Science*, **76**,491-501.

Lee, M.R.F., Winters, A.L., Scollan, N.D., Dewhurst, R.J., Theodorou, M.K. and Minchin, F.R. (2004) Plant mediated lipolysis and proteolysis in red clover with different polyphenol oxidase activities. *Journal of the Science of Food and Agriculture*, **84**, 1639-1645.

Lee, M.R.F., Connelly, P.L. Tweed, J.K.S., Dewhurst, R.J., Merry, R.J. and Scollan, N.D. Effects of mixing ryegrass and red clover silages on rumen function in beef steers. II. Lipid metabolism. *Journal of Animal Science* (submitted).

Lock, A.L., Pefield, J.W., Putnam, D.E. and Bauman, D.E. (2004) Evaluation of the degree of rumen inertness and bioavailability of trans-10, cis-12 CLA in a lipid encapsulated supplement. *Journal of Dairy Science*, **87 (Suppl. 1)**, 335 (abstract).

Lock, A.L., Corl, B.A., Bauman, D.E., Barbano, D.M. and Ip, C. (2004) The anticancer effects of vaccenic acid in milk fat are due to its conversion to conjugated linoleic acid via Δ^9-desaturase. *Journal of Dairy Science*, **87 (Suppl. 1)**, 425 (abstract).

Loor,J.J., Soriano, F.D., Lin, X., Herbein, J.H. and Polan, C.E. (2003) Grazing allowance after the morning or afternoon milking for lactating cows fed a total mixed ration (TMR) enhances trans11-18:1 and cis9,trans11-18;2 (rumenic acid) in milk fat to different extents. *Animal Feed Science and Technology*, **109**, 105-119.

Loyola, V.R., Murphy, J.J., O'Donovan, M., Devery, R., Oliveira, M.D.S. and Stanton, C. (2002) Conjugated linoleic acid (CLA) content of milk from cows on different ryegrass cultivars. *Journal of Dairy Science*, **85 (Suppl.1)**, (abstract).

McDonald, I.W. and Scott, T.W. (1977) Foods of ruminant origin with elevated content of polyunsaturated fatty acids. *World Review of Nutrition and Dietetics*, **26**, 144-207.

Mattos, R., Staples, C.R., Arteche, A., Wiltbank, M.C., Diaz, F.J., Jenkins, T.C.

and Thatcher, W.W. (2004).. The effects of feeding fish oil on uterine secretion of PGF$_{2\alpha}$, milk composition, and metabolic status of periparturient Holstein cows. *Journal of Dairy Science*, **87**, 921-932.

Mertens, D.R. (1997). Creating a system for meeting the fiber requirements of dairy cows. *Journal of Dairy Science*, **80**, 1463-1481.

Moate, P.J., Chalupa, W., Jenkins, T.C. and Boston, R.C. (2004) A model to describe ruminal metabolism and intestinal absorption of long chain fatty acids. *Animal Feed Science and Technology*, **112**, 79-105.

Offer, N.W., Marsden, M., Dixon, J., Speake, B.K. and Thacker, F.E. (1999) Effect of dietary fat supplements on levels of n-3 poly-unsaturated fatty acids, trans acids and conjugated linoleic acid in bovine milk. *Animal Science*, **69**, 613-625.

Parks, J.R. (1982) *A Theory of Feeding and Growth of Animals*. Berlin: Springer-Verlag. 322 pages.

Petit, H.V., Dewhurst, R.J., Scollan, N.D., Proulx, J.G., Khalid, M., Haresign, W., Twagiramungu, H. and Mann, G.E. (2002) Milk production and composition, ovarian function, and prostaglandin secretion of dairy cows fed omega-3 fats. *Journal of Dairy Science*, **85**, 889-899.

Phipps, R.H., Sutton, J.D. and Jones, B.A. (1995) Forage mixtures for dairy cows: the effect on dry-matter intake and milk production of incorporating either fermented or urea-treated whole-crop wheat, brewers' grains, fodder beet or maize silage into diets based on grass silage. *Animal Science*, **61**, 491-496.

Precht, D. and Molkentin, J. (1997) Effect of feeding on conjugated *cis*-9, *trans*-11,-octadecadienoic acid and other isomers of linoleic acid in bovine milk fats. *Nahrung*, **41**, 330-335.

Saito, T., Takadama, S., Kasuga, H. and Nakanishi, T, (1969) Effects on fatty acid composition of lipids in cows milk by grass and legume fed (VI) Differences of effects of district and seasons, on fatty acid composition of lipids in grass and legume. *Japanese Journal of Dairy Science*, **18**, 183-189. (In Japanese, with English Summary and Tables)

Schroeder, G.F., Delahoy, J.E., Vidaurreta, I., Bargo, F., Gagliostro, G.A. and Muller, L.D. (2003) Milk fatty acid composition of cows fed a total mixed ration or pasture plus concentrates replacing corn with fat. *Journal of Dairy Science*, **86**, 337-3248.

Stanton, C., Lawless, F., Kjellmer, G., Harrington, D., Devery, R., Connolly, J.F. and Murphy, J. (1997) Dietary influences on bovine milk *cis*-9, *trans*-11-conjugated linoleic acid content. *Journal of Food Science*, **62**, 1083-1086.

Stockdale, C.R., Walker, G.P., Wales, W.J., Dalley, D.E., Birkett, A., Shen, Z. and Doyle, P.T. (2003) Influence of pasture and concentrates in the diet of grazing dairy cows on the fatty acid composition of milk. *Journal of Dairy*

Research, **70**, 267-276.

Sutton, J.D. (1986) Milk composition. In: *Principles and Practice of Feeding Dairy Cows*. Eds. W.H. Broster, R.H. Phipps and C.L. Johnson. Pages 203-218. National Institute for Research in Dairying Technical Bulletin 8.

Sutton, J.D., Oldham, J.D. and Hart, I.C. (1980) Products of digestion, hormones and energy utilization in milking cows given concentrates containing varying proportions of barley or maize. In: *Energy Metabolism*. Ed. L.E. Mount. Pages 303-306. Butterworths: London.

Sutton, J.D., Broster, W.H., Napper, D.J. and Siviter, J.W. (1985) Feeding frequency for lactating cows: effects on digestion, milk production and energy utilization. *British Journal of Nutrition*, **53**, 117-130.

Sutton, J.D., Bines, J.A., Morant, S.V. and Givens, D.I. (1987) A comparison of starchy and fibrous concentrates for milk production, energy utilization and hay intake by Friesian cows. *Journal of Agricultural Science*, **109**, 375-386.

Thomas H. (1986) The role of polyunsaturated fatty acids in senescence. *Journal of Plant Physiology*, **123**, 97-105.

Thomas, H. and Smart, C.M. (1993) Crops that stay green. *Annals of Applied Biology*, **123**, 193-219.

Thomas, C., Aston, K. and Daley, S.R. (1985) Milk production from silage. 3. A comparison of red clover with grass silage. *Animal Production*, **41**, 23-31.

Thomson, N.A. and Van Der Poel, W. (2000) Seasonal variation of the fatty acid composition of milk fat from Friesian cows grazing pasture. *Proceedings of the New Zealand Society of Animal Production*, **60**, 314-317.

Timmen, H. and Patton, S. (1988) Milk fat globules: fatty acid composition, size and in vivo regulation of fat liquidity. *Lipids*, **23**, 685-689.

Wang, J.H., Song, M.K. Son, Y.S and Chang, M.B. (2002) Effect of concentrate level on the formation of conjugated linoleic acid and *trans*-octadecenoic acid by ruminal bacteria when incubated with oilseeds *in vitro*. *Asian-Australasian Journal of Animal Science*, **15**, 687-694.

Warensjö, E., Jansson, J-H., Bergkund, L., Boman, K., Ahrén, B., Weinehall, L., Lindahl, B., Hallmans, G. and Vessby, B. (2004). Estimated intake of milk fat is negatively associated with cardiovascular risk factors and does not increase the risk of a first acute myocardial infarction. A prospective case-control study. *British Journal of Nutrition*, **91**, 635-642.

White, S.L., Bertrand, J.A., Wade, M.P., Washburn, S.P., Green, J.T. and Jenkins, T.C. (2001) Comparison of fatty acid content of milk from Jersey and Holstein cows consuming pasture of total mixed ration. *Journal of Dairy Science*, **84**, 2295-2301.

Whiting, C.M., Mutsvangwa, T., Walton, J.P., Cant, J.P. and McBride, B.W. (2004) Effects of feeding either fresh alfalfa or alfalfa silage on milk fatty acid content in Holstein dairy cows. *Animal Feed Science and Technology*, **113**, 27-37.

Whitlock, L.A., Schingoethe, D.J., Hippen, A.R., Kalscheur, K.F., Baer, R.J., Ramaswamy, N. and Kasperson, K.M. (2002) Fish oil and extruded soybeans fed in combination increases conjugated linoleic acids in milk of dairy cows more than when fed separately. *Journal of Dairy Science*, **85**, 234-243.

Yu, S., Derr, J., Etherton, T.D. and Kris-Etherton, P.M. (1995). Plasma cholesterol-predictive equations demonstrate that stearic acid is neutral and monounsaturated fatty acids are hypocholesterolemic. *American Journal of Clinical Nutrition*, **61**, 1129-1139.

15

DEVELOPMENT OF A SIMPLE NUTRIENT BASED FEED EVALUATION MODEL FOR DAIRY COWS

HARMEN VAN LAAR[1*], ROBERT MEIJER[2], KARST MULDER[1], WILDRIK BUREMA[1], MARCEL BROK[2]
Nutreco Ruminant Research Centre, Veerstraat 38, P.O. Box 220, 5830AE, Boxmeer, The Netherlands; [2]Hendrix UTD, Botterweg 4, P.O. Box 537, 8000 AM, Zwolle, The Netherlands

Introduction

Current dairy farming is faced with ever-smaller margins, threatening individual farm survival, making economic decisions very important. Therefore, nutrition of the diary herd is, next to a nutrient based decision, more and more becoming an economic decision based on feed cost and expected milk response. An accurate prediction of milk production response to dairy diets can greatly support modern dairy farming in making this economic balance between cost and response. Current feed evaluation systems (energy and protein) are reported to be not fit for this task. To deal with the flaws of current energy and protein systems more dynamic, mechanistic and detailed feed evaluation systems have been proposed (BBSRC, 1988; Beever *et al.*, 2000). Over the past decades several models have been developed that describe dairy cow digestion and metabolism of individual nutrients in more detail (Baldwin *et al.*, 1987a; Baldwin *et al.*, 1987b; Danfear, 1990; Dijkstra *et al.*, 1993). These have given insight into actual nutrient flows in the digestive tract and metabolism of dairy cows, beyond just energy and protein. This has given rise to the concepts of glucogenic, ketogenic and aminogenic nutrients as tools to better understand dairy production responses.

Although recently developed models strive to describe digestion and metabolism processes accurately, conceptual differences between models can cause fairly large differences in outcomes (Bannink and De Visser, 1997). It is recognized that these types of model are very detailed and cannot be directly introduced for commercial routine feed evaluation, but have to be converted to suit this purpose (Boston *et al.*, 2000). Feed evaluation systems designed for routine feed evaluation should be as simple as possible (Kohn *et al.* 1998).

* Corresponding Author: Nutreco Ruminant Research Centre, Veerstraat 38, P.O. Box 220, 5830AE, Boxmeer, The Netherlands

Our business and our research group have realized the need for a prediction of dairy performance based on nutrient intake. However finding an optimal balance between the paradox of the complexity of modern mechanistic models and the need for simple models for routine feed evaluation is a continuous process that will evolve over time. The model described in this chapter is an initiative to find this balance in developing a relatively simple nutrient based feed evaluation model, incorporating some of the new insights on digestion mechanisms, yet remaining close to current more static models of feed evaluation.

Material and methods

NUTRIENT MODEL

Rumen fermentable components

To be able to calculate nutrients available to the metabolism of the dairy cow, intestinal fermentation and digestion kinetics of individual feed components have to be known. The basic chemical components taken into account in this nutrient model are, total organic matter (OM), crude protein (CP), neutral detergent fibre (NDF), acid detergent fibre (ADF), acid detergent lignin (ADL), starch, sugars, fat, and residual organic matter (ROM). ROM is calculated by subtracting CP, NDF, starch, sugars and fat from OM, and is thought to consist mainly of pectin and is therefore classified as part of the carbohydrates. ADF and ADL are used to calculate cellulose and hemicellulose. First, the model calculates the amount of rumen fermentable material for all components of all feedstuffs in the database (Equation 1). These calculations are based on nylon bag degradation characteristics of each individual component. For this purpose a database of nylon bag degradation characteristics (soluble (S in %), degradable (D in %), undegradable (U in %), and degradation rate (k_d in %/h)) of OM, CP, NDF and starch has been compiled for all feedstuffs, based on both literature values and research at our own laboratory. For ROM, degradation kinetics were calculated from degradation of the other components, relative to OM degradation. Additional degradation kinetics for sugar and fat had to be assumed for this purpose (S = 100 %, D= 0,U= 0, k_d = 300 %/h for sugar and S = 0, D= 80 %,U= 20 %, k_d= 5 %/h for fat), which were the same for all feedstuffs. Whereas for most feedstuffs nylon bag degradation characteristics were considered to be constant, for grass and maize silages (GS, MS) an exception was made. For GS and MS, linear regression equations between composition and nylon bag degradation were developed for OM, CP, NDF and starch, based on available datasets, in a way similar to Van Straalen *et al.* (1995). Passage rates of CP, and starch were considered to be 4.5 %/h for roughage and 6 %/h for concentrate feedstuffs (Tamminga *et al.*, 1994). For NDF, a lower passage rate of 3 %/h was assumed.

$$\text{Fermentable component}_i = \text{component}_i * (S_i + D_i * (k_{di} / k_{pi} + k_{di}))/100$$
$$\text{(Equation 1)}$$

Component$_i$	=	Content of component in feedstuff i (g/kg DM or product).
S$_i$	=	Washable fraction in nylon bag of component i (%).
D$_i$	=	Potentially degradable fraction of component i (%).
k$_{di}$	=	Fractional degradation rate of component i (%/h).
k$_{pi}$	=	Fractional passage rate of component i (%/h).
i	=	component being either, CP, NDF, starch, sugar, fat or ROM.

VFA production in the rumen

The amounts of rumen fermentable components form the basis for calculation of volatile fatty acid production (VFA) in the rumen. The coefficients of Bannink *et al.* (2000) were used (Table 15.1) to convert fermentable components into four main types of VFA; acetic, propionic, butyric and branched chain (BCFA) fatty acids. It was assumed that 2/3 of fermentable carbohydrates and protein were available for VFA production, and no VFA production was attributed to "fermentable" fat.

Table 15.1 Molar conversion rates of different fermentable components into four categories of volatile fatty acids (%).

	Acetic	*Propionic*	*Butyric*	*BCFA[1]*
Sugar	53	16	26	6
ROM	53	16	26	6
Starch	49	31	15	5
Hemicellulose	51	12	32	5
Cellulose	68	12	20	0
Protein	44	18	17	21

[1] BCFA: Branched chain fatty acids.
From Bannink *et al.* (2000)

Digestion in the small intestine

In this model the components that are digested in the small intestine of the dairy cow are CP, starch and fat, as detailed in subsequent paragraphs. Although fermentation in the large intestine can be considerable, for example approximately 10 to 15 % of total NDF digestion (Van Straalen, 1995), large intestinal fermentation was not incorporated into the current version of the model. However, initial calculations suggest that VFA production in the large intestines may be around 10 to 15% of VFA production.

Small intestinal protein digestion

Small intestinal digestible protein is calculated from three components, namely, digestible microbial protein, digestible bypass protein and endogenous losses, with methodology similar to Tamminga *et al.* (1994). The main difference between Tamminga *et al.* (1994) and the current model is that microbial protein production is estimated from total rumen fermentable carbohydrates as calculated by nylon bag degradation parameters (Equation 1). Tamminga *et al.* (1994) used the amount of fermentable organic matter as the basis for microbial protein production, which is calculated in a different way. Assumptions of small intestinal digestibility of microbial and bypass protein are based on recent CVB (2003) publications.

Small intestinal starch digestion

Small intestinal digestible starch was calculated from the amount of bypass starch (BS) and small intestinal digestibility of starch. For each feedstuff, bypass starch was calculated from nylon bag degradation characteristics of starch with Equation 2. In this equation 10% of the washable starch fraction was assumed to bypass the rumen (Nocek and Tamminga, 1991).

Bypass starch (%) =   Starch content * $(S*0.1 + D * (k_d / k_p + k_d))/100$

(Equation 2)

S =   Washable starch fraction in nylon bag (%).
D =   Potentially degradable starch fraction (%).
k_d =   Fractional degradation rate of starch (%/h).
k_p =   Fractional passage rate of starch (%/h).

The small intestinal digestibility of bypass starch was calculated according to Equation 3, as published by Nocek and Tamminga (1991).

| %DBS | = | 87.9 – (0.728 * (%BP)) | (Equation 3) |

%DBS   =   Digestible bypass starch (%)
%BP   =   Starch bypassing the rumen (%).

From the amount of bypass starch and its digestibility, the total amount of small intestinal digestible starch can be calculated. It has been suggested that there is a maximal level of starch that can be digested in the small intestine (Huntington, 1997) and thus utilization of bypass starch is not unlimited. To allow for this effect, utilization of bypass starch was arbitrarily reduced to 50, 25 and 12.5 % for dietary levels of DBS above, 850, 1050 and 1200 g/day.

Small intestinal fat digestion

Fat digested in the small intestine is derived from either dietary or microbial fats. Although nylon bag degradation parameters for fat are used in the model, dietary fat is not assumed to be fermented as such. Triglycerides are assumed to be hydrolysed and possibly hydrogenated in the rumen, after which fatty acids are passed to the small intestine and are subsequently digested. Digestibility of these fatty acids is assumed to be 80 %, which is the lower end of the range of fat digestibility for tallow and soap as reported by Palmquist *et al.* (1993a).

Microbial fat digested in the small intestine was estimated using 90% digestibility and a microbial fat production of 1.5 g/kg fermentable organic matter (Jenkins, 1993). Fermentable organic matter is the sum of total fermentable protein and carbohydrates.

Nutrient supply

The dairy cow receives nutrients from all parts of the gastrointestinal tract. Different nutrients supply the animal with energy which can be classified according to its final usage. In the model nutrients are distinguished according to their ability to produce milk protein, lactose or milk fat, and are called aminogenic, glucogenic and ketogenic nutrients respectively. Nutrients that become available in the gastro intestinal tract and the type of energy they supply are shown in Table 15.2. The large intestine has been omitted from this table because it is not taken into account in the present model.

Aminogenic nutrients

The aminogenic nutrient supply is equal to the amount of small intestinal digestible protein.

Table 15.2 Site of digestion of nutrients and nutrient type available to metabolism of the dairy cow.

Site of digestion	Nutrient	Energy type
Rumen	Acetic acid	Ketogenic
	Propionic acid	Glucogenic
	Butyric acid	Ketogenic
	Branched chain fatty acids	Gluco + Ketogenic
Small intestine	Fatty acids	Ketogenic
	Amino acids from protein	Aminogenic
	Glucose from starch	Glucogenic

Glucogenic nutrients

The dairy cow can derive glucose from propionic acid, part of the BCFA, intestinal digestible starch, and intestinal digestible protein that is not used for growth or milk protein synthesis, called unused protein. Of the propionic acid produced in the rumen, 7% is assumed to be metabolised within the rumen wall, and the rest is converted to glucose, forming 1 mole of glucose from 2 moles of propionic acid. As BCFA are only partly glucogenic, half of the ruminal BCFA production is assumed to be converted to glucose, at a rate of 1 mole glucose per mole of BCFA. Small intestinal digestible starch (DBS) is totally available as glucose. The potential amount of glucose formed from unused protein was assumed to be 60 g/ 100 g unused protein (Schreurs, personal communication), assuming that all protein is converted to glucose. Unused protein was estimated to be 40 % of total digestible protein intake, as milk protein is synthesised from digestible protein with an efficiency of approximately 60% (Subnel *et al.*, 1994). Glucose available from protein used for maintenance was not taken into account, as this is a relatively small source.

Ketogenic nutrients

The total ketogenic nutrient supply consists of acetic acid, butyric acid, half of the BCFA, digestible dietary fat and digestible microbial fat. The quantity of milk fatty acids that can be produced from VFA is calculated based on an average chain length of newly synthesised fatty acids in milk of 14.3 (Hanigan and Baldwin, 1994). Digestible dietary and microbial fat are assumed to be totally incorporated into milk fat.

Total net energy from nutrients

To compare total net energy supply, as calculated by the Dutch VEM system (CVB, 2003), with net energy supply from nutrients calculated by the model, energy values were assumed to be 17.6 MJ ME/kg for glucogenic nutrients, 18.4 MJ for aminogenic nutrients, and 38 MJ for ketogenic nutrients. To convert ME to NE, a k_f of 0.6 was assumed, which is also used in the VEM and UFL systems (CVB, 2003; Sauvant *et al.*, 2002).

Nutrient requirements

Aminogenic nutrients

Requirements for aminogenic nutrients for maintenance and milk protein production were calculated using the equations of CVB (2003).

Glucogenic nutrients

Requirements for glucose for maintenance and milk production were based on the work of Dijkstra (1996). Dijkstra (1996) suggests a glucose requirement of 4 g per kg metabolic body weight for maintenance. Glucose requirements for milk production are composed of a requirement for lactose production and a requirement for milk fat production. Milk lactose is assumed to be formed from glucose with an efficiency of 95%. Glucose requirements for milk fat are composed of a requirement for glucose to produce glycerol to form triglycerides, and a requirement for the processes of liponeogenesis in the mammary gland. Glycerol requires 0.12 g glucose per g milk fat, and 0.38 g glucose is required per g of fat synthesised from acetic acid.

Ketogenic nutrients

Maintenance requirements for ketogenic components are calculated by difference. First maintenance requirements for ME are estimated by assuming a requirement of 450 kJ per kg metabolic bodyweight. To estimate the ketogenic energy requirement, the energy required for maintenance from protein and glucose, and 7% of propionic acid production (see paragraph on glucogenic nutrients) are subtracted from total energy requirement for maintenance, assuming an ME value of 18.4 and 17.6 MJ/kg for protein and glucose and 1.536 kJ/mole for propionic acid. The resulting ketogenic maintenance requirements were assumed to be preferentially satisfied by butyric acid and BCFA, leaving an energy requirement to be filled by acetic acid. The acetic acid left after subtraction of the amount needed to fulfil ketogenic energy requirements was assumed to be available for milk fat synthesis.

DATABASES

To study the relationships between nutrient supply and milk response, and to develop a nutrient based milk prediction system, two datasets were used. Dataset 1 was used for initial comparisons of nutrients and production, and was used for calibration of the milk prediction system. Dataset 2 was an independent dataset used for validation of the milk prediction system. For an experiment to be incorporated into the dataset detailed information on the amount of individual raw materials and composition of roughages had to be available. From this information total ration composition was calculated, based on standard chemical composition for individual raw materials, and analysed composition for roughages.

Dataset 1

Dataset 1 consisted of data obtained from a trial by Van Duinkerken (2005). The trial was carried out with a herd of 55 to 57 Holstein-Friesian dairy cows, which were subjected to two factors, with each three levels. The first experimental factor was the level of fermentable protein above that of the requirement for microbial protein production (ruminal protein balance RPB), with levels of 0, 500 and 1000 g of surplus fermentable protein per cow per day (based on calculations of DVE/OEB system, Tamminga *et al.*, 1994). The second factor was the proportion of roughage fed as maize silage, with levels of 0, 50 and 100% of roughage fed as maize silage (100, 50 and 0 % of grass silage respectively). The experimental design was not a true factorial design, as the total heard was subjected to one treatment for a period of three weeks, after which a subsequent treatment was tested. The total number of cows was constant, throughout the trial, however animals did change during the trial, as the total trial lasted for 2 years. Each treatment was repeated either 3 or 4 times, leading to information on a total of 36 diets with accompanying milk production results. For analysis, data were averaged per cow and day. Table 15.3 shows some general statistics for this dataset.

Dataset 2

Dataset 2 was used for validation of milk production prediction, which had been calibrated on dataset 1. This dataset consisted of 14 separate experiments with a total of 38 treatments. This dataset was compiled from 4 experiments from the Animal Sciences Group of the Wageningen University and Research Centre (unpublished) and 10 experiments that were conducted at the Nutreco experimental dairy farm, the Kempenshof (unpublished results). Experiments differed in their objectives. Table 15.4 shows general statistics for dataset 2.

Table 15.3 General statistics for database 1.

	Average	*Minimum*	*Maximum*	*Std deviation*
DMI (kg/d)	20.9	17.7	23.2	1.3
NEL (MJ/kg DM)	6.59	6.15	6.97	0.21
DVE (g/kg DM)	87.2	80.7	101.2	4.7
OEB (g/d)	491	-94	1195	416
CP (g/kg DM)	166	141	195	18
Starch (g/kg DM)	137	42	269	71
Sugars (g/kg DM)	63	30	136	30
Milk (kg/d)	29.2	25.4	33.4	2.0
Milk protein (g/d)	1013	822	1180	88
Milk fat (g/d)	1337	1163	1468	82

Table 15.4 General statistics for database 2.

	Average	Minimum	Maximum	Std deviation
DMI (kg/d)	20.9	18.8	25.3	1.6
NEL (MJ/kg DM)	6.86	6.57	7.37	0.19
DVE (g/kg DM)	91.7	81.4	108.5	6.3
OEB (g/d)	544	220	959	206
CP (g/kg DM)	174	150	204	14
Starch (g/kg DM)	161	9	333	66
Sugars (g/kg DM)	78	37	169	27
Milk (kg/d)	30.9	22.7	38.0	3.0
Milk protein (g/d)	1061	820	1286	97
Milk fat (g/d)	1285	944	1677	190

CALCULATIONS AND STATISTICAL ANALYSIS

Dataset 1

The relationship between nutrient supply and milk production was studied by comparing the calculated amount of milk based on glucogenic supply, corrected for glucose needed for fat synthesis in the mammary gland and maintenance, with observed milk production. It was assumed that glucose requirement per kg of milk was 48 g plus the amount needed for fat production (see glucose requirements). Subsequently milk, milk fat, and milk protein production were compared with nutrient supply using single and multiple linear regression conducted with SAS for Windows version 8. From the analysis with multiple linear regression, equations predicting milk, milk fat, and milk protein supply were chosen to be validated with dataset 2.

Dataset 2

Dataset 2 was used to validate the multiple linear regression equations calibrated with dataset 1. This validation focused on predicting the change in milk production between control and different treatments within each experiment of dataset 2. This method was chosen to avoid problems with potential differences in genetic background of dairy cows and management level of farms in datasets 1 and 2. This would also be the preferred method of application of the model for on farm advice, because on farm milk production is influenced by many farm-specific factors.

Results and discussion

GENERAL RESULTS (DATASET 1)

Average nutrients supplied from different nutrient sources are given for diets in dataset 1 in Table 15.5. The average molar proportions of acetic, propionic, butyric and branched chain fatty acids are 0.62, 0.21, 0.12 and 0.04 respectively, which is within the ranges reported by Bannink *et al.* (2000). From Table 15.5, it is clear that for both glucogenic and ketogenic nutrients the largest nutrient source is VFA produced in the rumen. For glucogenic nutrients this has also been shown by Huntington (1997), even on high starch diets. A significant proportion (around 22%) of total glucogenic nutrients is derived from metabolism of protein. This is in agreement with Huntington (1997), who reported 23% of glucose supply to be from metabolism of amino acids. The potential of protein to substantially contribute to glucose availability was also suggested by Overton *et al.* (1999). Average glucose production from protein (27% of gluconeogenesis) is within the range of 20 to 30% of glucose formed in the liver from amino acids, as suggested by Overton and Waldron (2004). Total glucose supply is almost 2400 g/day, varying from 1745 to almost 3000 g. This is considerably lower than Huntington (1997), who reported a total glucose entry of 4589 g/day for a cow on an "American-type" diet producing 32 kg milk/day. However, it is between the values of glucose supply of 2000 and 3500 g/day reported by Overton (2001) and Overton (2003) respectively. In conclusion, the current level of glucose supply found in dataset 1 may be somewhat low compared with other sources, although not outside the range reported in literature.

Table 15.5 General statistics for the production of VFA, and availability for metabolism of total glucogenic (GN), ketogenic (KN) and aminogenic (AN) nutrients from different sources in dairy cows of dataset 1.

	Average	*Minimum*	*Maximum*	*Std deviation*
HAc (mol/d)	42	36	47	2
HPr (mol/d)	14	11	17	2
HBu (mol/d)	8	7	10	1
HBc (mol/d)	3	2	4	0
GN VFA (g/d)[1]	1470	1117	1723	155
GN B-starch (g/d)	547	97	1515	384
GN protein (g/d)	545	464	655	52
KN microbial (g/d)	152	129	171	8
KN dietary (g/d)	563	442	731	60
KN VFA (g/d)	2108	1815	2382	118
GN total (g/d)	2389	1745	2999	369
AN total (g/d)[2]	1887	1561	2353	213
KN total (g/d)[3]	2823	2416	3208	167

[1] glucogenic nutrients expressed as g glucose.
[2] aminogenic nutrients expressed as g protein.
[3] ketogenic nutrients expressed as g trigliceride (based on milk fatty acid profile).

Figure 15.1 shows the relationship between net energy supply calculated using the Dutch VEM system (CVB, 2003) and net energy calculated from supply of nutrients, both expressed in kJ per kg of DM in the diet. The r^2 of this relationship is 0.54, which represents a weak relationship between energy values calculated by these two very different models. Net energy calculated from nutrient supply is approximately 20% lower than that calculated using the VEM system. Part of this discrepancy might be explained by the absence of large intestinal fermentation in the current nutrient model. Approximately 10 to 15 % of NDF (Van Straalen, 1995) and around 4% (+/- 6 % std) of starch (Offner and Sauvant, 2004), might be fermented in the large intestine. This fermentation could still contribute considerably to total nutrient supply. Thus, the lack of a large intestinal module in the model could partly explain the low net energy supply calculated directly from nutrients, and also part of the relatively low total glucose supply as compared to literature.

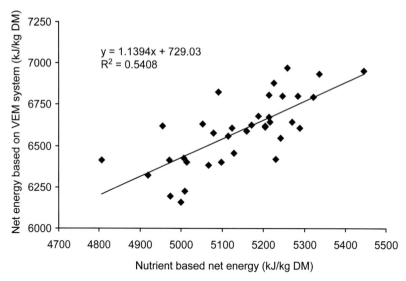

Figure 15.1 Relationship between nutrients based net energy content and net energy according to the Dutch VEM system, for diets in dataset 1.

DIRECT PREDICTION OF MILK PRODUCTION (DATASET 1)

The ultimate goal in development of nutrient based feed evaluation systems is to predict milk and milk component production, which is of great economic importance (Beever *et al.*, 2000). For the current model, the relationship between predicted and observed milk production, based on supply of total glucogenic nutrients, is shown in Figure 15.2. Milk production based on glucogenic nutrients was calculated by correcting total glucogenic nutrient supply with requirements for maintenance and milk fat production (see material and methods), assuming a glucose requirement of 48 g/kg milk. This requirement is based on an assumed lactose content in milk of 46 g/kg and a 0.95 efficiency of glucose conversion to lactose (Dijkstra *et al.*, 1996). The r^2 of the relationship between predicted and observed milk production is fairly weak (0.51). Milk production was under predicted at low milk production, and drastically over predicted at high milk production.

   To investigate possibilities for improvement of the relationship between observed and predicted milk production, two correlation analyses were performed. Firstly the correlation of the difference between observed and predicted milk production with feed intake and feed composition was calculated. Secondly the correlation of the residual of observed milk production to the regression line with chemical composition of the diet was calculated. The difference between observed and predicted milk production was highly correlated (r=0.8) with total DM intake, showing large over prediction at higher production levels. However, a stronger correlation was found (r=0.9) with starch content of the diet, suggesting a possible overestimation of the glucose supply from starch at high production levels or a decreased efficiency of glucose use at high glucose supplies.

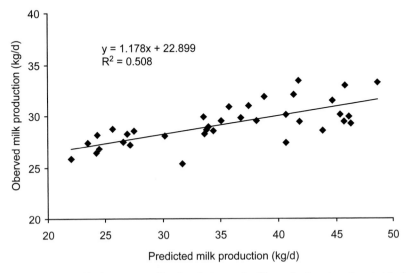

Figure 15.2 Relationship between predicted and observed milk production, based on total glucose supply.

The residual of observed milk production to the regression equation of observed and predicted milk production did not correlate well with any characteristic of diet composition. For composition traits, the highest correlation of residuals was with the content of rapidly fermentable carbohydrates (r=0.37). This might indicate that rumen fermentation was directed towards a more glucogenic nutrient (propionic acid) at higher ruminal fermentation rates (lower rumen pH), which is a know effect (Pitt *et al.*, 1996). Interactions like these, between fermentation level and proportion of propionic acid in the rumen from the same amount and type of fermentable components, are not incorporated into the present additive model.

CALIBRATION OF THE MODEL (DATASET 1)

It can be concluded that a good absolute prediction of observed milk production, based on glucogenic nutrients, cannot be obtained with the current model. Therefore, multiple linear regression analysis was used to relate milk production to nutrient supply and thus to calibrate a milk (component) prediction model on actual milk (component) production and nutrient supply as calculated in dataset 1. Table 15.6 shows the best single and multiple linear regression equations for milk, fat and protein production based on supply of total net energy (VEM), glucogenic, ketogenic and aminogenic energy supply. Although net energy (VEM) intake was the best predictor of total milk production, it was little better than glucose supply (r^2 0.59 vs 0.51). It is surprising that both fat and protein production were also best predicted by net energy intake. Protein intake should be highly correlated with aminogenic energy intake, as shown for the DVE system by Van Straalen *et al.* (1994). It is possible that higher energy (starch) diets have a larger ruminal production of microbial protein (Oba and Allen, 2003), thus explaining part of the strong correlation between milk protein and net energy.

To empirically improve the prediction of milk production and milk component production, multiple linear regressions were applied. The multiple linear regression equations relative to simple linear regression (Table 15.6) drastically improved the r^2 for the relationship with milk fat production (from 0.32 to 0.55) and marginally improved the r^2 for milk (from 0.59 to 0.61) and milk protein production (from 0.81 to 0.83).

Table 15.6 Best single and multiple linear regression equations for milk (kg/d), fat (kg/d) and protein (g/d) production. Equations shown are those with the best statistical fit (b) and those chosen for the final model (c).

	Intercept	*VEM*	*GN*	*KN*	*AN*	R^2
Milk	15.8	0.000916				0.59
Fat	932	0.0276				0.32
Protein	322	0.0472				0.81
Milk b	16.8	0.00058	0.00173	0.00092	-0.0003	0.61
Fat b	634	0.0300	-0.155	0.144	0.211	0.55
Protein b	339	0.0800	-0.0779	-0.105	-0.123	0.83
Milk c	16.7	0.00054	0.00174	0.001017		0.61
Fat c	694	0.0574	-0.159	0.0824		0.52
Protein c	4978		0.0844		0.20174	0.68

According to the best multiple linear regressions aminogenic nutrients are supposed to have a negative effect (negative coefficient) on milk and a positive effect on fat production. Furthermore, milk protein production was negatively influenced by all nutrients except total net energy. These effects might be statistically correct for this dataset, but are difficult to explain and to relate to dairy practice because they do not concur with current views of the effects of different nutrients. As the final goal of our research is practical application of the milk prediction model, it was undesirable to incorporate effects that are hard to explain. Therefore, multiple linear regression equations were chosen in which the direction of the effect of all coefficients could be explained by conventional knowledge regarding the mode of action of nutrients.

The final equations are displayed in Table 15.6 and describe a positive effect of net energy, glucogenic nutrients and ketogenic nutrients on milk production. The effect of net energy and glucogenic nutrients is self-evident, both stimulating milk production. Ketogenic energy might improve milk production by supplying intact fatty acids through the diet. As these can be efficiently incorporated into milk fat (Dijkstra *et al.*, 1996) this could decrease the need for glucogenic nutrients for liponeogenesis in the mammary gland, which would spare glucose for lactose production. Also Onnetti and Grummer (2004) conclude from their review that addition of fat to diets generally results in a moderately positive milk production response.

The equation for fat production contains a positive relationship with net energy and total ketogenic nutrients, which is logical. Glucogenic nutrients are negatively correlated with total fat production. This is thought to be caused by two possible modes of action. Firstly a higher glucogenic nutrient supply is correlated with both total energy and starch level in the diet (r = 0.71 and 0.99 respectively). High energy, high starch diets are generally shown to decrease fat production due to a lower ruminal acetate production with low rumen pH (Palmquist *et al.*, 1993b). The other possible mode of action lies in the cow's hormonal regulation. An increased glucose supply might increase insulin, thus creating an anabolic effect. Because of this effect, extra fatty acids might be used for body fat synthesis, and are therefore not available for milk fat synthesis (Lesmosquet *et al.*, 1997, Rigout *et al.*, 2002).

The equation for milk protein production contains only glucogenic and aminogenic nutrients, both positively related. The background of the positive relationship for aminogenic nutrients is clear. The positive relationship between glucogenic nutrients and milk protein production could be explained by the amino acid sparing effect of glucogenic energy. This effect might be due to the use of glucose by the small intestine as an energy source, which replaces the use of amino acids, which are thus spared; this, however, is still a contended hypothesis (Oba *et al.*, 2004). Another effect of increased glucose supply could be that conversion of amino acids to glucose, to relieve a glucose shortage, is reduced, again sparing amino acids for milk protein production as suggested by Clark *et al.* (1977).

VALIDATION OF THE MODEL (DATASET 2)

The second dataset was used to study the accuracy of the milk prediction model. As mentioned in the material and methods, it was attempted to predict the change in milk production between control and treatment within each experiment, rather than absolute milk production. To compare predicted and observed change in milk production and composition the distribution of the difference between predicted and observed (predicted – observed) is shown in Figure 15.3. Figure 15.3 shows that for milk yield, fat concentration and protein concentration most predictions are within respectively 1 kg milk, 1 g/kg fat and 1.5 g/kg protein of the observed response.

The average of the difference between predicted and observed change was -0.1 kg for milk, -1 g/kg for milk fat and 0 g/kg for milk protein showing that there was no systemic over or under prediction. The change in fat concentration tended to be under predicted. The average of the *absolute* difference between predicted and observed change was 0.95 kg for milk, 1.6 g/kg for milk fat and 0.9 g/kg for milk protein. The average difference between predicted milk production response and actual response is therefore quite high, almost 1 kg. Although this is lower than that found by Kohn *et al.* (1998), other validations of milk production prediction by CNCPS have shown better results (Kolver *et al.*, 1998). Part of the explanation for the error in predicting milk responses are the characteristics of dataset 2. Some experiments that studied the effect of glucogenic energy (more starch) did not yield results that would normally be expected. Milk production did not respond positively, as predicted, which increased errors in prediction. Other overriding factors within these experiments could have been the cause for this.

For practical application the model can be used to indicate theoretical milk production and milk component responses to dietary changes. Care has to be taken when interpreting predicted changes smaller than the error of the model.

Conclusions

A simple practical milk production and milk component prediction model has been developed on the basis of estimated supply of glucogenic, ketogenic and aminogenic nutrients. Variation still exists between actual and predicted changes in milk production and composition. However, the model does indicate theoretical responses in milk production and composition as effected by changes in nutrient composition with different diets.

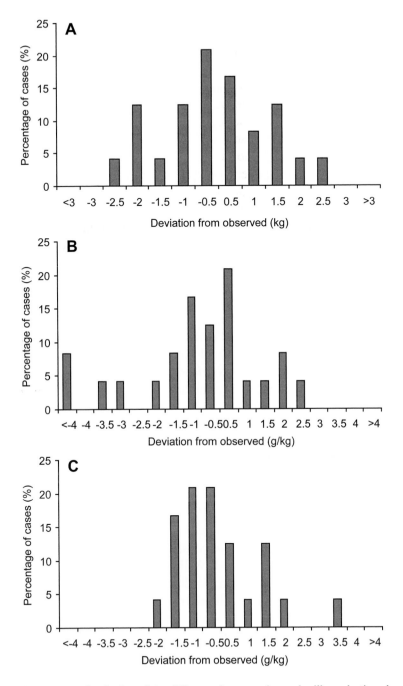

Figure 15.3 Frequency distribution of the difference between observed milk production change and predicted milk production change after a dietary change (predicted change - observed change) for milk production (A), fat concentration (B) and protein concentration (C).

Acknowledgements

Ir. Gert van Duinkerken of the animal sciences group of the Wageningen University and Research Centre is highly acknowledged for providing background data on the experiment used in dataset 1.

References

Baldwin, R.L., France, J. and Gill, M. (1987a) Metabolism of the lactating cow. I. Animal elements of a mechanistic model. *Journal of Dairy Research*, **54**, 55-105.

Baldwin, R.L, Thornley, J.H.M. and Beever, D.E. (1987b) Metabolism of the lactating cow. II. Digestive elements of a mechanistic model. *Journal of Dairy Research*, **54**, 107-131.

Bannink, A. and De Visser, H. (1997) Comparison of mechanistic rumen models on mathematical formulation of extramicrobial and microbial processes. *Journal of Dairy Science*, **80**, 1296-1314.

Bannink, A., Kogut, J., Dijkstra, J., France, J., Tamminga, S., and Van Vuuren, S. (2000) Modelling production and portal appearance of volatile fatty acids in dairy cows. In *Modelling Nutrient Utilization in Farm Animals – 2000*, pp 87-102. Edited by J.P. McNamara, J. France and D.E. Beever. CAB International, Wallingford.

Boston, R.C., Fox, D.G., Sniffen, C., Janczewski, E., Munson, R. and Chalupa, W. (2000) The conversion of a scientific model describing dairy cow nutrition and production to an industry tool: the CPM diary project. In *Modelling Nutrient Utilization in Farm Animals – 2000*, pp 87-102. Edited by J.P. McNamara, J. France and D.E. Beever. CAB International, Wallingford.

BBSRC, Biotechnology and Biological Sciences Research Council (1998) *Responses in the yield of milk constituents to the intake of nutrients by diary cows*. CAB International, Wallingford.

Beever, D.E., France, J. and Alderman, G. (2000) Prediction of response to nutrients by ruminants through mathematical modeling and improved feed characterization. In *Feedings Systems and Feed Evaluation Models – 2000*, pp 275-297. Edited by MK Theodorou and J. France. CAB International, Wallingford.

Clark, J.H., Spires, H.R., Derrig, R.G. and Bennink, M.R. (1977) Milk production, nitrogen utilization and glucose synthesis in lactating cows infused post ruminally with sodium caseinate and glucose. *Journal of Nutrition*, **107**, 631-644.

CVB (2003) *Veevoedertabel 2003, gegevens over chemische samenstelling, verteerbaarheid en voederwaarde van voedermiddelen.* [in dutch] CVB, Lelystad.

Danfaer, A. (1990) *A dynamic model of nutrient digestion and metabolism in lactating dairy cows.* Ph.D. Dissertation, National institute for Animal Science, Foulum, Denmark.

Dijkstra, J. (1993) Mathematical modeling and integration of rumen fermentation processes. Ph.D. Dissertation. Wageningen Agricultural University, Wageningen, The Netherlands.

Dijkstra, J., France, J., Assis, A.G., Neal, H.D.ST.C, Campos, O.F. and Aroeira, L.J.M. (1996) Simulation of digestion in cattle fed sugar cane: prediction of nutrient supply for milk production with locally available supplements. *Journal of Agricultural Science, Cambridge,* **127**, 247-260.

Jenkins, T.C. (1993) Lipid metabolism in the rumen. *Journal of Dairy Science,* **76**, 3851-3863.

Hanigan, M.D. and Baldwin, R.L. (1994) A mechanistic model of mammary gland metabolism in the lactating cow. *Agricultural Systems,* **45**, 396-419.

Huntington, G.B. (1997) Starch utilization by ruminants: From basics to bunk. *Journal of Animal Science,* **75**, 852-867.

Kohn, R.A., Kalscheur, K.F. and Hanigan, M. (1998) Evaluation of models for balancing the protein requirements of dairy cows. *Journal of Dairy Science,* **81**, 3402-3414.

Kolver, E.S., Muller, L.D., Barry, M.C. and J.W. Penno (1998) Evaluation and application of the Cornell net carbohydrate and protein system for dairy cows fed diets based on pasture. *Journal of Dairy Science,* **81**, 2029-2039.

Lesmosquet, S., Rideau, N., Rulquin, H., Faverdin, P., Simon, J. and Verite, R. (1997) Effects of a duodenal glucose infusion on the relationship between plasma concentrations of glucose and insulin in dairy cows. *Journal of Dairy Science,* **80**, 2854-2865.

Nocek, J.E. and Tamminga, S. (1991) Site of digestion of starch in the gastrointestinal tract of dairy cows and its effect on milk yield and composition. *Journal of Dairy Science,* **74**, 3598-3629.

Oba, M. and Allen, M.S. (2003) Effects of diet fermentability on efficiency of microbial nitrogen production in lactating dairy cows. *Journal of Dairy Science,* **86**, 195-207.

Oba, M., Baldwin VI, R.L. and Bequette, B.J. (2004) Oxidation of glucose, glutamate, and glutamine, by isolated ovine enterocytes in vitro is decreased by the presence of other metabolic fuels. *Journal of Animal Science,* **82**, 479-486.

Onetti, S.G. and Grummer, R.R. (2004) Response of lactating cows to three supplemental fat sources as affected by forage in the diet and stage of lactation: a meta-analysis of literature. *Animal Feed Science and Technology,* **115**, 65-82.

Offner, A. and Sauvant, D. (2004) Prediction of in vivo starch digestion in cattle from in situ data. *Animal Feed Science and Technology,* **111**, 41-56.

Overton, T.R., Drackley, J.K., Otteman-Abbamonte, C.J., Beaulieu, A.D., Emmert, L.S. and Clark, J.H. (1999) Substrate utilization for hepatic glucogenesis is altered by increased glucose demand in ruminants. *Journal of Animal Science,* **77**, 1940-1951.

Overton, T.R. (2001) Transition cow programs – The good, the bad, and how to keep them from getting ugly. *Advances in dairy technology,* **13**, 17-26.

Overton, T.R. (2003) Managing the metabolism of transition cows. *Proceedings of the 6th Western Dairy Management Conference March 12-14, Reno NV,* 7-16.

Overton, T.R. and Waldron, M.R. (2004) Nutritional management of transition dairy cows: Strategies to optimize metabolic health. *Journal of Dairy Science (E. Supplement),* **87**, E105-E119.

Palmquist, D.L., Weisbjerg, M.R. and Hvelplund, T. (1993a) Ruminal intestinal and total digestibilities of nutrients in cows fed diets high in fat and undegradable protein. *Journal of Dairy Science,* **76**, 1353-1364.

Palmquist, D.L., Beaulieu, A.D. and Barbano, D.M. (1993b) Feed and animal factors influencing milk fat composition. *Journal of Dairy Science,* **76**, 1753-1771.

Pitt, R.E., Van Kessel, J.S., Fox, D.G., Pell, A.N., Barry, M.C. and Van Soest, P.J. (1996) Prediction of ruminal volatile fatty acids and pH within the net carbohydrate and protein system. *Journal of Animal Science,* **74**, 226-244.

Rigout, S., Lemosquet, S., Bach, A., Blum, J.W. and Rulquin, H. (2002) Duodenal infusion of glucose decreases milk fat production in grass silage fed dairy cows. *Journal of Dairy Science,* **85**, 2541-2550.

Sauvant, D., Perez, J.M. and Tran, G. (2002) *Tables de composition et de valeur nutritive des matières premières destinées aux animaux d'élevage.* [in French] INRA, Paris.

Subnel, A.P.J., Meijer, R.G.M., Van Straalen, W.M. and Tamminga, S. (1994) Efficiency of milk protein production in the DVE protein evaluation system. *Livestock Production Science,* **40**, 139-155.

Tamminga, S., Van Straalen, W.M., Subnel, A.P.J., Meijer, R.G.M., Steg, A., Wever, C.G.J., and Blok, M.C. (1994) The Dutch protein evaluation system: the DVE/OEB-system. *Livestock Production Science,* **40**, 139-155.

Van Duinkerken, G., André, G., Smits, M.C.J., Monteny, G.J. and Sebek, L.B.J (2005) Effect of rumen degradable protein balance and forage type on bulk milk urea concentration and emission of ammonia from dairy cow houses. *Journal of Dairy Science,* **88**, 1099-1112.

Van Straalen, W.M., Salaün, C., Veen, W.A.G., Rijpkema, Y.S., Hof, G. and Boxem, T.J. (1994) Validation of protein evaluation systems by means of milk production experiments with dairy cows. *Netherlands Journal of Agricultural Science,* **42**, 89-104.

Van Straalen, W.M. (1995) *Modelling of nitrogen flow and excretion in dairy cows.* Ph.D. Dissertation, Wageningen Agricultural University, Wageningen, The Netherlands.

LIST OF PARTICIPANTS

The thirty-eighth University of Nottingham Feed Conference was organised by the following committee:

MR N.J. CHANDLER *(National Renderers Association)*
MRS M. GOULD *(Volac International)*
MR J.R. PICKFORD
DR I.H. PIKE *(IFOMA)*
MR J. TWIGGE *(Trouw Nutrition Ltd)*
DR R. TEN DOESCHATE *(ABNA Ltd)*
DR M.A. VARLEY *(Provimi Ltd)*

DR J.M. BRAMELD
PROF P.J. BUTTERY
DR P.C. GARNSWORTHY *(Secretary)*
DR T. PARR   } *University of*
DR A.M. SALTER   } *Nottingham*
PROF R. WEBB
DR J. WISEMAN *(Chairman)*

The conference was held at the University of Nottingham Sutton Bonington Campus, 14-16 September 2004. The following persons registered for the meeting.

Allen, Dr J	Frank Wright Ltd, Blenheim House, Blenheim Rd, Ashbourne DE6 1HA, UK
Allen, Mrs U	CABI Publishing, Nosworthy Way, Wallingford, Oxfordshire OX10 8DE, UK
Appleby, Mr G	Elanco Animal Health, Eli Lilley and Co Ltd, Basingstoke RG21 6XA, UK
Arhin, Mr R	Midland Poultry Processing Ltd, Post Office Box 513, Kumasi, Ghana
Aronen, Dr I P	Raiso Nutrition Ltd, P O Box 101, Raiso 21201, Finland
Atherton, Dr D	Thompson & Joseph Ltd, 119 Plumstead Road, Norwich NR1 4JT, UK
Bardsley, Dr R	University of Nottingham, Sutton Bonington Campus, Loughborough, Leics LE12 5RD, UK
Bartram, Dr C	Pye Bibby Agriculture, Lansil Way, Lancaster LA1 3QY, UK
Bayles, Mr T	Trouw Nutrition GB, Wincham Lane, Wincham, Northwich CW9 6DF, UK
Beaumont, Mr D	Soda Feed Ingredients Ltd, c/o Maple Lodge, Ryknield Hill, Denby DE5 8NW, UK
Bedford, Dr M	Zymetrics, Chestnut House, Marlborough SN8 1QJ, UK
Bennett, Mr R	Adisseo France S.A.S, 42 Avenue Aristide Briand, Antony 92160, France
Best, Mr P	Feed International, 18 Chapel Street, Petersfield GU32 3DZ, UK
Blake, Dr J	Dietcheck Ltd, Highfield, Little London, Andover SP11 6JE, UK

Bone, Mr P	Telsol Ltd, 39 Stratton Heights, Cirencester GL7 2RH, UK
Boyd, Dr J	BOCM Pauls Ltd, Tucks Mill, Burston, Diss IP23 5TJ, UK
Boydell, Miss A	BOCM Pauls Ltd, First Ave, Royal Portbury Dock, Portbury BS20 7XS, UK
Brameld, Dr J M	University of Nottingham, Sutton Bonington Campus, Loughborough, Leics LE12 5RD
Brown, Mr G	DSM Nutritional Products, Heanor Gate, Heanor, Derbyshire DE75 7SG, UK
Buttery, Prof P	University of Nottingham, Sutton Bonington Campus, Loughborough, Leics LE12 5RD, UK
Chandler, Mr N J	NRA, 52 Packhorse Rd, Gerrards Cross SL9 8EF, UK
Chijiiwa, Mr A	Kyodo Shiryo Co Ltd, 5-12 Takashima 2 Chome, Nishi-Ku, Yokohama 220 - 0011, Japan
Chuffart, Mr M	Crina S.A, Rue de la Combe 15, Gland 1196, Switzerland
Clarke, Dr E	University of Nottingham, Sutton Bonington Campus, Loughborough, Leics LE12 5RD, UK
Clarke, Mr N	Britphos Ltd, Rawden House, Green Lane, Yeadon, Leeds LS19 7BY, UK
Clay, Mr J	Alltech UK Ltd, Alltech House, Ryall Rd, Peterborough PE9 1TZ, UK
Cooke, Dr B	Consultant, 1 Jenkins Orchard, Wick St Lawrence, Weston-Super-Mare BS22 7YP, UK
Cooksley, Mr R	Cooksley & Co, Portbury House, Sheepway, Portbury BS20 7TE, UK
Cottrill, Dr B	ADAS, Woodthorne, Wergs Road, Wolverhampton WV6 9BG, UK
David, Mr F	AUSMOZ, Australia Mozambique Farm Holding, CP384, Chimio, Mozambique
Davies, Mr B	Dugdale Nutrition, Bellman Mill, Clitheroe, Lancs BB7 1QW, UK
Davies, Mr T	Kite Consulting, Rodbaston College, Rodbaston, Penkridge, Staffs ST19 5PH, UK
Davies, Dr Z E	DEFRA, Rm 701 Cromwell House, Dean Stanley St London, UK
Dewhurst, Dr R J	IGER, Plas Gogerddan, Aberystwyth SY23 3EB, UK
Doppenberg, Dr J	VVM, p o Box 2121, Deventee 7420AC, The Netherlands
Ewing, Dr W	Context Publications, 53 Mill St, Packington, Ashby de la Zouch LE65 1WN, UK
Falkowski, Dr J	Univ of Warmia & Mazury in Olsztyn, Kortowo BL 37, Olsztyn 10-719, Poland
Farley, Mr R	Trouw Nutrition GB, Wincham Lane, Wincham, Northwich CW9 6DF, UK
Fjermedal, Dr A	Fiska Molle, Grooseveien 94, Grimstad 4879, Norway

Fornos, Mr J	Cooperation D'Artesa de Segre, Av Eduard Maluquer 9, Artesa de Segre 25730, Spain
Fouladi, Dr A	University of Nottingham, Sutton Bonington Campus, Loughborough, Leics LE12 5RD
Fullarton, Mr P	Forum Products Ltd, 41-51 Brighton Road, Redhill Surrey, RH1 6YS
Garland, Mr P W	March House, Kettleburgh, Woodbridge IP13 7J2, UK
Garnsworthy, Dr P C	University of Nottingham, Sutton Bonington Campus, Loughborough, Leics LE12 5RD, UK
Gibson, Mr J H	Parnutt Foods Ltd, Hadley Road, Woodbridge Ind Est, Sleaford, Lincs NG34 7ES, UK
Gilbert, Mr R	Grain & Feed Milling Technology, 214 Prestbury Rd, Cheltenham, Gloucs GL52 3ER, UK
Givens, Prof D I	University of Reading, Earley Gate, School of Agric, Policy, & Development, Reading, Berks RG6 6AR, UK
Godfrey, Miss S-J	ABN, ABN House, P O Box 250, Oundle Road, Peterborough PEL 9QF, UK
Golds, Mrs S	University of Nottingham, Sutton Bonington Campus, Loughborough, Leics LE12 5RD, UK
Gould, Mrs M	Volac International, Volac House, Orwell, Royston, Herts SG8 5QX, UK
Granero-Rosell, Mr M A	Commission European, D G Health & Consumer Protection, Animal Nutrition Unit, Belgium
Green, Mrs K	Fishmeal Information Network, 4 The Forum, Minerva Business Park, Peterborough PE2 6FT, UK
Green, Dr S	Adisseo, Berkhamsted House, 121 High St, Berkhamsted HP4 2DJ, UK
Gutierrez, Dr C	University of Nottingham, Sutton Bonington Campus, Loughborough, Leics LE12 5RD, UK
Hammond, Dr L	British United Turkeys, Hockenhull Hall, Tarvin, Chester CH3 8LE, UK
Hegeman, Mr F	Avebe Feed Department, Transportweg 11, Veendam 9640AA, The Netherlands
Higginbotham, Dr J D	Tate & Lyle Europe, Gibraltar House, 1st Avenue, Burton-on-Trent, Staffs DE14 2WE, UK
Hill, Dr R	University of Idaho, P O Box 442330, Moscow ID 83844, USA
Hooley, Mrs E	University of Nottingham, Sutton Bonington Campus, Loughborough, Leics LE12 5RD, UK
Howells, Dr D	Advanced Liquid Feeds, Alexander House, Regent Rd, Liverpool L20 1ES, UK
Hurdidge, Mr L	Lallemand, P O Box 190, Hereford HR4 7YD, UK
Hyam, Mr J M	Qualivet, Avda Reyes Vatolicos 6; of 16A, Majadahonda 28220, Spain

Ingham, Mr R W	Kemira Growhow UK Ltd, Ince, Chester, Cheshire CH2 42B, UK
Jordan, Mr G	Regal Processors, 2 Silverwood Industrial Estate, Craigavon, Co Armagh BT66 6LN, N Ireland
Keane, Mr N	Countrywest, Underlane, Holsworthy, Devon EX22 6EE, UK
Keller, Dr T	BASF, Carl Bosch St 38, Ludwigshafen 67056, Germany
Kelley, Mr J	AIC (Agric Industries Confederation), Confederation House, East of England Showground Peterborough, PE2 6XE, UK
Kenyon, Dr R	Forum Products Ltd, 41-51 Brighton Road, Redhill RH16 6YS, UK
Kocher, Dr A	Alltech Biotechnology Centre, Sarney Summerhill Rd, Dundayne, Meath, Ireland
Lee, Miss E	University of Nottingham, Sutton Bonington Campus, Loughborough, Leics LE12 5RD, UK
Leek, Dr A	Sun Valley Foods, Tram Inn, Allensmore, Hereford HR2 9AW, UK
Lima, Mr S	Felleskjopet Rogalandagder, Box 208, Stavanger 4001, Norway
Llanes, Miss N	Cooperativa D'Ivars, Plaça Bisbe Coll 9, Ivars D'Urgell 25260, Spain
Lowe, Dr J	Dodson and Horrell, Ringstead, Kettering, Northants NN14 4BY, UK
Lucey, Mr P	Dairygold Co-operative Society Ltd, Lambardstown, Co Cork, Ireland
Macmillan, Miss S	Farmers Weekly, Quadrant House, The Quadrant, Sutton SM2 5AS, UK
Mafo, Mr A	Hi Peak Feeds Ltd, Hi Peak Feed Mills, Sheffield Rd, Killamarsh S21 1ED, UK
Martyn, Mr S	BFI Innovations Ltd, 1 Telford Court, Chester Gates, Dunkirk Lea, Chester CH1 6LT, UK
McAdoo, Mr S	Regal Processors, 2 Silverwood Industrial Estate, Craigavon, Co Armagh BT66 6LN,
McGrane, Mr M	Trouw Ireland, 206 Ryevale Lawns, Leixlip, Co Kildare , Ireland
McGuire, Mr P	Ceva Animal Health, 90 The Broadway, Chesham HP5 1EG, UK
McIlmoyle, Dr A	Nutrition Consultants, 20 Young St, Lisburn BT275 5EB, Antrim
Meeusen, Dr A	Kemin Europa N V, Toekomstlaan 42, Herentals 2200, Belgium
Meggison, Dr P A	Aust-Asia Business Solutions, 6 Brunskill Rd, New South Wales 2650, Australia
Mellor, Miss Sarah	Feed Mix, Reed Business Information, Hanzestraat 1, Doetinchem 7006 RH, Netherlands
Millar, Mr K	Food Standards Agency, Aviation House, 125 Kingsway, London, UK
Montagu, Miss J	W & H Marriage & Sons Ltd, Chelmer Mills, New St, Chelmsford, Essex CM1 1PN, UK

Mounsey, Mr A	Pentlands Publishing Ltd, Station Road, Great Longstone, Bakewell DE45 1TS, UK
Mudd, Dr A J	Alpharma, Landfall, Curdridge Lane, Waltham Chase, Southampton SO32 2LD, UK
Murray, Dr I	SAC, Ferguson Building, Craibstone, Aberdeen AB21 9YA, UK
Nam, Mr I	University of Nottingham, Sutton Bonington Campus, Loughborough, Leics LE12 5RD, UK
Nelson, Miss J	AIC (Agric Industries Confederation), Confederation House, East of England Showground Peterborough, PE2 6XE, UK
Nordang, Dr L	Fellskjopet Forutvikling, Bromstadveien 57, Trondheim N-7005, Norway
Olupona, Mr J A	Inst of Agricultural Research & Training, (IAR&T) PMB 5029, Ibadan , Nigeria
Overend, Dr M	Forum Bioscience Ltd, Redhill, Surrey , UK
Overton, Prof T	Cornell Univesity, 272 Morrison Hall, Ithaca, NY 14853 , USA
Packington, Mr A J	DSM Nutritional Products, Heanor Gate, Heanor, Derbyshire DE75 7SG, UK
Parker, Dr D	Novus Europe, 200 Ave Marcel Thiry, Building D, Fourth Floor, Brussels B-1200, Belgium
Parr, Dr T	University of Nottingham, Sutton Bonington Campus, Loughborough, Leics LE12 5RD, UK
Pass, Mr R	Diageo, Carsebridge Road, Alloa, Clackmannanshire FK10 3LT, Scotland
Pickford, Mr R	Bocking Hall, Bocking Church St, Braintree, Essex CM7 5JY, UK
Pike, Dr I H	IFFO, 2 College Yard, Lower Dagnall Street, St Albans AL3 4PA, UK
Pine, Dr A	Premier Nutrition Products Ltd, The Levels, Rugeley, Staffs WS15 1RD, UK
Probert, Miss L	Danisco Animal Nutrition, P O Box 777, Marlborough, Wiltshire SN8 1XN, UK
Rafael, Mr J	Nuri & Espadaler SL, Cantonigros 8 (Malloles), Vic 08500, Spain
Ratcliffe, Mrs J	Frank Wright Ltd, Blenheim House, Blenheim Rd, Ashbourne DE6 1HA, UK
Rhodes, Mrs A	Dairy Farmer Magazine, Low Laithes Farm, Ardsley, Barnslwy S71 5HB, UK
Richards, Dr S	Provimi Ltd, SCA Mill, Dalton Airfield Industrial Estate, Dalton, Thirsk YO7 3HE, UK
Robert, Mr J-C	Adisseo France, 42 Avenue, Aristide Briand, Antony 92160, France
Robinson, Mr S	Nottingham University Press, Manor Farm, Thrumpton, Nottingham NG11 0AX, UK
Rogers, Mr M	Volac International, Volac House, Orwell, Royston, Herts SG8 5QX, UK

Rosen, Dr G	Pronutrient Services, 66 Bathgate Rd, London SW19 5PH, UK
Russell, Dr D	Masterfoods, Mill Street, Melton Mowbray, Leics LE13 1BB, UK
Russell, Mrs H	University of Nottingham, Sutton Bonington Campus, Loughborough, Leics LE12 5RD, UK
Salter, Dr A	University of Nottingham, Sutton Bonington Campus, Loughborough, Leics LE12 5RD, UK
Sanchez, Dr W K	Diamond V Mills, 14145 SW 97th Place, Tigard, Oregon OR 97224, USA
Serrano, Dr X	Cotecnica, CTRA NII KM 494.5, Bellpuig 25250, Spain
Shilton, Mrs E	Wheyfeed Ltd, Hill Farm, Melton Rd, Stanton-on-the-Wolds, Notts NG12 5PJ, UK
Short, Dr F	ADAS, ADAS Gleadthorpe, Meden Vale, Mansfield, Notts NG20 9PF, UK
Sinclair, Dr K	University of Nottingham, Sutton Bonington Campus, Loughborough, Leics LE12 5RD, UK
Slatterman, Mrs M	Svenska Lantmannen, Steffen Sohst Gatan, Ahus 29632, Sweden
Solvas, Mr X	Qualivet, Mallorca 186 1, Barcelona 08036, Spain
Stebbens, Dr H	Trouw Nutrition GB, Wincham Lane, Wincham, Northwich CW9 6DF, UK
Steinbock, Mr M	Forum Bioscience, 41-51 Brighton Rd, Redhill, Surrey RH1 6YS, UK
Sterk, Miss A-R	CCL BV, NCB-IAAN 52, Veghel 5462 6E, Netherlands
Stevenson, Miss Z	Park Tonks Ltd, 48 North Road, Great Abington, Cambridge CB1 6AS, UK
Swanson, Dr K	University of Illinois, 162 Aninmal Sciences Lab, 1207 W Gregory Dr Urbabna, IL 61801, USA
Taylor-Pickard, Dr J	Alltech Europe Ltd, Tudor Oaks, 26 Milner Drive, Whitton, Twickenham TW2 7PJ, UK
ten Doeschate, Dr R	ABNA Ltd, ABN House, P O Box 250, Oundle Rd, Peterborough PE2 9QF, UK
Tesfa, Dr A	EELA, Box 45 Hameentie 57, Helsinki FIN 00581, Finland
Thompson, Mr R	Shire Consulting, Shirebrook, Whinney Brow, Forton, Preston, Lancs PR3 OAE, UK
Toplis, Mr P	Primary Diets Ltd, Melmerby Industrial Estate, Melmerby, Ripon HG4 5HP, UK
Toyomaki, Mr K	Kyodo Shiryo Co Ltd, 5-12 Takashima 2 Chrome, Nishi-Ku, Yokohama 220-0011, Japan
Tuck, Mr K	Alltech, Sarney, Summerhill Rd, Dunboyne Co Meath, Ireland
Tucker, Prof G	University of Nottingham, Sutton Bonington Campus, Loughborough, Leics LE12 5RD,
Tucker, Dr L	Alltech, c/o The Old Wainhouse, Canon Frome, Ledbury HR8 2TE, UK

Twigge, Mr J	Nutreco Ruminant Research Centre, Trouw Nutrition, Wincham Lane, Wincham, Northwich CW9 6DF, UK
Van Den Bighelaar, Mr H	BFI Innovations Ltd, 1 Telford Court, Chester Gates, Dunkirk Lea, Chester CH1 6LT, UK
van Hees, Mr H	Nutreco Swine Research Centre, Boxmeer se weg 30, St Anthonis 5845 ET, The Netherlands
Van Laar, Dr H	Nutreco RRC, Veerstraat 308, Boxmeer 5830AE, Netherlands
Varley, Dr M	Provimi Ltd, Dalton, SCA Mill, Thirsk, N Yorks YO7 3HE, UK
Vik, Mr K-R	Fiska Molle, Box 300, Sandane 6821 , Norway
Vinyeta, Mrs E	ESPORC, Ctra Barcelona-Puigcerda, KMGO, Tona 08551, Spain
Watanabe, Mr K	Ajinomoto Eurolysine, 153 Rue de Courcelles, 75817, Paris Cedex 17, France
Webb, Prof R	University of Nottingham, Sutton Bonington Campus, Loughborough, Leics LE12 5RD, UK
West, Mr A	Victam International, P O 411, Redhill RH1 6WE, UK
White, Mr G	University of Nottingham, Sutton Bonington Campus, Loughborough, Leics LE12 5RD, UK
Wilde, Mr D	Alltech UK Ltd, Alltech House, Ryhall Rd, Stamford PE9 1TZ, UK
Williams, Dr P	Syngenta International AG, WRO 1002 13 63, Schwarzwaldallee 215, CH 4058 Basel, Switzerland
Wiseman, Prof J	University of Nottingham, Sutton Bonington Campus, Loughborough LE12 5RD, UK
Wright, Mr I	Hyde House, Fossil Bank, Upper WR13 6PJ, UK
Wynn, Dr R	ABNA, P O Box 259, Oundle Rd, Peterborough PE2 9QF, UK
Yeo, G	Premier Nutrition Products Limited, The Levels, Rugeley, Staffs WS15 1RD, UK

INDEX